Radome Electromagnetic Theory and Design

Radome Electromagnetic Theory and Design

Reuven Shavit
Department of Electrical and Computer Engineering
Ben-Gurion University of the Negev, Israel

WILEY

IEEE PRESS

This edition first published 2018

Registered Office(s)
John Wiley & Sons, Inc., 111 River Street, Hoboken, NJ 07030, USA
John Wiley & Sons Ltd, The Atrium, Southern Gate, Chichester, West Sussex, PO19 8SQ, UK

Editorial Office
The Atrium, Southern Gate, Chichester, West Sussex, PO19 8SQ, UK

For details of our global editorial offices, customer services, and more information about Wiley products visit us at www.wiley.com.

Wiley also publishes its books in a variety of electronic formats and by print-on-demand. Some content that appears in standard print versions of this book may not be available in other formats.

Library of Congress Cataloging-in-Publication Data:

Names: Shavit, Reuven, 1949- author.
Title: Radome electromagnetic theory and design / by Reuven Shavit.
Description: Hoboken, NJ : John Wiley & Sons, 2018. | Includes bibliographical references and index. |
Identifiers: LCCN 2017057991 (print) | LCCN 2018006051 (ebook) | ISBN 9781119410829 (pdf) | ISBN 9781119410843 (epub) | ISBN 9781119410799 (cloth)
Subjects: LCSH: Radomes.
Classification: LCC TK6590.R3 (ebook) | LCC TK6590.R3 S43 2018 (print) | DDC 621.3848/3–dc23
LC record available at https://lccn.loc.gov/2017057991

Cover Design: Wiley
Cover Image: © ra-photos/Getty Images;
Background: © rionm/Getty Images

Set in 10/12pt WarnockPro by SPi Global, Chennai, India

Printed and bound by CPI Group (UK) Ltd, Croydon, CR0 4YY

10 9 8 7 6 5 4 3 2 1

To my wife, Liora, and my daughters: Sarit, Liat, Shirley, and Libby, without whose constant support this book would not have become a reality.

Contents

Preface

Thirty years ago, I started working for ESSCO company in the United States while being on leave of absence from my job in Israel, and for the first time I was exposed to the subject of large radomes and their design. I found this subject very interesting and enlightening, especially because it includes almost all disciplines of analytics and numerical analysis in electromagnetics (EM), such as scattering, array theory, numerical analysis of integral equations and differential equations, circuit analysis modeling, and measurements (near and far field). It is quite amazing that such a niche topic in applied EM involves so many disciplines in electromagnetics. At ESSCO, I was lucky to meet Al Cohen, the CEO and founder, who tried successfully to couple his enthusiasm to the subject with his engineers and with me in particular, and Professor R. Mittra from University of Illinois at Champaign–Urbana, from which I learned a lot on computational electromagnetics. He served at that time as the company's advisor, and with his students developed many of the numerical codes used at ESSCO. I also would like to mention my colleagues at ESSCO, A. Smolski, C. Cook, E. Ngai, T. Wells, J. Sangiolo, T. Monk, A. Mantz, M. Naor, and Y. Hozev, from which I have learned a lot and have been enriched on the subject throughout numerous interactions.

It took me a long time to come to the conclusion that such a book is necessary. Over the years some books on the subject have been published, but most of them are practical books for radome design engineers. The book compiles in a unified manner the major theoretical and numerical tools needed for radomes design, which are currently dispersed over many scientific papers. The book is intended for graduate students and electromagnetic engineers working in the field who are interested in the theoretical and numerical aspects of the subject beyond running simulations with commercial EM software such as CST and HFSS. Enough detail is given such that readers can write their own computer programs to test new numerical algorithms or designs.

September 2017 *Reuven Shavit*

Acknowledgments

I would like to thank Al Cohen from ESSCO (now a subsidiary of L3 Communications) for introducing me to the subject through my association with ESSCO, Joe Sangello my direct supervisor who gave me full support and encouragement in my stay at ESSCO, and Professor R. Mittra, advisor to ESSCO, from whom I learned a lot on computational electromagnetics.

1

Introduction

The word *radome,* is an acronym of two words "radar" and "dome" and is a structural, weatherproof enclosure that protects the enclosed radar or communication antenna. The main objective of the radome is to be fully transparent to the electromagnetic energy transmitted/received by the enclosed antenna, and in this sense its objective is similar to that of a glass window for light in the optics spectrum. Radomes protect the antenna surfaces from weather and, in contrast to a glass window, can also conceal the antenna electronic equipment from the outside radome observer. Another benefit for using a radome is that it enables use of a low-power antenna rotating systems and weaker antenna mechanical design, followed by a significant price reduction, since the enclosed antenna is not exposed to the harsh outside weather. Radomes can be constructed in several shapes (spherical, geodesic, planar, etc.), depending on the particular application using various construction materials (e.g., fiberglass, quartz, polytetrafluoroethylene (PTFE)-coated fabric, closed cell foam (rohacell), honeycomb). The radomes are assembled on aircrafts, ships, cars, and in fixed ground-based installations. In case of high-speed moving platforms like aircrafts, another important consideration is related to the streamline shape of the radome to reduce its drag force.

The materials used to construct radomes are often used to prevent ice and freezing rain (sleet) from accumulating directly on its external surface to avoid extra losses of the communication link. In case of a spinning radar dish antenna, the radome also protects the antenna from debris and rotational problems due to wind. For stationary antennas, excessive ice accumulation on the radome surface can de-tune the antenna, causing extra losses and internal reflections, which may

Radome Electromagnetic Theory and Design, First Edition. Reuven Shavit.

go back to the transmitter and cause overheating. A good designed radome prevents that from happening by covering the exposed parts with a sturdy' weatherproof material like PTFE, which keeps debris or ice away from the antenna. One of the main driving forces behind the development of fiberglass as a structural material was the need for radomes during World War II. Sometimes radomes may be internally heated to melt the accumulated ice on their exterior surface. The most common shape of ground-based radomes is spherical because of the rotational symmetry such a radome offers. Large ground-based radomes are made of sandwich panels interconnected by seams or beams, which may affect the enclosed antenna radiation pattern as described in Chapter 6. Small or medium-sized radomes are usually made of one molded piece. In this case, only the transmission loss and boresight error caused by the radome need to be considered in the design as explained in Chapter 4.

Static electricity caused by air friction on the radome surface can present a serious shock hazard. Thin antistatic coatings are used to neutralize static charge by providing a conducting path to attached structures. Lightning strikes to aircraft are common, so metallic lightning-diverter strips are used to minimize structural damage to the radome. Diverters cause some increase in sidelobe levels; this effect can be estimated using the computational tools described in Chapter 5.

The US Air Force Aerospace Defense Command operated and maintained dozens of air defense radar stations in the United States, including Alaska, during the Cold War. Most of the radars used at these ground stations were protected by rigid or inflatable radomes. The radomes were typically at least 15 m (50 ft) in diameter, and the radomes were attached to standardized radar tower buildings that housed the radar transmitter, receiver, and antenna. Some of these radomes were very large. The CW-620 was a rigid space frame radome with a maximum diameter of 46 m (150 ft), and a height of 26 m (84 ft). This radome consisted of 590 panels and was designed for winds of up to 240 km/h (150 mph). The total radome weight was 92,700 kg (204,400 lb) with a surface area of 3680 m^2 (39,600 ft^2). The CW-620 radome was designed and constructed by Sperry-Rand Corp. for the Columbus Division of the North American Aviation. This radome was originally used for the FPS-35 search radar at Baker Air Force Station in Oregon. Two typical airborne radomes are shown in Fig. 1.1 and Fig. 1.2. Both of them are ogive type, but with different contours.

Figure 1.1 Tejas aircraft (India) radome picture.

Figure 1.2 Norton radome B787 dreamliner picture.

For maritime satellite communication service, radomes are widely used to protect dish antennas, which are continually tracking fixed satellites while the ship experiences pitch, roll, and yaw movements. Large cruise ships and oil tankers may have radomes over 3 m in diameter covering antennas for broadband transmissions for television, voice, data, and the internet, while recent developments allow similar services from smaller installations, such as the 85 cm motorized dish used in the ASTRA2 Connect Maritime Broadband system. Small private yachts may use radomes as small as 26 cm in diameter for voice and low-speed data transmission/reception.

1.1 History of Radome Development

The first radomes appeared in United States in 1940 with the introduction of the radar during the World War II when radars were installed on aircrafts and aerodynamics considerations were imposed to cover the radar antennas to reduce the drag forces on a high speed aircraft. The first reported aircraft radomes used simple, thin-wall designs. In 1941, the first in-flight radome was a hemispherical nose radome fabricated from plexiglass [1, 2]. It protected an experimental S-band, Western Electric radar flown in a B-18A aircraft. Beginning 1943, production airborne radars used plywood radomes [1].

In this period, plywood radomes also appeared on Navy PT boats and blimps, as well as in ground installations. Because plywood has moisture absorption problems and does not easily bend into doubly curved shapes, new radome construction techniques and materials were introduced. In 1944, the MIT Radiation Laboratory developed the three-layer *A-sandwich,* which used dense skins and a low-density core material. The skins were fabricated from fiberglass and the core from polystyrene. Since World War II, radome materials have developed in the following areas: ceramics for high-speed missile radomes, quartz, fiberglass, honeycomb, and foam for sandwich composites radomes. Today, the majority of aircraft radomes use sandwich-wall designs. Fig. 1.3 shows some typical radomes installed on ships.

Various authors have contributed to the literature describing the evolution, design, and manufacture of radomes. Cady [3] describes the electrical design of normal and streamlined radomes and their installation, together with the theory of reflection and transmission

Figure 1.3 Typical radomes installed on ships.

of electromagnetic waves through dielectric materials. The focus is on airborne radomes. Hansen [4] describes large ground radomes, their environmental, structural and design problems. Walton [5] describes advanced airborne radomes. Skolnik [6] and Volakis [7] gave theoretical electrical characteristics for sandwich panels and typical requirements for airborne and ground radomes. Chapters describing the radome theory and design rules can also be found in [8] and [9] and in books like [10].

The major electrical parameters that determine the radome performance are:

- Insertion loss (IL) due to the presence of the radome.
- Increased antenna sidelobe levels of the radiation pattern.
- Depolarization radiation pattern increase.
- Boresight error (BSE) and boresight error slope (BSES).

Insertion loss is a reduction in the signal strength as the electromagnetic wave propagates through the radome wall. Part of the loss is due to the reflection from the air/dielectric interface. Additional losses are due to internal and external diffraction, refraction effects, and polarization shift. The remainder is due to the dissipation within the dielectric layers. The computation of these parameters in multilayer radomes, including analysis in frequency selective surfaces (FSS), are described in Chapters 2 and 3.

The reflection and scattering from the radome also causes changes of the main beam-shape and increases the radiation pattern sidelobes. The scattering mechanism from a single beam in a space structure radome is described in Chapter 5 and the total effect of all beams array in front of the antenna is described in Chapter 6.

Depolarization is an energy diverted from the primary antenna polarization to the orthogonal polarization. This phenomenon occurs as a result of the radome wall curvature and the difference in complex transmission coefficient between orthogonal polarized vectors. Depolarization can be problematic, particularly with satellite communication (SATCOM) ground terminals that utilize frequency reuse with orthogonal polarization. In this application, two independent signals are transmitted or received within the same frequency channel, but in opposite polarization senses. This topic is discussed in Chapters 4 and 6.

BSE is caused by the distortion of the electromagnetic wavefront as it propagates through the radome wall and bends the angle of arrival of the received signal relative to its actual angle of arrival. For monopulse antennas, it is the shift that the radome causes in the direction of the difference mode pattern's deepest minimum, or the shift that is obtained by comparing the phases in a pair of antennas. The BSES is defined as the rate of change of BSE with respect to the angle between axes of the radome and the antenna. Radome's BSES can cause severe degradation for modern guidance systems as well as for classical proportional navigation systems. This topic is discussed in Chapter 4.

1.2 Types of Radomes

We will now introduce and briefly discuss some forms of the various radome types in order to familiarize ourselves with the types of radomes discussed in the rest of the textbook.

1.2.1 Solid Laminate

The solid laminate radome is made of doubly curved solid fiberglass panels in which for small radomes (less than 1 m in diameter) are made from one piece and in case of larger sizes are made of panels arranged in neat vertical and horizontal rows. A typical solid laminate radome is shown in Fig. 1.4.

Figure 1.4 Solid laminate radome of L-3 Communications-ESSCO, Concord, Massachusetts.

The radome exhibits excellent performance below 3 GHz or at higher frequencies when wall thickness is tuned for narrow bandwidths.

1.2.2 Inflatable

The inflatable radome is actually a truncated spherical balloon made of strong fabric, which keeps its shape since it is highly pressurized. In terms of electromagnetic performance, it is superior to other radome types due to its highly transmittance feature from low to high frequencies. Its drawback is its pressurizing system that is vulnerable to electricity breakdowns and the entire radome may collapse over the enclosed antenna. This deficiency may be ameliorated by backing up the main power supply by an uninterruptible power supply (UPS) system. The inflatable radome can't sustain extreme environment conditions like high winds. A typical inflatable radome is shown in Fig. 1.5.

Figure 1.5 A typical inflatable radome.

1.2.3 Sandwich

The sandwich radome is a multilayer structure doubly curved, which may be fabricated in one piece for small size radomes or as a multipanel geodesic form for larger size radomes with polygonal panels bolted together to form the radome. The radome shell is made of highly developed composites to obtain panel consistency and strength. The fiberglass skins fully enclose each panel core to make the panels weather tight. The sandwich radome exhibits very good performance over relatively narrow frequency bands or potentially at multiple discrete frequencies. A typical ground-based multipanel sandwich radome is shown in Fig. 1.6.

1.2.4 Metal Space Frame

The metal space frame (MSF) radome is made of triangular frames quasi-randomly oriented in space and bolted together to form a geodesic dome. The frames are usually made of metal aluminum extrusion. Thin membranes made of materials with low dielectric constant and low losses are bonded into the frames. To avoid rain and snow accumulation, a thin laminate based on PTFE is permanently bonded on the membranes. A typical MSF radome is shown in Fig. 1.7.

Figure 1.6 Multipanel sandwich radome of L-3 Communications-ESSCO, Concord, Massachusetts.

Figure 1.7 Metal space frame radome of L-3 Communications-ESSCO, Concord, Massachusetts.

The operational frequency of the MSF radomes is 1–20 GHz with an insertion loss lower than 0.6 dB.

1.2.5 Dielectric Space Frame

The dielectric space frame (DSF) radome is made of panel geometries available in both regular and quasi-randomized. It is similar in structure to the MSF with the difference being that its beams are made of pultruded fiberglass beams instead of metal. The DSF panels may be flat or double curved, yielding a faceted geodesic or spherical smooth appearance. This type of radome extends the operational frequency band of the MSF to lower frequencies than 1 GHz. For frequencies above 1 GHz, the DSF insertion loss is oscillating and increasing, making the MSF the correct choice for a radome. Fig. 1.8 shows a typical DSF from AFC, Florida.

1.3 Organization of the Book

The book is organized into seven chapters. Chapter 2 describes the analogy in the analysis of a multilayer sandwich radome and transmission line theory. This analogy enables an efficient analysis and use of all the tools developed in transmission line theory like matching techniques, Smith chart, scattering matrix, and more. It also describes and brings design data and graphs for various types of sandwich radomes, such as A-sandwich; B-sandwich; and C-sandwich.

Chapter 3 describes analysis techniques of an arbitrary type FSS element geometry using Floquet modes and numerical solution of the current distribution on the FSS unit cell element. It also describes simulation results of the transmission and reflection coefficients through an FSS as a function of frequency, incident angle, and polarization for different type of elements, such as square patch, circular patch, crossed dipole, Jerusalem cross slot, double square loops, and more. It analyzes the benefits and drawbacks of the various type of element geometries. In addition, it describes the scattering analysis techniques for multilayer FSS structures. It also describes a metamaterial-inspired radome with a narrow passband to allow the enclosed antenna transmission

Figure 1.8 Dielectric space frame radome of AFC, Ocala, Florida.

and reception and a wide band above the passband with maximum absorption.

Chapter 4 describes various techniques for the analysis of airborne radomes with conformal shape like ogive and conical. The basic analysis for relatively large aperture antennas enclosed in conformal radomes is based on ray tracing combined with physical optics to determine the radiation pattern of the antenna enclosed in the radome. The ray-tracing method is not accurate near tips and discontinuities in the radome. In these cases and for small antennas enclosed in conformal radomes, integral equations (surface or volume) are formulated and solved numerically with the method of moments. The alternative numerical solution of integral equations is a solution using the finite element method (FEM).

Chapter 5 describes various methods to compute scattering from infinite cylinders with arbitrary cross section, made of conductive, dielectric and composite mixture of dielectric material and conductive strips. This information is critical in the evaluation of the radiation pattern of the enclosed antenna in a ground-based space frame radome. The scattering analysis can be performed by solution of an integral equation (volume or surface) or by a differential equation formalism. The integral equation can be solved numerically by the method of moments, while the differential equation can be solved numerically by the FEM. The surface integral equation is numerically efficient but is limited to homogeneous and conductive cylinders analysis. On the other hand, the volume integral equation enables analysis of inhomogeneous cylinders but is more numerically involved.

Chapter 6 describes the computation of the scattering parameters of a ground-based space frame radome as a function of the scattering from a single beam and the array factor of all the beams present in front of the antenna. Design considerations and the trade-offs made in terms of minimal optical blockage to the antenna, amount of scattering of a single beam (tuned or untuned), and the beam geometry randomness are also described.

Chapter 7 describes various methods to measure the electrical parameters (insertion loss and reflection) of sandwich panels and the scattering parameters (forward scattering and scattering radiation pattern) of dielectric (tuned and untuned) and conductive cylinders with arbitrary cross section using near field and far field techniques.

References

1 Baxter, JP. Scientists against time. Boston: Little Brown and Company, 1952.
2 Tice, TE. Techniques for Airborne radome design. Air Force Avionics Laboratory, Wright Patterson AFB, Ohio, 1966.
3 Cady, W, Karelitz, M, and Turner, L. Radar scanners and radomes. New York: McGraw-Hill, 1948.
4 Hansen, RC. Microwave scanning antennas. New York: Academic Press, 1964.

5 Walton, JD. Radome engineering handbook. New York: Marcel DeKker, 1976.
6 Skolnik, M. Radar handbook. New York: Wiley, 1990.
7 Volakis, JL. Antenna engineering handbook. New York: McGraw-Hill, 2007.
8 Lo, YT, and Lee, SW. Antenna handbook: theory, applications, and design. New York: Van Nostrand Reinhold, 1988.
9 Rudge, AW, Milne, K, Olver, AD, and Knight, P. The handbook of antenna design. London: Peter Peregrinus, 1986.
10 Kozakoff, DJ. Analysis of radome-enclosed antennas. Boston: Artech House, 1997.

2

Sandwich Radomes

Radomes are used as electromagnetic (EM) windows and are made of multilayer structures commonly called *sandwich radomes*. Usually, their geometrical shape is conformal and designed to protect the enclosed antenna from environmental hazards and be transparent to electromagnetic energy transmitted or received through them. The shape of a radome is determined by a combination of considerations, which include mechanical, thermal, aerodynamics, environmental, and electromagnetics. However, if the radome curvature is large enough in terms of wavelength, the interaction problem between the antenna fields and the radome surface can be locally considered as a problem of an electromagnetic plane wave with arbitrary polarization and oblique incident angle illuminating a planar multilayer structure extending to infinity in the x-y plane and being finite in z direction.

Section 2.1 illustrates the analogy between the analysis of a multilayer sandwich radome and its transmission line model. This analogy is important because it enables us to use all developed tools and concepts in transmission line analysis for the sandwich radome design. In section 2.2, the analysis is extended to a multilayer structure, and in sections 2.3 to 2.6, the analysis focuses on particular cases such as single layer slab, A-sandwich, B-sandwich, and C-sandwich, respectively.

Radome Electromagnetic Theory and Design, First Edition. Reuven Shavit.

2.1 Transmission Line Analogy

The propagation direction of the incident plane wave with respect to the z-axis is θ and its propagation constants are (k_x, k_y, k_z), as shown in Fig. 2.1.

We begin by decomposing the propagating fields along the z-axis into longitudinal $(e_z(x,y)e^{-jk_z z}, h_z(x,y)e^{-jk_z z})$ and transverse $(\underline{e}_t(x,y)e^{-jk_z z}, \underline{h}_t(x,y)e^{-jk_z z})$ components propagating through a medium with dielectric constant $\varepsilon = \varepsilon_0\varepsilon_r$ and permeability $\mu = \mu_0\mu_r$:

$$\underline{E}(x,y,z) = (\underline{e}_t(x,y) + e_z(x,y)\hat{z})e^{-jk_z z}$$
$$\underline{H}(x,y,z) = (\underline{h}_t(x,y) + h_z(x,y)\hat{z})e^{-jk_z z} \tag{2.1}$$

The z propagation constant is given by $k_z = \sqrt{k_0^2\varepsilon_r\mu_r - k_x^2 - k_y^2}$. Substituting (2.1) into the first time harmonic $\{e^{j\omega t}\}$ Maxwell equation,

$$\nabla \times \underline{E} = -j\omega\mu\underline{H} \tag{2.2}$$

and if we define the operator $\nabla = \nabla_t - jk_z\hat{z}$ in which $\nabla_t = \frac{\partial}{\partial x}\hat{x} + \frac{\partial}{\partial y}\hat{y}$ results in two equations:

$$\nabla_t \times \underline{e}_t = -j\omega\mu h_z\hat{z}$$
$$\hat{z} \times (\nabla_t e_z - jk_z\underline{e}_t) = j\omega\mu\underline{h}_t \tag{2.3}$$

In a similar fashion, we obtain from the second Maxwell equation

$$\nabla \times \underline{H} = j\omega\varepsilon\underline{E} \tag{2.4}$$

the following equations,

$$\nabla_t \times \underline{h}_t = j\omega\varepsilon e_z\hat{z}$$
$$\hat{z} \times (\nabla_t h_z - jk_z\underline{h}_t) = -j\omega\varepsilon\underline{e}_t \tag{2.5}$$

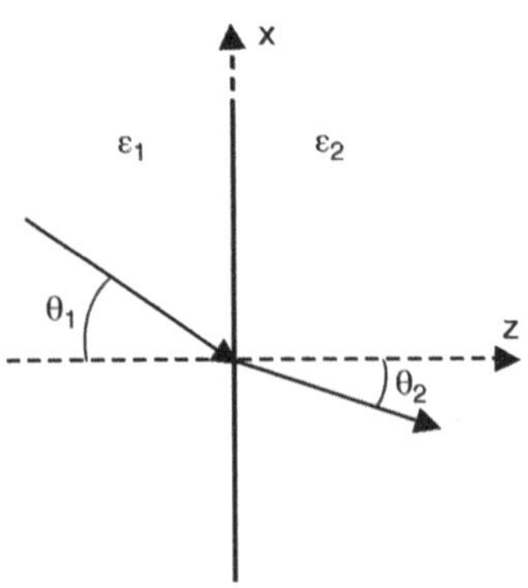

Figure 2.1 The geometry of the incident plane wave.

Inspection of eqs. (2.3)b and (2.5)b reveals that in the transverse electric (TE) case ($e_z = 0$) and transverse magnetic (TM) case ($h_z = 0$), the ratio of the transverse electric and magnetic components is constant. This ratio is called the medium characteristics impedance η^{TE} and η^{TM}, respectively. Thus,

$$\eta^{TE} = \frac{|\underline{e}_t|}{|\underline{h}_t|} = \frac{\omega\mu}{k_z}; \quad TE \ case$$

$$\eta^{TM} = \frac{|\underline{e}_t|}{|\underline{h}_t|} = \frac{k_z}{\omega\varepsilon}; \quad TM \ case \tag{2.6}$$

Thus, in analogy to transmission line theory, the propagation of a plane wave through a planar dielectric layer is similar to the propagation through a transmission line with characteristic impedances given by (2.6), depending on the polarization, propagation constant k_z, voltage wave e_t, and current wave h_t. This analogy enables to use all tools developed for transmission lines like matching techniques and Smith chart analysis to optimize the radome performance. Once the optimization process is performed using the transmission line theory, we can switch back to EM fields using the relations (2.3) and (2.5).

2.2 Multilayer Analysis

The geometry of an m multilayer sandwich radome is shown in Fig. 2.2.

In a multilayer radome problem, we assume that the input and output media is free space; therefore, for convenience, we define for the ith

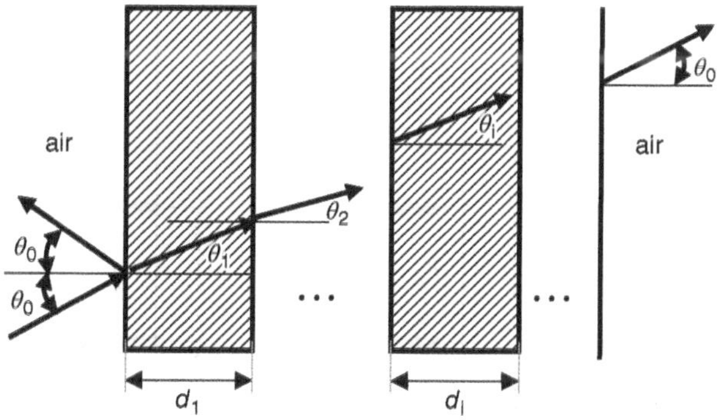

Figure 2.2 The geometry of a multilayer sandwich radome.

layer normalized to free space (input and output layers) characteristic impedances $\overline{\eta}_i^{TE}$ for the TE and $\overline{\eta}_i^{TM}$ for TM cases, such that [1]

$$\overline{\eta}_i^{TE} = \frac{\mu_{ri} k_{z0}}{k_{zi}} = \frac{\overline{\eta}_i \cos\theta_0}{\cos\theta_i} \quad ; \quad TE\ case$$

$$\overline{\eta}_i^{TM} = \frac{k_{zi}}{\varepsilon_{ri} k_{z0}} = \frac{\overline{\eta}_i \cos\theta_i}{\cos\theta_0} \quad ; \quad TM\ case \tag{2.7}$$

in which $k_{zi} = k_0\sqrt{\varepsilon_{ri}\mu_{ri}}\cos\theta_i$, $k_{z0} = k_0\cos\theta_0$, $\overline{\eta}_i = \sqrt{\frac{\mu_{ri}}{\varepsilon_{ri}}}$. The equivalent transmission line circuit to represent the multilayer structure is shown in Fig. 2.3.

The ABCD matrix describing a single layer is [1]:

$$\begin{bmatrix} A & B \\ C & D \end{bmatrix}\Bigg|_i = \begin{bmatrix} \cos\xi_i & j\overline{\eta}_i^{TM,TE}\sin\xi_i \\ j\frac{1}{\overline{\eta}_i^{TM,TE}}\sin\xi_i & \cos\xi_i \end{bmatrix} \tag{2.8}$$

with

$$\xi_i = k_{zi}d_i = \frac{2\pi d_i\sqrt{n_i^2 - \sin^2\theta_0}}{\lambda_0} \quad ; \quad \text{Snell's law- } \sin\theta_0 = n_1\sin\theta_1 = \ldots\ldots = n_m\sin\theta_m \tag{2.9}$$

in which $n_i = \sqrt{\varepsilon_{ri}\mu_{ri}}$ with $\varepsilon_{ri} = \varepsilon'_{ri} - j\varepsilon''_{ri}$, $\mu_{ri} = \mu'_{ri} - j\mu''_{ri}$ and θ_0 being the incident and output angles in free space of the EM plane wave. Substitution of (2.9) into (2.7) yields

$$\overline{\eta}_i^{TE} = \frac{\overline{\eta}_i\cos\theta_0}{\sqrt{1 - \frac{\sin^2\theta_0}{n_i^2}}} \quad ; \quad \overline{\eta}_i^{TM} = \frac{\overline{\eta}_i}{\cos\theta_0}\sqrt{1 - \frac{\sin^2\theta_0}{n_i^2}} \tag{2.10}$$

Moreover, the transverse EM field components at the boundaries of the ith and $i+1^{th}$ layers are related through [1]:

$$\begin{pmatrix} E_t^i \\ H_t^i \end{pmatrix} = \begin{pmatrix} \cos\xi_i & j\overline{\eta}_i^{TM,TE}\sin\xi_i \\ j\frac{1}{\overline{\eta}_i^{TM,TE}}\sin\xi_i & \cos\xi_i \end{pmatrix}\begin{pmatrix} E_t^{i+1} \\ H_t^{i+1} \end{pmatrix} \tag{2.11}$$

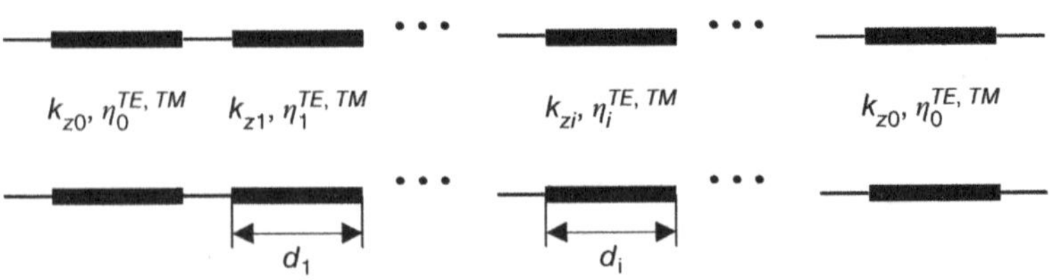

Figure 2.3 Equivalent transmission line circuit of the multilayer structure.

and the relationship between the normal components at the input and output boundaries are:

$$E_z^1 = E_z^m \rightarrow \underline{E}^i = E_t^i \hat{t} + E_z^i \hat{z} \tag{2.12}$$

Cascading the ABCD matrices of all layers results in the total ABCD matrix of the multilayer structure, which is the product of all m layers matrices such that

$$\begin{bmatrix} A & B \\ C & D \end{bmatrix} = [1][2][3]....[m] \quad layers \tag{2.13}$$

Once the multilayer structure matrix is known, the transmission T and reflection R coefficients of the transverse electric and magnetic fields can be computed [1] using these identities:

$$T = \frac{2}{A+B+C+D}; \quad R = \frac{A+B-C-D}{A+B+C+D} \tag{2.14}$$

In many instances, parallel PEC strips or wires with circular cross section (radius a) and spacing d [1] are used as interim layers to match and optimize the performance of a multilayer radome. Such layers can also be represented by ABCD matrices [1], depending on the electric field polarization relative to the wire axis. The equivalent circuit for a normal incident electric field with parallel polarization to the PEC wires screen is shown in Fig. 2.4.

The $ABCD$ matrix for such a wire grating for parallel polarization is given by [1]:

$$\begin{pmatrix} A & B \\ C & D \end{pmatrix} = \begin{pmatrix} 1 - \dfrac{\overline{X}_b}{\overline{X}_a} & j\overline{X}_b\left(\dfrac{\overline{X}_b}{\overline{X}_a} - 2\right) \\ -j\dfrac{1}{\overline{X}_a} & 1 - \dfrac{\overline{X}_b}{\overline{X}_a} \end{pmatrix} \tag{2.15}$$

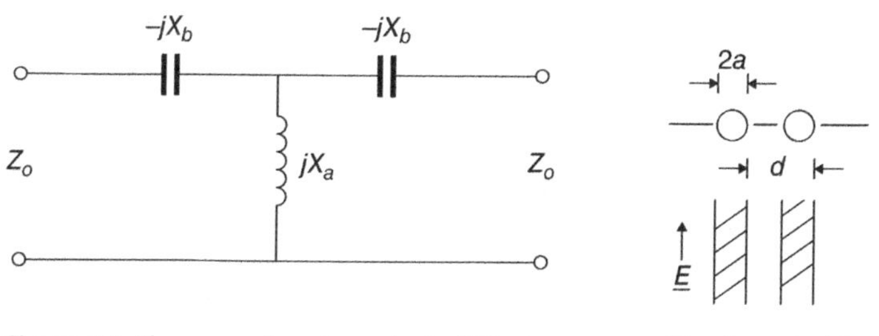

Figure 2.4 The geometry of a periodic PEC wire structure illuminated with parallel polarization and its circuit model.

in which the normalized inductances $\overline{X}_{a,b} = \frac{X_{a,b}}{Z_0}$ can be computed through [1] and are derived using modal analysis. The wire effect in this case is inductive.

$$\overline{X}_a = \frac{d}{\lambda}\left\{\log\left(\frac{d}{2\pi a}\right) + \frac{1}{2}\sum_{m=-\infty}^{\infty}\left[\left(m^2 - \frac{d^2}{\lambda^2}\right)^{-\frac{1}{2}} - \frac{1}{|m|}\right]\right\} \tag{2.16}$$

and

$$\overline{X}_b = \frac{d}{\lambda}\left(\frac{2\pi a}{d}\right)^2 \tag{2.17}$$

In case of a normal incident electric field and perpendicular polarization to the PEC wires with circular cross section (radius a) and spacing d [1], the equivalent circuit of the wire layer is shown in Fig. 2.5, and its effect is mainly capacitive.

The ABCD matrix in this case is given by

$$\begin{pmatrix} A & B \\ C & D \end{pmatrix} = \begin{pmatrix} 1 + \dfrac{\overline{B}_a}{\overline{B}_b} & -j\dfrac{1}{\overline{B}_b} \\ j\overline{B}_a\left(\dfrac{\overline{B}_a}{\overline{B}_b} + 2\right) & 1 + \dfrac{\overline{B}_a}{\overline{B}_b} \end{pmatrix} \tag{2.18}$$

in which the normalized reactances are $\overline{B}_{a,b} = \frac{B_{a,b}}{Y_0}$ and can be computed through [1] and derived by modal analysis. The wire effect in this case is capacitive.

$$\overline{B}_a = \frac{2\pi^2 a^2}{\lambda d}\left\{1 - \frac{\pi^2 a^2}{\lambda^2}\left[\frac{11}{2} + 2\log\left(\frac{2\pi a}{d}\right)\right] - \frac{\pi^2 a^2}{6d^2}\right\} \tag{2.19}$$

$$\overline{B}_b = \frac{d}{\lambda}\left[\frac{3}{4} - \log\left(\frac{2\pi a}{d}\right)\right] + \frac{d\lambda}{2\pi^2 a^2} \tag{2.20}$$

Periodic PEC strips embedded in a dielectric slab can help matching at the operating frequency without making any change in the thickness

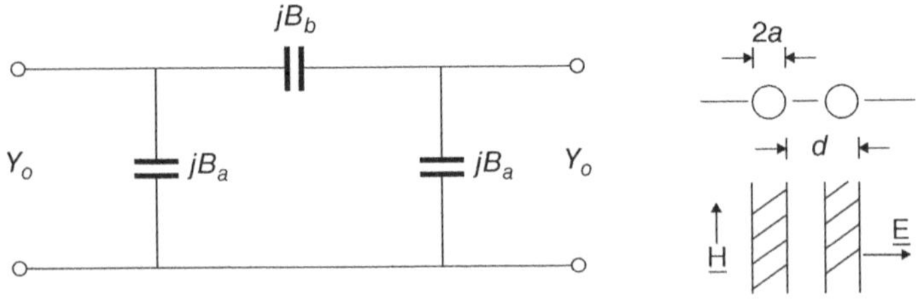

Figure 2.5 The geometry of a periodic wire structure illuminated with perpendicular polarization and its circuit model.

and the electrical parameters of the slab. To demonstrate this effect let us consider a dielectric slab with thickness of $h = 0.4$ in. and $\varepsilon_r = 4.6$. At the operating frequency of 9 GHz the dielectric slab is not matched and its insertion loss is high. To improve its matching, two grids of PEC strips are embedded in the dielectric slab, as shown in Fig. 2.6. The PEC strips width is $t = 0.062$ in. and their horizontal d and vertical h/2 spacings are 0.277 in. and 0.2 in., respectively.

Fig. 2.7 shows simulation results of the transmission as a function of frequency of the plain dielectric slab and the slab with the PEC grids at different incident angles for parallel polarization to the PEC strips.

Figure 2.6 A dielectric slab with two grids of PEC strips embedded.

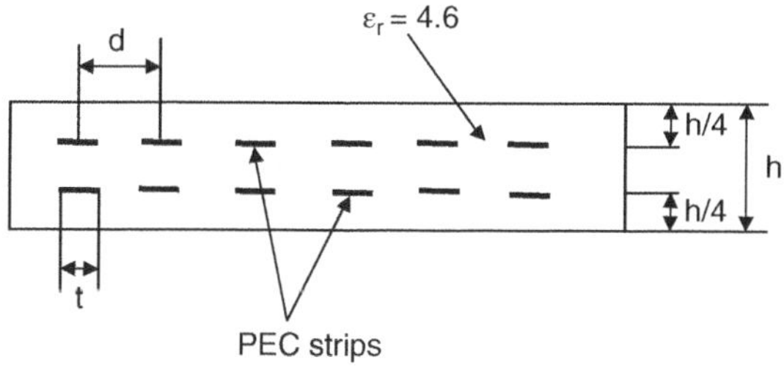

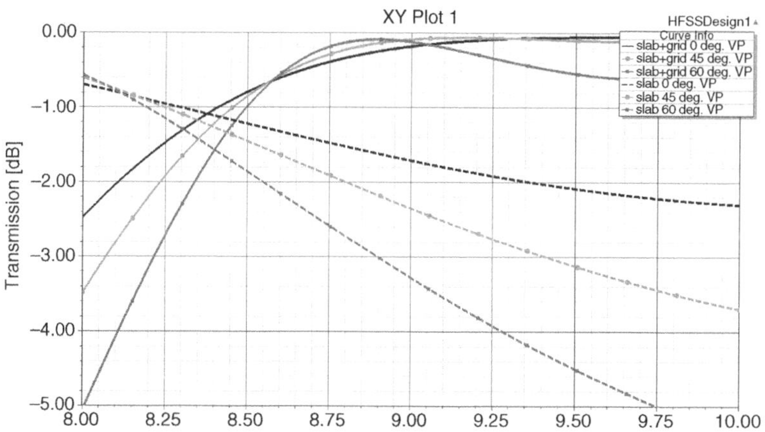

Figure 2.7 Comparison of the transmission loss through a dielectric slab and the dielectric slab with two PEC grids as a function of frequency at different incident angles for parallel polarization to the PEC strips. All simulations performed with HFSS.

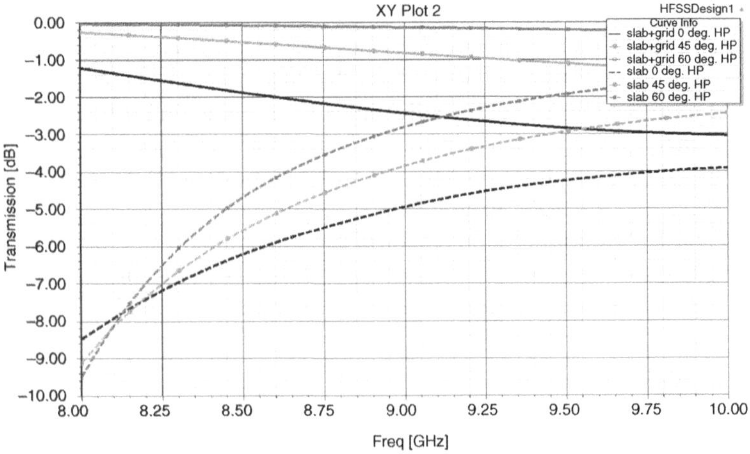

Figure 2.8 Comparison of the transmission loss through a dielectric slab and the dielectric slab with two PEC grids as a function of frequency and different incident angles for perpendicular polarization to the PEC strips. All simulations performed with HFSS.

One can observe that at the operational frequency of 9 GHz, the transmission of the plain dielectric slab is varying between −1.8 dB and −3.3 dB as a function of the incident angle. The addition of the PEC grids increases the transmission to −0.2 dB for incident angles 0–60 deg. Fig. 2.8 shows the same comparison, but for perpendicular polarization to the PEC strips.

In this case, the transmission of the slab with the conductive grids is higher than that of the plain dielectric slab but not as much as in the case of the parallel polarization.

2.3 Single Layer

Initially, a single layer with relative electrical parameters (ε_r, μ_r) shown in Fig. 2.9 will be considered. For both the input and output, the medium is air with electrical parameters (ε_0, μ_0). The incident plane is defined by the normal to the slab and the incident plane wave propagation direction. The incident electric field can be decomposed into a component normal to the incident plane (TE case) and a component in the incident plane (TM case). The TE case may also be called the perpendicular polarization, and the TM case called

Figure 2.9 The geometry of a single layer illuminated by an oblique plane wave.

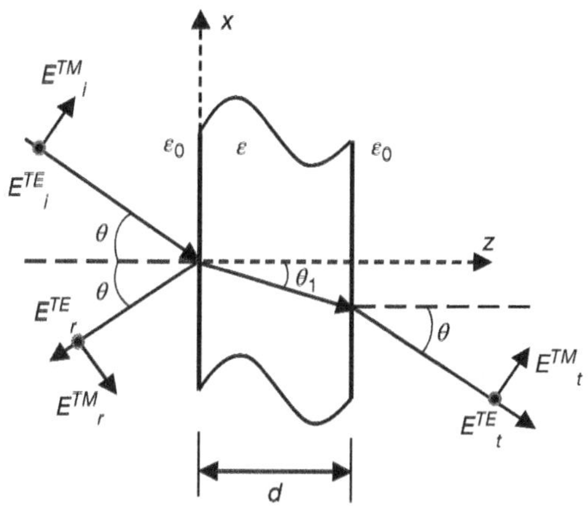

the parallel polarization. The incident angle is θ. By using boundary conditions on the front and back slab surfaces, it is easy to show that for a slab with parallel surfaces, the output angle is also θ, as shown in Fig. 2.9. The incident electric field is $E_i^{TE,TM}$ and the reflected, transmitted electric fields are $E_r^{TE,TM}$ and $E_t^{TE,TM}$, respectively. The reflected $S_{11}^{TM,TE}$ and transmitted $S_{21}^{TM,TE}$ coefficients from the slab can be computed using transmission line theory [1].

$$S_{11}^{TM,TE} = \frac{E_r^{TM,TE}}{E_i^{TM,TE}} = \frac{\Gamma^{TM,TE}(1 - e^{-j2k_z d})}{1 - (\Gamma^{TM,TE})^2 e^{-j2k_z d}} \tag{2.21}$$

and

$$S_{21}^{TM,TE} = \frac{E_t^{TM,TE}}{E_i^{TM,TE}} = \frac{(1 - (\Gamma^{TM,TE})^2)e^{-jk_z d}}{1 - (\Gamma^{TM,TE})^2 e^{-j2k_z d}} \tag{2.22}$$

where

$$\Gamma^{TM,TE} = \frac{\bar{\eta}^{TM,TE} - 1}{\bar{\eta}^{TM,TE} + 1} \tag{2.23}$$

with

$$\bar{\eta}^{TM} = \bar{\eta}\frac{\cos\theta_1}{\cos\theta} = \frac{\bar{\eta}\sqrt{1 - \sin^2\theta/n^2}}{\cos\theta} \tag{2.24}$$

in which θ_1 can be evaluated by Snell's law (2.9) and is shown in Fig. 2.9. Similarly,

$$\bar{\eta}^{TE} = \bar{\eta}\frac{\cos\theta}{\cos\theta_1} = \frac{\bar{\eta}\cos\theta}{\sqrt{1 - \sin^2\theta/n^2}} \tag{2.25}$$

in which $n = \sqrt{\varepsilon_r \mu_r}$; $\overline{\eta} = \sqrt{\frac{\mu_r}{\varepsilon_r}}$ and $k_z = k_0 n \cos\theta_1 = k_0 n\sqrt{1 - \sin^2\theta / n^2}$. One can observe that for $k_z d = m\pi$, $S_{11} = 0$ (no-reflection). The lowest order solution $m = 1$ for a no-reflection layer is possible for a slab thickness, $d = \frac{\lambda_0 / n}{2\sqrt{1 - \frac{\sin^2\theta}{n^2}}} = \frac{\lambda_0}{2\sqrt{n^2 - \sin^2\theta}}$.

The geometry of a single-layer slab illuminated by an oblique plane wave with amplitude E_0, linear polarized at an angle γ with respect to the incident plane is shown in Fig. 2.10. In this case the incident electric field is decomposed in a component in the incident plane and the other orthogonal to it. Each component is multiplied by the corresponding transmission coefficient $S_{21}^{TM} \triangleq T_{\|} e^{j\phi_1}$ for the in-plane (TM) component and $S_{21}^{TE} \triangleq T_{\perp} e^{j\phi_2}$ for the orthogonal (TE) component. Moreover, the transmitted field can be decomposed in a component in the original polarization direction (copol-plane) and orthogonal to the original polarization direction (crosspol-plane).

In case that the incident field is circular polarized right-hand circular polarization (RHCP) or left hand circular polarization

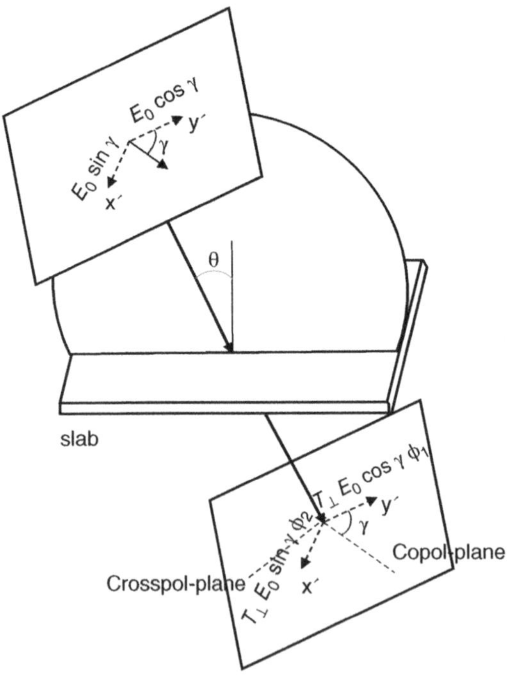

Figure 2.10 The geometry of a single layer slab illuminated by an oblique linear polarized plane wave.

(LHCP) the polarization of the transmitted field would be elliptical. The total transmitted elliptical polarized field can be expressed as a superposition of two orthogonal circular polarization fields—one LHCP with amplitude A and one RHCP with amplitude B. In case of an LHCP incident wave we obtain,

$$T_{\perp}e^{j\phi_2}\hat{x}' + T_{\parallel}e^{j\phi_1}e^{j\frac{\pi}{2}}\hat{y}' = A\left(\hat{x}' + e^{j\frac{\pi}{2}}\hat{y}'\right) + B\left(\hat{x}' - e^{j\frac{\pi}{2}}\hat{y}'\right) \tag{2.26}$$

Comparison of the $\hat{x}'$ and $\hat{y}'$ components results in two identities:

$$T_{\perp}e^{j\phi_2} = A + B \quad ; \quad T_{\parallel}e^{j\phi_1} = A - B \tag{2.27}$$

Evaluation of A and B results for LHCP incident wave

$$(\hat{x}' + j\hat{y}') \Rightarrow A = \frac{T_{\perp}e^{j\phi_2} + T_{\parallel}e^{j\phi_1}}{2} \quad ; \quad B = \frac{T_{\perp}e^{j\phi_2} - T_{\parallel}e^{j\phi_1}}{2} \tag{2.28}$$

A and B coefficients can be also written in the form,

$$|A| = \frac{T_{\parallel}}{2}\sqrt{1 + k^2 + 2k\cos(\phi_2 - \phi_1)} \quad ;$$

$$k \triangleq \frac{T_{\perp}}{T_{\parallel}} \text{ and } |B| = \frac{T_{\parallel}}{2}\sqrt{1 + k^2 - 2k\cos(\phi_2 - \phi_1)}$$

Similarly, for RHCP incident wave:

$$(\hat{x}' - j\hat{y}') \Rightarrow A = \frac{T_{\perp}e^{j\phi_2} - T_{\parallel}e^{j\phi_1}}{2}; \quad B = \frac{T_{\perp}e^{j\phi_2} + T_{\parallel}e^{j\phi_1}}{2} \tag{2.29}$$

Fig. 2.11 shows the transmission coefficient through a dielectric slab ($\varepsilon_r = 3.8$, $\tan\delta = 0.015$) as a function of the thickness d normalized to free space wavelength for perpendicular and parallel polarizations and different incident angles.

At any given angle of incidence, the insertion phase delay is the phase difference at the transmission surface between an unimpeded plane wave and the wave in the presence of the layer. This is given by

$$\psi_{\parallel,\perp} = \phi_{1,2} - \frac{2\pi}{\lambda_0}d\cos\theta \tag{2.30}$$

Fig. 2.12 shows the phase delay through the dielectric slab.

One can observe two peaks in the transmission at a very low thickness $d < \frac{\lambda_0}{10\sqrt{\varepsilon_r}}$ and at a thickness $d \approx \frac{\lambda_0}{2\sqrt{\varepsilon_r}}$. Operation with a thin layer radome is beneficial in terms of transmittance performance (high transmittance, wide frequency bandwidth, and insensitivity to the incident angle and polarization), but its major drawback is its mechanical strength. However, in many applications in which the radome can perform as a tight membrane, as in the inflatable radome and radome panel designs, this is a very good EM solution.

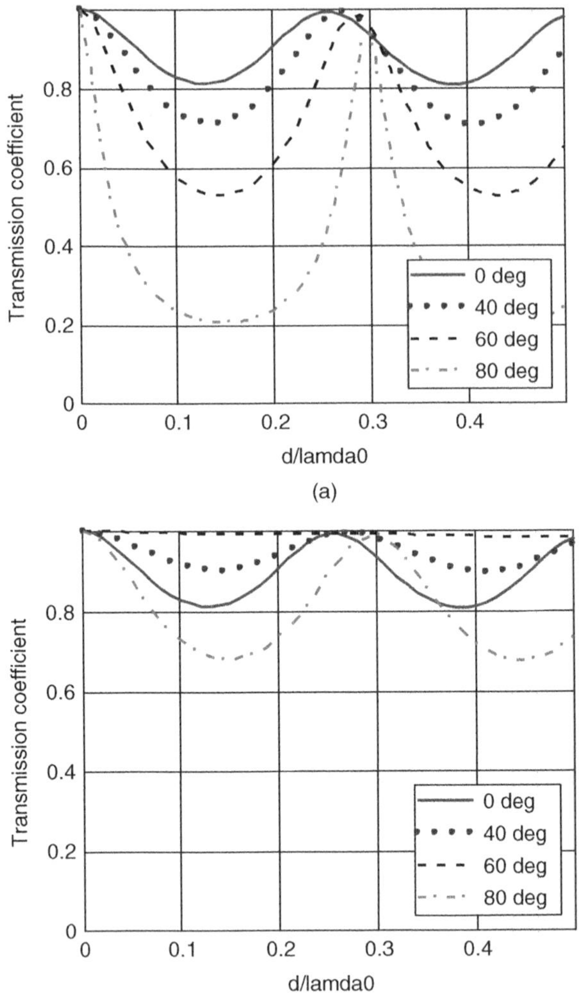

Figure 2.11 Transmission coefficient through dielectric slab ($\varepsilon_r = 3.8$, $\tan\delta = 0.015$) (a) perpendicular polarization, (b) parallel polarization.

The second option for maximum transmittance occurs at $d \approx \frac{\lambda_0}{2\sqrt{\varepsilon_r}}$ or multiples of this value. The transmittance in this case is not 100% and decreases as the thickness increases because of the loss tangent of the slab. At $\lambda_g \sim \frac{\lambda_0}{4\sqrt{\varepsilon_r}}$ the transmittance is minimal. Moreover, the transmittance is better for parallel polarization than for perpendicular

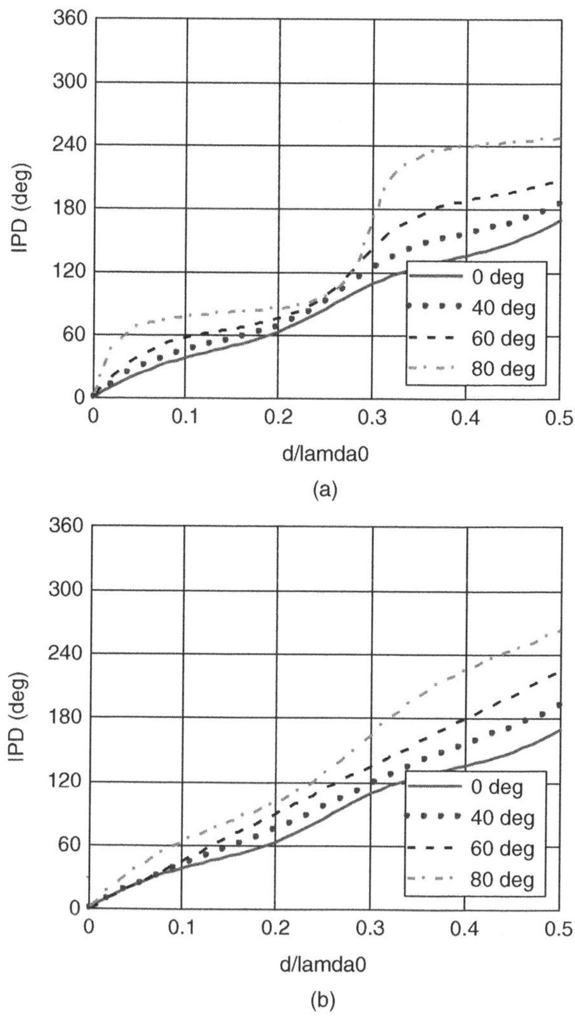

Figure 2.12 Phase delay through dielectric slab ($\varepsilon_r = 3.8$, $\tan\delta = 0.015$): (a) perpendicular polarization, (b) parallel polarization.

polarization because of the existance of Brewster angle [1] in the parallel polarization case. One can also observe that the frequency bandwidth around the optimal transmittance is narrower as the incident angle increases. Moreover, the phase delay through the slab is almost linear as a function of its thickness for normal incidence and becomes nonlinear as the incident angle increases and again the perpendicular polarization is more sensitive.

2.4 A-Sandwich

The relative narrow frequency bandwidth and the sensitivity of the transmittance to the incident angle in a single layer radome have driven the scientific community to look for alternative solutions. The A-sandwich is an improved solution, and the price paid is a multilayer structure with increased thickness and additional cost for fabrication. The three-layer A-sandwich consists of a low dielectric constant core with two skins of higher dielectric constant on top and bottom, as shown in Fig. 2.13.

In this case, $\mu_r = 1$ for all layers and the *ABCD* matrix based on eqs. (2.8) and (2.13) simplifies to:

$$\begin{bmatrix} A & B \\ C & D \end{bmatrix} = \begin{bmatrix} \cos\xi_1 & j\bar{\eta}_1^{TM,TE}\sin\xi_1 \\ j\frac{1}{\bar{\eta}_1^{TM,TE}}\sin\xi_1 & \cos\xi_1 \end{bmatrix} \begin{bmatrix} \cos\xi_2 & j\bar{\eta}_2^{TM,TE}\sin\xi_2 \\ j\frac{1}{\bar{\eta}_2^{TM,TE}}\sin\xi_2 & \cos\xi_2 \end{bmatrix} \times \begin{bmatrix} \cos\xi_1 & j\bar{\eta}_1^{TM,TE}\sin\xi_1 \\ j\frac{1}{\bar{\eta}_1^{TM,TE}}\sin\xi_1 & \cos\xi_1 \end{bmatrix} \tag{2.31}$$

in which

$$\xi_i = \frac{2\pi d_i}{\lambda_0}\sqrt{\varepsilon_{ri} - \sin^2\theta}; \quad \bar{\eta}_i^{TE} = \frac{\cos\theta}{\sqrt{\varepsilon_{ri} - \sin^2\theta}}; \quad \bar{\eta}_i^{TM} = \frac{\sqrt{\varepsilon_{ri} - \sin^2\theta}}{\varepsilon_{ri}\cos\theta}; \quad i = 1,2$$

A double-frequency design strategy is usually implemented such that at the higher frequency, the parameters of the skins are chosen to obtain a total phase delay of 90 deg, while the parameters of the core are chosen to obtain no reflection performance at a second frequency.

A typical A-sandwich radome consists of a foam or honeycomb core with dielectric constant near 1.07 and loss tangent 0.002 and resin glass-fiber skins with dielectric constant 3.8 and loss tangent of 0.015. Fig. 2.14 shows the percentage of the transmission power for parallel and perpendicular polarizations as a function of the core thickness normalized to the free space wavelength for different incident angles.

Fig. 2.15 shows the insertion phase delay through this radome.

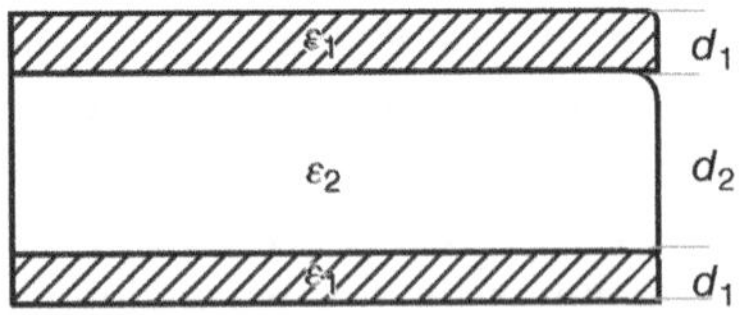

Figure 2.13 A sandwich panel geometry.

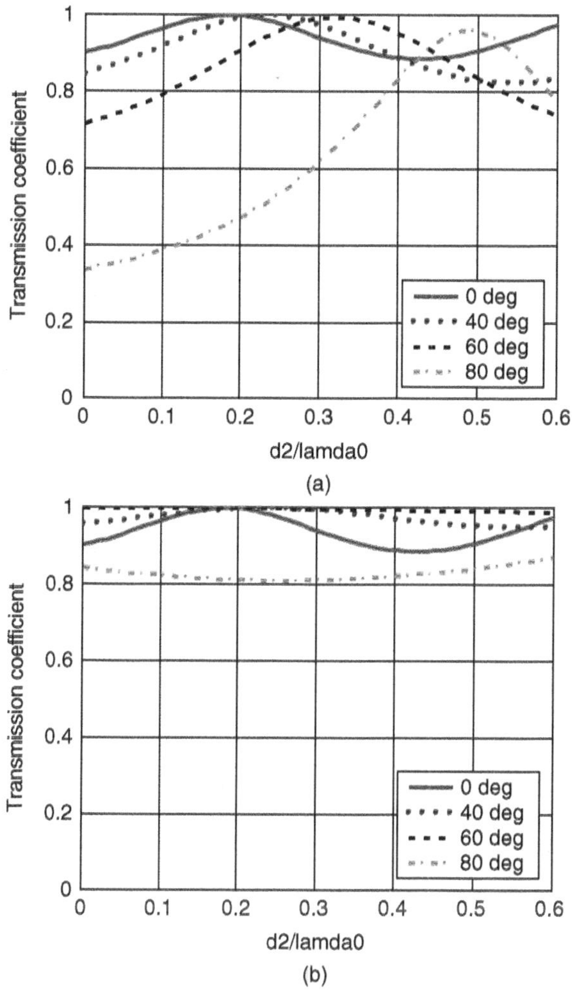

Figure 2.14 Transmission coefficient through A sandwich with skins dielectric constant ($\varepsilon_r = 3.8$, $\tan\delta = 0.015$) and thickness/wavelength (free space) ratio 0.03 and with core dielectric constant ($\varepsilon_r = 1.07$, $\tan\delta = 0.002$) (a) perpendicular polarization, (b) parallel polarization.

One can observe that the frequency bandwidth is wider compared to its counterpart single-layer radome for incident angles up to 60 deg, especially for parallel polarization. Its phase delay is also almost linear as a function of the core thickness up to 60 deg incident angle.

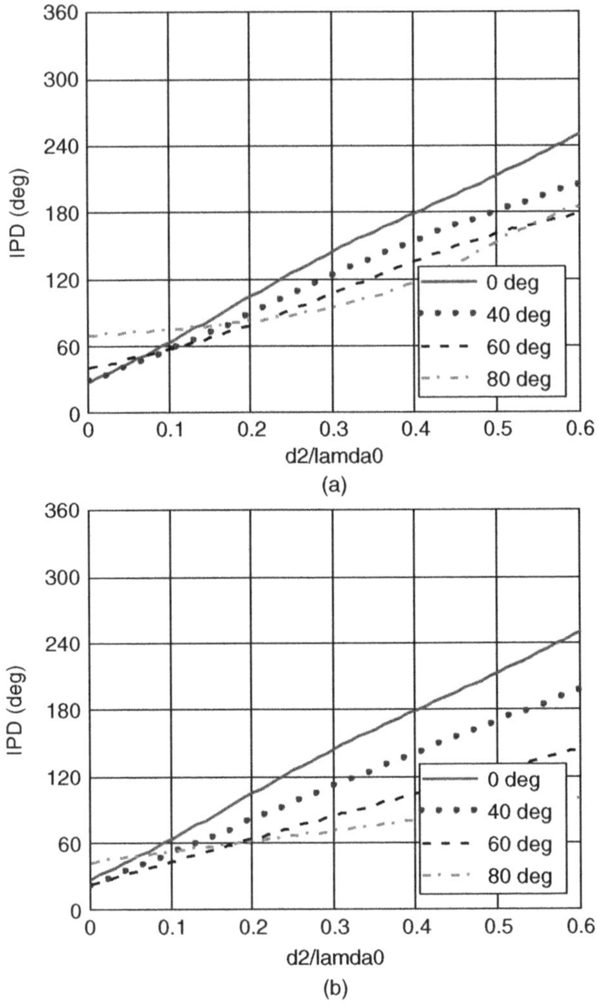

Figure 2.15 Insertion phase delay as a function of the core thickness of A sandwich with skins dielectric constant ($\varepsilon_r = 3.8$, $\tan\delta = 0.015$) and thickness/wavelength ratio 0.03 and with core dielectric constant ($\varepsilon_r = 1.07$, $\tan\delta = 0.002$) (a) perpendicular polarization, (b) parallel polarization.

The three-layer structure with the lighter core compared to the skins gives the A-sandwich good mechanical properties in addition to a relatively lightweight structure compared to the half wavelength single-layer radome.

A possible way to reduce the insertion-phase delay variation as a function of frequency and incident angle is to design the radome around a zero-phase delay point. The insertion phase delay in this case is given by

$$\psi_{\parallel,\perp} = \phi_{1,2} - \frac{2\pi}{\lambda_0}(2d_1 + d_2)\cos\theta \tag{2.32}$$

For this to be made zero, it is intuitively obvious that the center medium must be a "phase advance" material if the outer layers of dielectric are *phase retarding*. Such a material exists in the form of the two-dimensional metamaterial with double negative permeability and permittivity [2, 3]. The obvious advantage of such a structure is in those applications where the insertion phase of the radome is a major cause of beam distortion or displacement. The result shows that a very broadband radome results and that over the larger part of this band and for a wide range of incidence angles, the insertion phase delay can be kept within ±1 radian.

2.5 B-Sandwich

A variation of the A-sandwich radome is also designed with three layers, but with the skins being with a low dielectric constant and a core with the higher dielectric constant. This is the B-sandwich, shown in Fig. 2.16.

The design in this case is also based on a two-frequency concept. The parameters of the skins are chosen such that at the higher frequency they make a $\lambda_g/4$ matching layer to the core layer. This implies that $\varepsilon_{r,1}^2 \sim \varepsilon_{r,2}$. The core parameters are chosen to obtain a $\lambda_g/2$ phase delay at the lower frequency. The advantage of this design is that it has more freedom in the choice of the core thickness to satisfy electrical and structural properties and also fits for multiband operation.

Fig. 2.17 shows the transmission coefficient as a function of frequency for four incident angles (0, 40, 60, and 80 deg) and perpendicular/parallel polarizations for a typical B-sandwich with skins

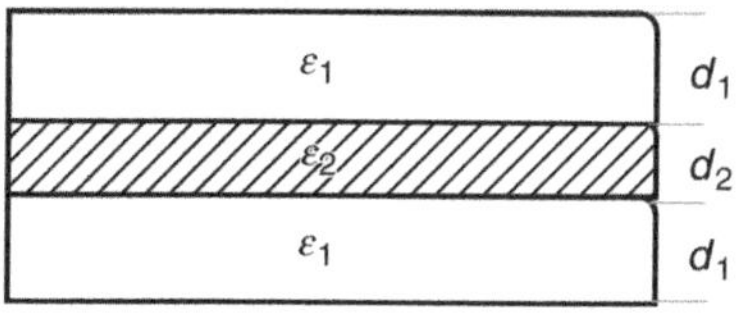

Figure 2.16 B-sandwich panel geometry.

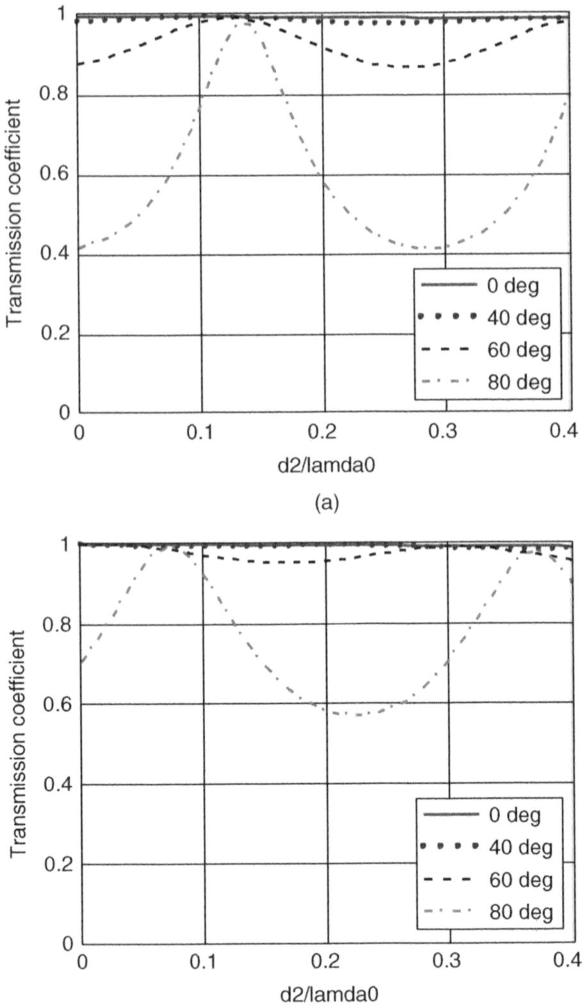

Figure 2.17 Transmission coefficient for (a) perpendicular and (b) parallel polarizations through B-sandwich with skins thickness $0.18\lambda_0$ and electrical parameters ($\varepsilon_r = 1.95$, $\tan\delta = 0.002$) and core parameters ($\varepsilon_r = 3.8$, $\tan\delta = 0.015$) for angle of incidence (0°, 40°, 60°, and 80°).

dielectric parameters ($\varepsilon_r = 1.95$, $\tan\delta = 0.002$) and thickness $0.18\lambda_0$ and with core dielectric parameters ($\varepsilon_r = 3.8$, $\tan\delta = 0.015$).

One can observe that the radome is tuned at several frequencies, and its transmission coefficient sensitivity to the incident angle up to 60 deg is relatively low. Fig. 2.18 shows the insertion phase delay through this radome.

2.6 C-Sandwich

The C-sandwich is an extension of the A-sandwich and comes to deliver a larger frequency bandwidth for two orthogonal polarizations. It consists of a five-layer sandwich radome. The five-layer C-sandwich may be considered as two A- sandwiches back to back, as shown in Fig. 2.19.

The radome can produce maximum transmission at the incident angle where the individual A-sandwiches have zero reflections, and also at the angle where the reflection for each A-sandwich is of opposite

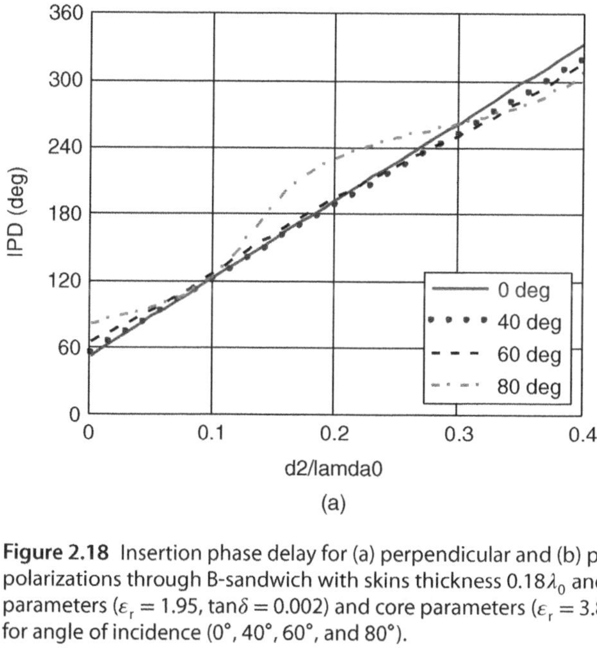

Figure 2.18 Insertion phase delay for (a) perpendicular and (b) parallel polarizations through B-sandwich with skins thickness $0.18\lambda_0$ and electrical parameters ($\varepsilon_r = 1.95$, $\tan\delta = 0.002$) and core parameters ($\varepsilon_r = 3.8$, $\tan\delta = 0.015$) for angle of incidence (0°, 40°, 60°, and 80°).

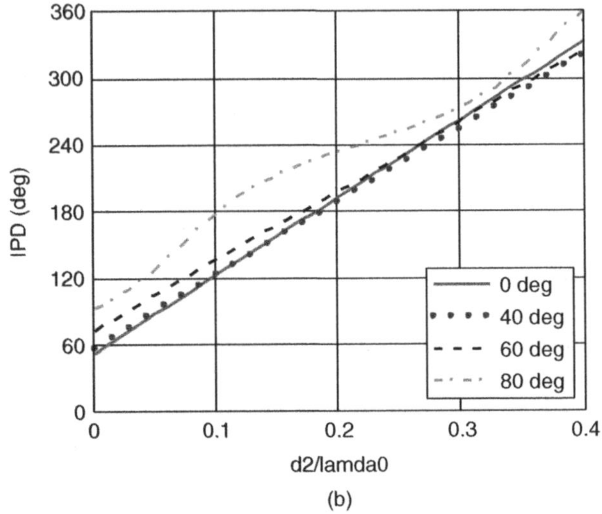

Figure 2.18 (*Continued*)

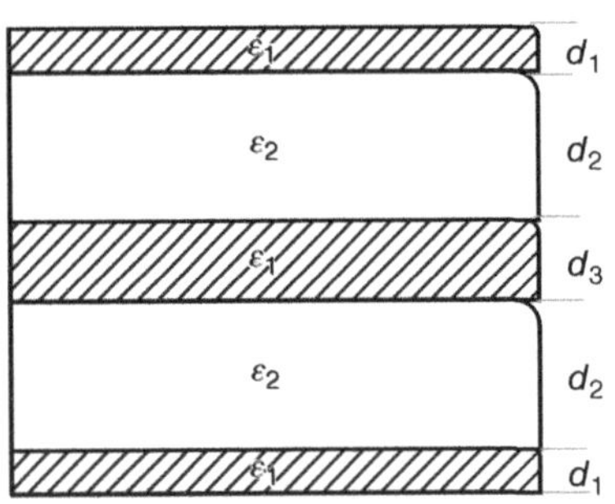

Figure 2.19 C sandwich panel geometry.

phase. If the individual A-sandwiches give maximum transmission at one angle of incidence and the combined sandwich at another, then wide-band coverage over a reasonable range of incidence angles can be obtained. In consequence, the C-sandwich has found application on high-incidence streamlined airborne radomes.

Fig. 2.20 and Fig. 2.21 show the transmission coefficients and the phase delay as a function of the core thickness normalized to the free space wavelength for different incident angles and two orthogonal polarizations for a typical C sandwich with skins dielectric parameters

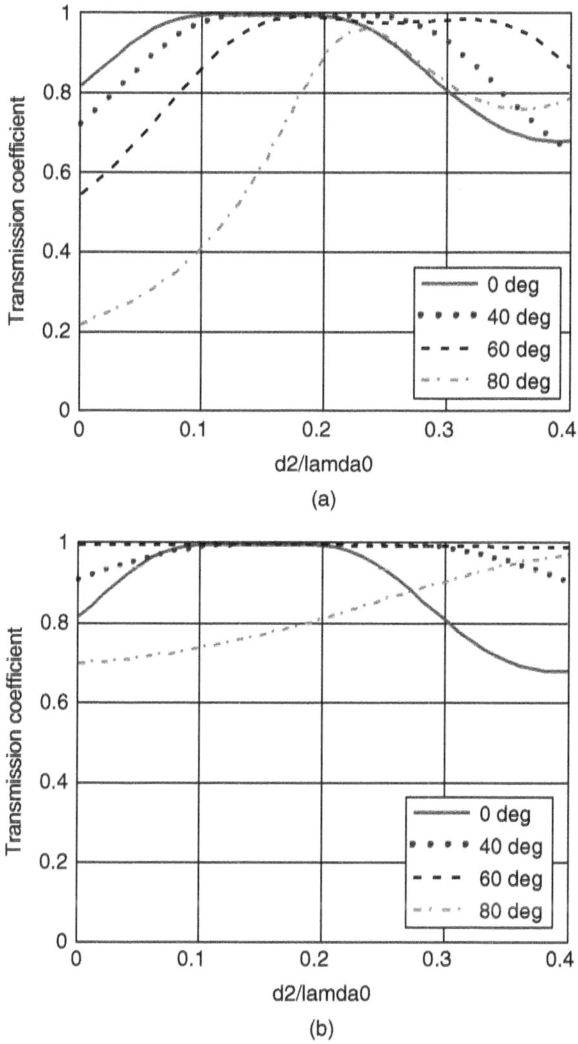

Figure 2.20 Transmission coefficients through a C sandwich with skins dielectric constant ($\varepsilon_r = 3.8$, $\tan\delta = 0.015$), thickness/wavelength ratio = 0.03 outer, 0.06 center, and with core dielectric constant ($\varepsilon_r = 1.07$, $\tan\delta = 0.002$) at various angles of incidence: (a) perpendicular polarization, (b) parallel polarization.

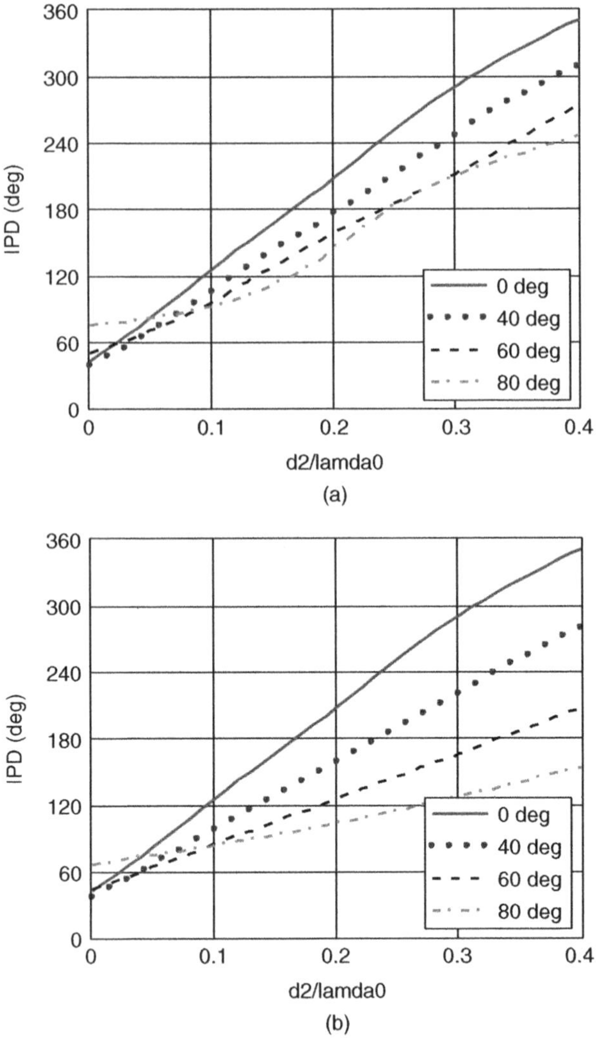

Figure 2.21 Insertion phase delay as a function of core thickness through a C sandwich with skins dielectric constant ($\varepsilon_r = 3.8$, $\tan\delta = 0.015$), thickness/wavelength ratio = 0.03 outer, 0.06 center, and with core dielectric constant ($\varepsilon_r = 1.07$, $\tan\delta = 0.002$) at various angles of incidence: (a) perpendicular polarization), (b) parallel polarization.

($\varepsilon_r = 3.8$, $\tan\delta = 0.015$), thickness/wavelength ratio = 0.03 outer, 0.06 center, and with core dielectric parameters ($\varepsilon_r = 1.07$, $\tan\delta = 0.002$).

One can observe a high transmission coefficient out to high angles of incidence for both polarizations for core thickness up to 0.2λ. However, phase differences of some 40 deg in the phase delay for different incidence angles are likely to produce significant aberrations, which might affect the radome performance, especially for conformal geometries.

Sandwich constructions consisting of more than five layers have not been widely reported. Calculations have shown that by using thin skin layers and low dielectric constant cores, useful performance can be obtained, particularly for low angles of incidence.

References

1 Cornbleet, S. Microwave optics. New York: Academic Press, 1976.
2 Pendry, J.B, Holden, AJ, Stewart, WJ, and Youngs, I. Extremely low frequency plasmons in metallic mesostructures. *Phys. Rev. Lett.*, 76(25), 4773–4776, 1996.
3 Pendry, JB, Holden, AJ, Robbins, DJ, and Stewart, WJ. MAgnetism from conductors and enhanced nonlinear phenomena. *IEEE Trans. Microw. Theory Tech.*, 47, 2075–2084, 1999.
4 Rudge, AW, Milne, K, Olver, AD, and Knight, P. The handbook of antenna design. London: Peter Peregrinus, 1986.

Problems

P2.1 Prove eq. (2.14) of the transmission T and reflection R coefficients in terms of A, B, C, D parameters.

P2.2 Design a planar radome made of parallel PEC wires (as shown in Fig. 2.4), with radius a and periodicity d operating at 5 GHz with transmission loss 0.5dB for parallel polarization to the wires and normal incidence. Use the layer equivalent circuit to compute its optimum a and d. Plot the transmission T and reflection R coefficients as a function of frequency in the frequency band 3–7 GHz. Repeat the computation of T and R coefficients using a commercial EM simulator and compare to the simulation results based on the equivalent circuit model. What kind of conclusion can you draw based on the comparison?

P2.3 Use the radome design in problem P2.2 to compute the transmission and reflection coefficients for parallel polarization to the wires in the frequency band 3–7 GHz and incident angles $\theta = 20, 40, 60,$ and 80 deg. Compare results to simulation results based on EM simulator. Discuss the results.

P2.4 Repeat the design in P2.2, but for perpendicular polarization.

P2.5 Repeat the simulations in problem P2.3 for perpendicular polarization.

P2.6 Design an A-sandwich to be optimal (insertion loss less than 1 dB) in the frequency range 5–6 GHz, parallel and perpendicular polarizations, incident angles $\theta = 0, 20, 40,$ and 60 deg. Use for the skins polyester material with $\varepsilon_r = 3.6$, $\tan\delta = 0.018$ and for the core honeycomb material with $\varepsilon_r = 1.07$, $\tan\delta = 0.005$. The skins can be fabricated with thickness increment of 15 mils and the honeycomb can be cut at any thickness.

P2.7 Repeat P2.6 and try to extend the radome's bandwidth for parallel polarization, keeping the insertion loss less than 1 dB in the pass-band by inserting one or two PEC wire layers in the A-sandwich designed in P2.6. Plot the insertion loss with and without the PEC wire layers.

P2.8 Design a C-sandwich to be optimal (insertion loss less than 1 dB) in the frequency range 5–6 GHz, parallel and perpendicular polarizations, incident angles $\theta = 0, 20, 40,$ and 60 deg. Use for the skins polyester material with $\varepsilon_r = 3.6$, $\tan\delta = 0.018$ and for the cores honeycomb material with $\varepsilon_r = 1.07$, $\tan\delta = 0.005$. The skins can be fabricated with thickness increment of 15 mils and the honeycomb can be cut to any thickness. Compare performance to that of the A-sandwich designed in P2.6.

3

Frequency Selective Surfaces (FSS) Radomes

Frequency selective surfaces are 2D periodic structures that transmit perfectly in certain frequency bands and reflect perfectly in others in case of incident waves with arbitrary polarization and different incident angles. The unit cell size of the FSS periodic structure is approximately $\lambda/2$. Historically, the FSS evolved from the investigation of diffraction gratings in optics. This filtering process, as well as the diffraction grating itself, was discovered by D. Rittenhouse, as documented by a scientific exchange between Hopkinson and Rittenhouse published in 1786 [1]. In this paper, Hopkinson reports about a street lamp he could see through his handkerchief stretched through his hands. There are many applications to FSS planar structures. Among them, the most common is for hybrid radomes in which conformal pass-band radomes are used to reduce the radar cross section (RCS) of the enclosed antennas in the radome outside their operating frequency band [2] by diverting the incident wave in another direction from the incident wave direction. This concept is shown in Fig. 3.1.

At the operating frequency of the enclosed antenna, the FSS is fully transparent, while for out-of-band frequency, the FSS is fully reflective or absorptive. In case of a fully reflective FSS, almost all reflected energy is redirected in a different direction from the incoming wave, while if the FSS is fully absorptive [3], the incident wave energy is dissipated in the radome. Thus, in both cases, out of band the monostatic RCS is reduced significantly. The FSS can be also used to stabilize the optimal transmittance through a $\lambda_g/2$ dielectric radome, with λ_g being the wavelength in the radome material and to widen its operational frequency bandwidth [2]. Another important utilization of FSS is as a dichroic surface [2], in a two-reflector antenna system with two feeds

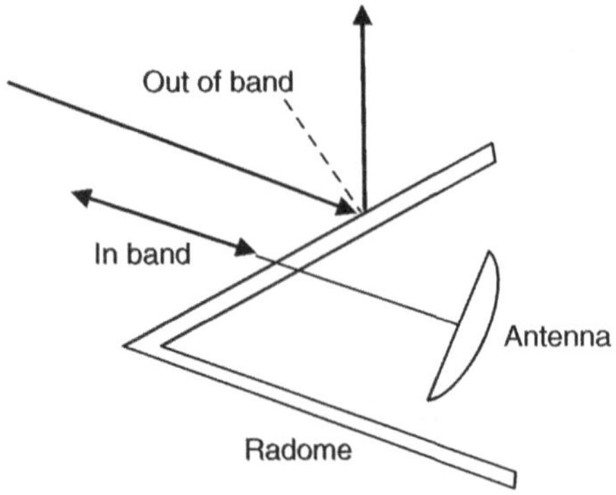

Figure 3.1 The concept of using FSS to reduce antenna RCS out of band.

operating in two frequency bands. The main reflector is parabolic, with the subreflector being hyperbolic in shape, and is made of FSS, which operates in two modes. In the FSS pass-band frequency, the FSS subreflector performs in its full transmittance mode and enables operation of the antenna system as a parabolic primary feed antenna, while in the second frequency band the FSS operates as a full reflector transforming the main reflector into a Cassegrain system antenna. An additional application of FSS is as an absorber [2], in which an FSS made of resistive material is placed at a distance of approximately $\lambda/4$ from a perfect conductor and absorbs the plane wave incident upon the periodic structure, while in the enclosed antenna operating frequency the FSS radome is transparent. FSS structures can be also used as polarizers for different incident angles transforming linear polarization into circular polarization or rotating the incident linear polarization [2].

This chapter is organized as follows. Section 3.1 describes the scattering analysis of a planar FSS using its integral equation formulated with Floquet harmonics and numerically solved with MoM. Section 3.2 extends the scattering analysis to multilayer FSS structures. Section 3.3 describes metamaterials applications for radome's through an example of a radome, which allows transmission and reception of an enclosed antenna in the pass-band and absorbs the EM energy out of the pass-band.

3.1 Scattering Analysis of Planar FSS

There are basically two approaches to determine the scattering from a periodic structure like FSS: the mutual impedance approach described in detail in [2] and the plane wave expansion or the spectral approach presented in [4] and [5]. The mutual impedance approach is especially advantageous for finite size FSS, while the spectral approach fits well for infinite size cases. The spectral approach is more computational efficient and therefore adopted in this case. In the analysis presented, we follow the derivation presented in [4].

Fig. 3.2 shows the geometry of an FSS with conductive elements arranged in a rectangular grid with the unit cell dimensions $a \times b$.

Some typical elements of FSS periodic structures are shown in Fig. 3.3.

The time harmonic $\{e^{j\omega t}\}$ scattered electric and magnetic fields $(\underline{E}^s, \underline{H}^s)$ from the conductive elements of the FSS surface can be expressed in terms of the vector magnetic potential $\underline{A}$ defined by

Figure 3.2 Geometry of FSS structure.

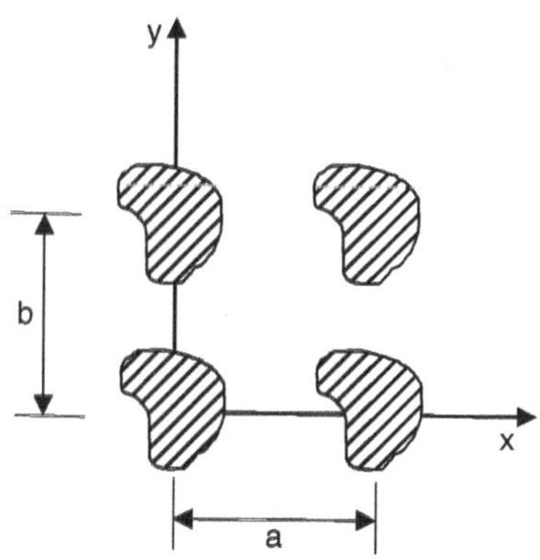

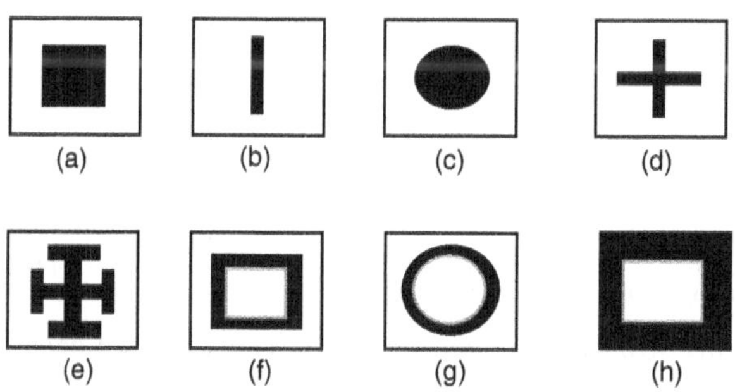

Figure 3.3 Some FSS unit cell geometries: (a) Square patch; (b) Dipole; (c) Circular patch; (d) Cross dipole; (e) Jerusalem cross; (f) Square loop; (g) Circular loop; (h) Square aperture.

$\underline{H}^s \triangleq \nabla \times \underline{A}$. If we introduce the scalar electric potential Φ and use the Lorentz gauge $\nabla \cdot \underline{A} = -j\omega\varepsilon_0\Phi$, the scattered electric field $\underline{E}^s$ can be expressed by

$$\underline{E}^s = \frac{1}{j\omega\varepsilon_0}[\nabla(\nabla \cdot \underline{A}) + k_0^2\underline{A}] \tag{3.1}$$

in which $k_0 = \omega\sqrt{\mu_0\varepsilon_0}$ and ε_0, μ_0 are the permittivity and permeability in free space. The induced electric currents $\underline{J}$ of the planar FSS and the vector magnetic potential $\underline{A}$ have only x and y components. These two vectors are related through the Green's function. Initially, we consider the case of a perfectly conducting FSS in free space positioned at $z = 0$. Later, the formulation will be modified to consider an FSS on a dielectric substrate and the Green's function will change correspondingly. To set the integral equation of the problem, we are only interested in the transverse components of the scattered electric fields at $z = 0$. Accordingly, the relationship between the transverse vector magnetic potential components and the transverse electric field components is

$$\begin{pmatrix} A_x(x,y) \\ A_y(x,y) \end{pmatrix} = \overline{\overline{G}}(x,y) * \begin{pmatrix} J_x(x,y) \\ J_y(x,y) \end{pmatrix} \tag{3.2}$$

in which $*$ denotes convolution [6]. $\overline{\overline{G}}(x,y) = \frac{e^{-jk_0r}}{4\pi r}\overline{\overline{I}}$ with $r = \sqrt{x^2+y^2}$ and $\overline{\overline{I}}$ is the identity tensor. Substitution of (3.2) into (3.1) yields

$$\begin{pmatrix} E_x^s \\ E_y^s \end{pmatrix} = \frac{1}{j\omega\varepsilon_0}\begin{pmatrix} \frac{\partial^2}{\partial x^2} + k_0^2 & \frac{\partial^2}{\partial x\partial y} \\ \frac{\partial^2}{\partial y\partial x} & \frac{\partial^2}{\partial y^2} + k_0^2 \end{pmatrix}\begin{pmatrix} A_x(x,y) \\ A_y(x,y) \end{pmatrix} \tag{3.3}$$

Next, we introduce the Fourier transform of the scattered fields,

$$\begin{aligned} \tilde{E}_{x,y}^s &= \int_{-\infty}^{\infty}\int_{-\infty}^{\infty} E_{x,y}^s e^{-jk_xx}e^{-jk_yy}\,dx\,dy \\ E_{x,y}^s &= \frac{1}{4\pi^2}\int_{-\infty}^{\infty}\int_{-\infty}^{\infty} \tilde{E}_{x,y}^s e^{jk_xx}e^{jk_yy}\,dk_x dk_y \end{aligned} \tag{3.4}$$

which transforms eq. (3.3) into

$$\begin{pmatrix} \tilde{E}_x^s \\ \tilde{E}_y^s \end{pmatrix} = \frac{1}{j\omega\varepsilon_0}\begin{pmatrix} k_0^2 - k_x^2 & -k_xk_y \\ -k_xk_y & k_0^2 - k_y^2 \end{pmatrix}\begin{pmatrix} \tilde{A}_x(k_x,k_y) \\ \tilde{A}_y(k_x,k_y) \end{pmatrix} \tag{3.5}$$

The Fourier transform of the Green's function can be found based on [6] as

$$\Im\left\{\frac{e^{-jk_0r}}{4\pi r}\right\} = \frac{1}{2j}\frac{e^{-jz\sqrt{k_0^2-k_x^2-k_y^2}}}{\sqrt{k_0^2-k_x^2-k_y^2}}; \quad r = \sqrt{x^2+y^2+z^2} \tag{3.6}$$

Accordingly, substitution of Fourier transform of eqs. (3.2) and (3.6) at $z = 0$ into (3.5) yields

$$\begin{pmatrix} \tilde{E}_x^s \\ \tilde{E}_y^s \end{pmatrix} = \frac{1}{j\omega\varepsilon_0} \begin{pmatrix} k_0^2 - k_x^2 & -k_x k_y \\ -k_x k_y & k_0^2 - k_y^2 \end{pmatrix} \tilde{\bar{\bar{G}}}(k_x, k_y) \begin{pmatrix} \tilde{J}_x(k_x, k_y) \\ \tilde{J}_y(k_x, k_y) \end{pmatrix} \tag{3.7}$$

in which $\tilde{\bar{\bar{G}}}(k_x, k_y) = \frac{1}{2j} \frac{1}{\sqrt{k_0^2 - k_x^2 - k_y^2}} \bar{\bar{I}}$. Next, we make the assumption that the FSS is periodic in two dimensions and only consider the fields in a unit cell $a \times b$, as shown in Fig. 3.2. The fields can be described as a superposition of Floquet harmonics $\Psi_{mn}(x, y) = \frac{1}{\sqrt{a \cdot b}} e^{j(k_{xm}x + k_{yn}y)}$ with discrete propagation coefficients k_{xm} and k_{yn}. The explicit expressions for these propagation coefficients are

$$\begin{aligned} k_{xm} &= k_0 \sin\theta^i \cos\phi^i + \frac{2\pi m}{a} = k_x^i + \frac{2\pi m}{a} \\ k_{yn} &= k_0 \sin\theta^i \sin\phi^i + \frac{2\pi n}{b} = k_y^i + \frac{2\pi n}{b} \end{aligned} \tag{3.8}$$

where θ^i and ϕ^i are the propagation angles of the incident plane wave (E_x^i, E_y^i) and m,n are integers. Enforcing the boundary condition that the total tangential electric field (incident and scattered) on the FSS PEC surface at $z = 0$ is equal to zero and expressing the scattered field in terms of Floquet modes results into

$$\begin{aligned} \begin{pmatrix} E_x^i(x, y) \\ E_y^i(x, y) \end{pmatrix} &= -\frac{1}{j\omega\varepsilon_0} \sum_m \sum_n \begin{pmatrix} k_0^2 - k_{xm}^2 & -k_{xm} k_{yn} \\ -k_{xm} k_{yn} & k_0^2 - k_{yn}^2 \end{pmatrix} \tilde{\bar{\bar{G}}}(k_{xm}, k_{yn}) \begin{pmatrix} \tilde{J}_x(k_{xm}, k_{yn}) \\ \tilde{J}_y(k_{xm}, k_{yn}) \end{pmatrix} \Psi_{mn}(x, y) \\ &= \sum_m \sum_n \begin{pmatrix} \tilde{G}_{xx}^e & \tilde{G}_{xy}^e \\ \tilde{G}_{yx}^e & \tilde{G}_{yy}^e \end{pmatrix} \begin{pmatrix} \tilde{J}_x(k_{xm}, k_{yn}) \\ \tilde{J}_y(k_{xm}, k_{yn}) \end{pmatrix} \Psi_{mn}(x, y) \end{aligned} \tag{3.9}$$

in which

$$\begin{pmatrix} \tilde{G}_{xx}^e & \tilde{G}_{xy}^e \\ \tilde{G}_{yx}^e & \tilde{G}_{yy}^e \end{pmatrix} = -\frac{1}{j\omega\varepsilon_0} \begin{bmatrix} \dfrac{k_0^2 - k_{xm}^2}{2j\sqrt{k_0^2 - k_{xm}^2 - k_{yn}^2}} & \dfrac{-k_{xm} k_{yn}}{2j\sqrt{k_0^2 - k_{xm}^2 - k_{yn}^2}} \\ \dfrac{-k_{xm} k_{yn}}{2j\sqrt{k_0^2 - k_{xm}^2 - k_{yn}^2}} & \dfrac{k_0^2 - k_{yn}^2}{2j\sqrt{k_0^2 - k_{xm}^2 - k_{yn}^2}} \end{bmatrix} \tag{3.10}$$

and $\Psi_{mn}(x, y) = \frac{1}{\sqrt{a \cdot b}} e^{j(k_{xm}x + k_{yn}y)}$.

The unknown induced currents $\underline{J}$ on the FSS can be computed using the MoM as described in the following. If the FSS has finite conductivity, the problem needs to be formulated in terms of the surface currents on the conducting elements of the FSS.

In case of an FSS with aperture type elements, we can apply the concepts of duality to eq. (3.9) by replacing $\underline{E}$ with $\underline{H}$ and ε_0 with μ_0. Furthermore, invoking the imaging of the magnetic currents $\underline{J}_{m} = -\hat{z} \times \underline{E}^a$ into the PEC FSS, and realizing that the transverse scattered magnetic field is equal to the transverse incident field at $z = 0$, we obtain the following boundary condition for the FSS elements with the aperture field $\underline{E}^a$ as unknown:

$$\begin{pmatrix} H_x^i(x,y) \\ H_y^i(x,y) \end{pmatrix} = -\frac{2}{j\omega\mu_0} \sum_m \sum_n \begin{pmatrix} k_0^2 - k_{xm}^2 & -k_{xm}k_{yn} \\ -k_{xm}k_{yn} & k_0^2 - k_{yn}^2 \end{pmatrix}$$

$$\tilde{\tilde{G}}(k_{xm}, k_{yn}) \begin{pmatrix} \tilde{E}_y^a(k_{xm}, k_{yn}) \\ -\tilde{E}_x^a(k_{xm}, k_{yn}) \end{pmatrix} \Psi_{mn}(x,y)$$

$$= \sum_m \sum_n \begin{pmatrix} \tilde{G}_{xx}^h & \tilde{G}_{xy}^h \\ \tilde{G}_{yx}^h & \tilde{G}_{yy}^h \end{pmatrix} \begin{pmatrix} \tilde{E}_y^a(k_{xm}, k_{yn}) \\ -\tilde{E}_x^a(k_{xm}, k_{yn}) \end{pmatrix} \Psi_{mn}(x,y) \tag{3.11}$$

in which

$$\begin{pmatrix} \tilde{G}_{xx}^h & \tilde{G}_{xy}^h \\ \tilde{G}_{yx}^h & \tilde{G}_{yy}^h \end{pmatrix} = -\frac{2}{j\omega\mu_0} \begin{bmatrix} \dfrac{k_0^2 - k_{xm}^2}{2j\sqrt{k_0^2 - k_{xm}^2 - k_{yn}^2}} & \dfrac{-k_{xm}k_{yn}}{2j\sqrt{k_0^2 - k_{xm}^2 - k_{yn}^2}} \\ \dfrac{-k_{xm}k_{yn}}{2j\sqrt{k_0^2 - k_{xm}^2 - k_{yn}^2}} & \dfrac{k_0^2 - k_{yn}^2}{2j\sqrt{k_0^2 - k_{xm}^2 - k_{yn}^2}} \end{bmatrix} \tag{3.12}$$

In this case, the unknown aperture distribution fields of the FSS elements can be computed using the MoM in a similar fashion to the induced currents computation in the case of FSS conductive elements.

Fig. 3.4 shows a practical extension of the FSS in free space to an FSS embedded between a dielectric superstrate and substrate.

The dielectric constants of the superstrate and substrate are ε_1 and ε_2, and their thicknesses are t_1 and t_2, respectively.

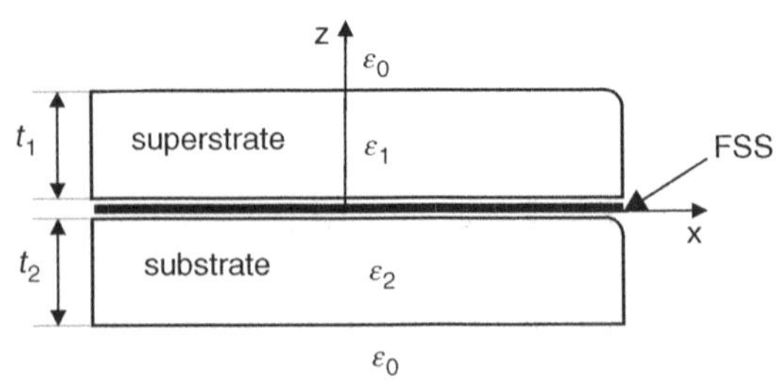

Figure 3.4 The FSS embedded between a superstrate and substrate dielectric media.

In this case, the Green's function changes and needs to be replaced with a new Green's function, which takes into consideration the superstrate and substrate of the FSS. To accomplish this task, it is convenient to consider a revised spectral Green's function using the spectral domain immitance approach as described in [7]. We begin by decomposing the scattered fields into longitudinal (e^s_{zmn}, h^s_{zmn}) and transverse ($\underline{e}^s_{tmn}$, $\underline{h}^s_{tmn}$) components:

$$\underline{E}^s(x,y,z) = \sum_m \sum_n a_{mn} \left(\underline{e}^s_{tmn}(x,y) + e^s_{zmn}(x,y)\hat{z}\right) e^{jk_{zmn}z}$$
$$\underline{H}^s(x,y,z) = \sum_m \sum_n a_{mn} \left(\underline{h}^s_{tmn}(x,y) + h^s_{zmn}(x,y)\hat{z}\right) e^{jk_{zmn}z} \tag{3.13}$$

in which a_{mn} are the weighting coefficients of the Floquet modes and $k_{zmn} = \sqrt{k_0^2\varepsilon_r - k_{xm}^2 - k_{yn}^2}$. Substituting (3.13) into the first Maxwell equation,

$$\nabla \times \underline{E} = -j\omega\mu\underline{H} \tag{3.14}$$

and if we define the operator $\nabla = \nabla_t + jk_{zmn}\hat{z}$ in which $\nabla_t = \frac{\partial}{\partial x}\hat{x} + \frac{\partial}{\partial y}\hat{y}$ results in two equations:

$$\nabla_t \times \underline{e}^s_{tmn} = -j\omega\mu h^s_{zmn}\hat{z}$$
$$\hat{z} \times \left(\nabla_t e^s_{zmn} + jk_{zmn}\underline{e}^s_{tmn}\right) = j\omega\mu\underline{h}^s_{tmn} \tag{3.15}$$

In a similar fashion, we obtain from the second Maxwell equation

$$\nabla \times \underline{H} = j\omega\varepsilon\underline{E} \tag{3.16}$$

the following equations:

$$\nabla_t \times \underline{h}^s_{tmn} = j\omega\varepsilon e^s_{zmn}\hat{z}$$
$$\hat{z} \times \left(\nabla_t h^s_{zmn} + jk_{zmn}\underline{h}^s_{tmn}\right) = -j\omega\varepsilon\underline{e}^s_{tmn} \tag{3.17}$$

Inspection of eqs. (3.13) and (3.15) reveals that in the transverse electric (TE) case ($e_{zmn} = 0$) and in the transverse magnetic (TM) case ($h_{zmn} = 0$), the ratio of the transverse electric and magnetic components is constant:

$$Z^h_{mn} = \frac{|\underline{e}^s_{tmn}|}{|\underline{h}^s_{tmn}|} = \frac{\omega\mu}{k_{zmn}}; \quad TE \ case$$
$$Z^e_{mn} = \frac{|\underline{e}^s_{tmn}|}{|\underline{h}^s_{tmn}|} = \frac{k_{zmn}}{\omega\varepsilon}; \quad TM \ case \tag{3.18}$$

This result implies that the propagation for each Floquet mode can be modeled as propagation along a transmission line with characteristics impedance Z^h_{mn} or Z^e_{mn} for the TE and TM cases, respectively, and

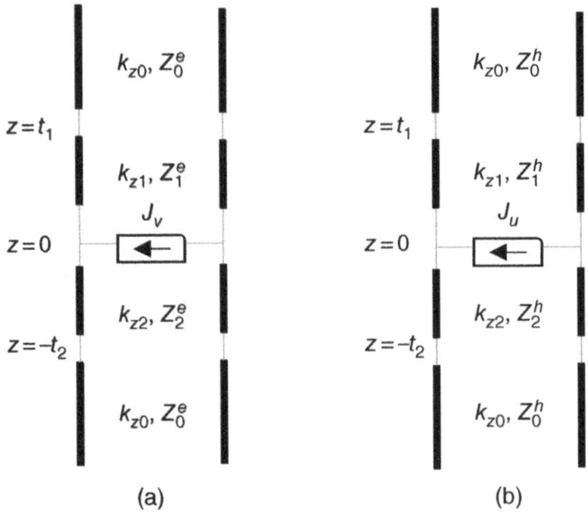

Figure 3.5 Equivalent circuit model of the FSS embedded between a superstrate and substrate dielectric media: (a) TM case; (b) TE case.

propagation constant k_{zmn}. Consequently, Fig. 3.5 shows the equivalent circuit of an elementary current element embedded between a substrate and a superstrate for the TM and TE cases.

Based on the transmission line analogy, the voltage wave on the transmission line describes the transverse electric field and the current wave describes the transverse magnetic field. As a first step in the evaluation of the Green's functions, which are the transverse electric fields at $z = 0$ due to an infinitesimal transverse current at $z = 0$, we compute the input impedance into the superstrate, $Z_{in}^{+e,h}$ and into the substrate $Z_{in}^{-e,h}$ (for simplicity we drop the index mn denoting each Floquet mode):

$$Z_{in}^{+e,h} = Z_1^{e,h} \frac{Z_0^{e,h} + jZ_1^{e,h} \tan k_{z1} t_1}{Z_1^{e,h} + jZ_0^{e,h} \tan k_{z1} t_1}$$

$$Z_{in}^{-e,h} = Z_2^{e,h} \frac{Z_0^{e,h} + jZ_2^{e,h} \tan k_{z2} t_2}{Z_2^{e,h} + jZ_0^{e,h} \tan k_{z2} t_2} \tag{3.19}$$

in which $Z_0^{e,h}$, $Z_1^{e,h}$ and $Z_2^{e,h}$ are the characteristics impedances of the free space, superstrate and substrate for TE and TM cases, respectively. The relative dielectric constants of the superstrate and substrate are ε_{r1} and ε_{r2} and their thicknesses are t_1 and t_2. The propagation constant

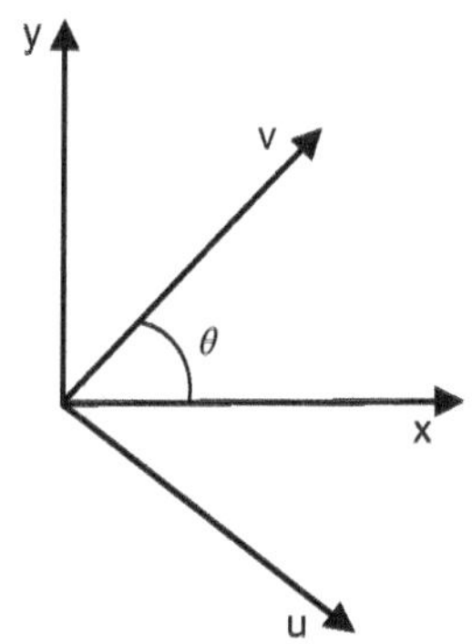

Figure 3.6 The coordinate system transformation.

is given by $k_{zi} = \sqrt{k_0^2 \varepsilon_{ri} - k_{xm}^2 - k_{yn}^2}$; $i = 1,2$. The voltage at $z = 0$ is equal to

$$\tilde{V}(z=0) = \tilde{i}_0 Z^{e,h}; \quad Z^{e,h} \triangleq \frac{Z_{in}^{+e,h} Z_{in}^{-e,h}}{Z_{in}^{+e,h} + Z_{in}^{-e,h}} = \frac{1}{Y_{in}^{-e,h} + Y_{in}^{+e,h}} \tag{3.20}$$

For solution simplicity, we introduce a coordinate system transformation as proposed in [7] and shown in Fig. 3.6 to realize current distributions, which generate TE and TM orthogonal modes.

In the new (u,v) coordinate system $\tilde{J}_v$ current creates only TM fields and $\tilde{J}_u$ current only TE fields. The coordinates (u,v) are related to (x,y) via

$$\begin{aligned} u &= x \sin\theta - y\cos\theta \\ v &= x\cos\theta + y\sin\theta \end{aligned} \tag{3.21}$$

Using (3.21), the TM and TE electric fields $\tilde{E}_v^s = Z^e \tilde{J}_v$ and $\tilde{E}_u^s = Z^h \tilde{J}_u$ can be mapped into the (x,y) coordinate system such that

$$\begin{aligned} \tilde{E}_x^s &= \tilde{E}_v^s \cos\theta - \tilde{E}_u^s \sin\theta = Z^e \tilde{J}_v \cos\theta - Z^h \tilde{J}_u \sin\theta \\ \tilde{E}_y^s &= \tilde{E}_v^s \sin\theta + \tilde{E}_u^s \cos\theta = Z^e \tilde{J}_v \sin\theta + Z^h \tilde{J}_u \cos\theta \end{aligned} \tag{3.22}$$

and

$$\begin{aligned} \tilde{J}_v &= \tilde{J}_x \cos\theta + \tilde{J}_y \sin\theta \\ \tilde{J}_u &= -\tilde{J}_x \sin\theta + \tilde{J}_y \cos\theta \end{aligned} \tag{3.23}$$

Substituting (3.23) into (3.22) yields

$$\begin{aligned} \tilde{E}_x^s &= \tilde{J}_x (Z^e \cos^2\theta + Z^h \sin^2\theta) + \tilde{J}_y \sin\theta\cos\theta (Z^e - Z^h) \\ \tilde{E}_y^s &= \tilde{J}_x \sin\theta\cos\theta (Z^e - Z^h) + \tilde{J}_y (Z^e \sin^2\theta + Z^h \cos^2\theta) \end{aligned} \tag{3.24}$$

Thus, the scattered fields E_x^s, E_y^s can be expressed in terms of Floquet modes $\Psi_{mn}(x,y) = \frac{1}{\sqrt{a \cdot b}} e^{j(k_{xm}x + k_{yn}y)}$ using (3.24) to obtain

$$
\begin{aligned}
E_x^s &= \sum_m \sum_n \Big[\left(Z_{mn}^e \cos^2\theta_{mn} + Z_{mn}^h \sin^2\theta_{mn}\right) \tilde{J}_x \\
&\quad + \left(Z_{mn}^e - Z_{mn}^h\right) \sin\theta_{mn} \cos\theta_{mn} \tilde{J}_y \Big] \Psi_{mn}(x,y) e^{\pm jk_{zmn}z} \\
E_y^s &= \sum_m \sum_n \Big[\left(Z_{mn}^e - Z_{mn}^h\right) \sin\theta_{mn} \cos\theta_{mn} \tilde{J}_x \\
&\quad + \left(Z_{mn}^e \sin^2\theta_{mn} + Z_{mn}^h \cos^2\theta_{mn}\right) \tilde{J}_y \Big] \Psi_{mn}(x,y) e^{\pm jk_{zmn}z}
\end{aligned}
\tag{3.25}
$$

Moreover, $\cos\theta_{mn} = \frac{k_{xm}}{\sqrt{k_{xm}^2 + k_{yn}^2}}$ and $\sin\theta_{mn} = \frac{k_{yn}}{\sqrt{k_{xm}^2 + k_{yn}^2}}$. Enforcing the boundary condition that the total tangential electric field (incident and scattered) on the FSS's PEC surface (z=0) is equal to zero, and using (3.25), yields

$$
\begin{pmatrix} E_x^i(x,y) \\ E_y^i(x,y) \end{pmatrix} = \sum_m \sum_n \begin{pmatrix} \tilde{G}_{xx}^e & \tilde{G}_{xy}^e \\ \tilde{G}_{yx}^e & \tilde{G}_{yy}^e \end{pmatrix} \begin{pmatrix} \tilde{J}_x(k_{xm}, k_{yn}) \\ \tilde{J}_y(k_{xm}, k_{yn}) \end{pmatrix} \Psi_{mn}(x,y) \tag{3.26}
$$

where

$$
\begin{pmatrix} \tilde{G}_{xx}^e & \tilde{G}_{xy}^e \\ \tilde{G}_{yx}^e & \tilde{G}_{yy}^e \end{pmatrix} = - \begin{bmatrix} Z_{mn}^e \cos^2\theta_{mn} + Z_{mn}^h \sin^2\theta_{mn} & (Z_{mn}^e - Z_{mn}^h) \sin\theta_{mn} \cos\theta_{mn} \\ (Z_{mn}^e - Z_{mn}^h) \sin\theta_{mn} \cos\theta_{mn} & Z_{mn}^e \sin^2\theta_{mn} + Z_{mn}^h \cos^2\theta_{mn} \end{bmatrix}
\tag{3.27}
$$

The next step in the solution of (3.26) is the determination of the incident transverse electric fields $E_x^i(x,y)$, $E_y^i(x,y)$ at $z = 0$ without the existence of the FSS conductive elements. This task can be accomplished using the derivation outlined in Chapter 2 for TM and TE cases. The *ABCD* matrix for this case (two dielectric layers) using (2.8) is given by

$$
\begin{bmatrix} A & B \\ C & D \end{bmatrix} = \begin{bmatrix} \cos\xi_1 & j\overline{Z}_1^{e,h} \sin\xi_1 \\ j\frac{1}{\overline{Z}_1^{e,h}} \sin\xi_1 & \cos\xi_1 \end{bmatrix} \begin{bmatrix} \cos\xi_2 & j\overline{Z}_2^{e,h} \sin\xi_2 \\ j\frac{1}{\overline{Z}_2^{e,h}} \sin\xi_2 & \cos\xi_2 \end{bmatrix} \tag{3.28}
$$

in which $\xi_i = k_{zi}t_i, \overline{Z}_i^h = \frac{k_{z0}}{k_{zi}}, \overline{Z}_i^e = \frac{k_{zi}}{\varepsilon_{ri}k_{z0}}$ (i=1,2). The transmission coefficient $T = \frac{2}{A+B+C+D}$ of the structure can be computed using (2.14) and the transverse fields (E_t, H_t) at $z = 0$ can be computed using (2.11) such that

$$
\begin{pmatrix} E_t \\ H_t \end{pmatrix}\Bigg|_{z=0} = \begin{pmatrix} \cos\xi_2 & j\overline{Z}_2^{e,h} \sin\xi_2 \\ j\frac{1}{\overline{Z}_2^{e,h}} \sin\xi_2 & \cos\xi_2 \end{pmatrix} \begin{pmatrix} E_t \\ H_t \end{pmatrix}\Bigg|_{z=-t_2} \tag{3.29}
$$

After some algebraic manipulations, one can show that the incident transverse electric and magnetic fields at the interface between the superstrate and substrate at $z = 0$ is given by [4]

$$\left.\begin{pmatrix} E_x^i \\ E_y^i \end{pmatrix}\right|_{z=0} = j(R_1 + R_2)e^{j(k_x^i x + k_y^i y)}\begin{pmatrix} -k_y^i \\ k_x^i \end{pmatrix}; \quad TE\ case \tag{3.30}$$

and

$$\left.\begin{pmatrix} E_x^i \\ E_y^i \end{pmatrix}\right|_{z=0} = \frac{jk_{z1}}{\omega\varepsilon_{r1}\varepsilon_0}(R_1 - R_2)e^{j(k_x^i x + k_y^i y)}\begin{pmatrix} k_x^i \\ k_y^i \end{pmatrix}; \quad TM\ case \tag{3.31}$$

where

$$R_1 = \frac{k_{z0}\left(e^{jk_{z0}t_1}/\sin(k_{z1}t_1)\right)\left[-\bar{k}_{z2}^2 - k_{z0}\bar{k}_{z1} + j(k_{z0} + \bar{k}_{z1})\bar{k}_{z2}\cot k_{z2}t_2\right]}{D}$$

$$R_2 = -\frac{k_{z0}\left(e^{jk_{z0}t_1}/\sin(k_{z1}t_1)\right)\left[-\bar{k}_{z2}^2 + k_{z0}\bar{k}_{z1} + j(k_{z0} - \bar{k}_{z1})\bar{k}_{z2}\cot k_{z2}t_2\right]}{D} \tag{3.32}$$

and

$$D = \left(jk_{z0} - \bar{k}_{z1}^2\right)\bar{k}_{z2}\cot(k_{z2}t_2) - \left(k_{z0}^2 + \bar{k}_{z2}^2\right)\bar{k}_{z1}\cot(k_{z1}t_1)$$
$$- jk_{z0}\left(\bar{k}_{z1}^2 + \bar{k}_{z2}^2\right) + 2jk_{z0}\bar{k}_{z1}\bar{k}_{z2}\cot(k_{z2}t_2)\cot(k_{z1}t_1)$$

$$\bar{k}_{zi} = \begin{cases} k_{zi} & TE\ case \\ k_{zi}/\varepsilon_{ri} & TM\ case \end{cases} \qquad i = 1,2 \tag{3.33}$$

Substitution of (3.30) and (3.31) into (3.26) for the incident fields yields the integral equation (3.26) to be solved for an FSS with substrates and superstrates. This integral equation is similar to (3.9) integral equation for the FSS in free space with modifications in the dyadic Green's functions and the incident fields. The solution of this integral equation can be done using MoM [8]. The basis functions can be entire domain functions or local functions. Usually, entire domain functions are more computationally efficient (smaller matrices to invert) but fit only canonical element shapes like rectangular and circular patches, dipoles, or slots and loops or aperture rings. In case of entire domain N basis functions $\underline{B}_j$'s the induced current can be expressed as

$$\underline{J} = \sum_{j=1}^{N} C_{xj}B_{xj}\hat{x} + C_{yj}B_{yj}\hat{y} \tag{3.34}$$

The Fourier transform of the basis function is denoted by $\underline{\tilde{B}}_j$. If we apply the Galerkin method [8] and choose the testing functions to be the same as the basis functions, apply on (3.9) scalar product with the complex conjugate testing function $\underline{B}_i^*$, and recognize that the scalar product on the right-hand side (RHS) of (3.9) results in the Fourier transform of the testing functions, we obtain

$$\begin{pmatrix} \int E_x^i(x,y)B_{xi}\,ds \\ \int E_y^i(x,y)B_{yi}\,ds \end{pmatrix} = \sum_j \sum_m \sum_n \begin{pmatrix} \tilde{B}_{xi}^* & 0 \\ 0 & \tilde{B}_{yi}^* \end{pmatrix} \begin{pmatrix} \tilde{G}_{xx}^e & \tilde{G}_{xy}^e \\ \tilde{G}_{yx}^e & \tilde{G}_{yy}^e \end{pmatrix}$$
$$\times \begin{pmatrix} \tilde{B}_{xj} & 0 \\ 0 & \tilde{B}_{yj} \end{pmatrix} \begin{pmatrix} C_{xj} \\ C_{yj} \end{pmatrix}; \quad i,j = 1..N \tag{3.35}$$

Eq. (3.35) is a matrix equation with the unknown coefficients vector $\begin{pmatrix} C_{xj} \\ C_{yj} \end{pmatrix}$. The numerical efficiency with which the elements of the matrix in (3.35) can be computed depends on the asymptotic behaviors of the Fourier transforms of the basis functions which, in turn, determine the number of terms needed for the convergence of the doubly infinite summations appearing in the expressions for the matrix elements. Local functions are more versatile and can fit more general element geometries. Moreover, the Fourier transforms of the local basis functions do not decay rapidly enough until large m and n in k_{xm} and k_{yn}. Therefore, more Floquet harmonic terms are needed for the double summation to converge. One way to accelerate the convergence is by using fast Fourier transform algorithm (FFT). In cases of large matrix sizes for inversion, we may resort to iterative techniques like conjugate gradient method (CGM), as described in [9, 10] and in Appendix D.

One of the most frequently used local basis functions is the rooftop [11, 12], which has a triangular or piecewise-linear dependence in the direction of the current and a pulse or stepwise-constant dependence in the orthogonal direction, as shown in Fig. 3.7. In this case, the unit

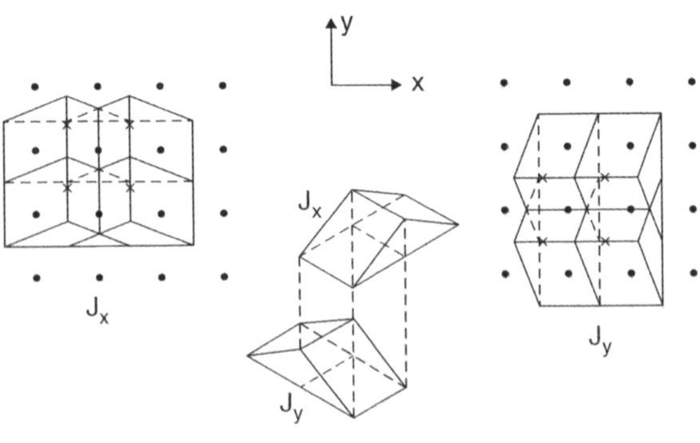

Figure 3.7 Rooftop basis functions.

cell is divided into an $N \times N$ grid and employ equal size rooftop basis functions.

The representations for the induced currents J_x and J_y take the form

$$J_x = \sum_{p=-N/2}^{N/2-1} \sum_{q=-N/2}^{N/2-1} C_x(p,q) B_x(p,q)$$

$$J_y = \sum_{p=-N/2}^{N/2-1} \sum_{q=-N/2}^{N/2-1} C_y(p,q) B_y(p,q) \tag{3.36}$$

The current basis functions are described by the following equations:

$$B_x(p,q) = \Lambda\left(p + \frac{1}{2}\right)\Pi(q)$$

$$B_y(p,q) = \Lambda(p)\Pi\left(q + \frac{1}{2}\right) \tag{3.37}$$

in which the rooftop functions are

$$\Pi(q) = \begin{cases} 1, & |y - q\Delta y| < \dfrac{\Delta y}{2} \\ 0, & elsewhere \end{cases} ; \quad q = 1 \ldots N$$

$$\Lambda(p) = \begin{cases} 1 - \dfrac{|x - p\Delta x|}{\Delta x}, & |x - p\Delta x| < \Delta x \\ 0 \quad , & elsewhere \end{cases} ; \quad p = 1 \cdots N \tag{3.38}$$

where $\Delta x = a/N$ and $\Delta y = b/N$ with a,b being the unit cell dimensions. Using Galerkin version of the MoM enables us to rewrite the MoM matrix equation after rearranging the order of summations in the form

$$-\begin{bmatrix} E_{x0}\tilde{B}_x^*(0,0)P^*\left(p + \frac{1}{2}, q\right) \\ E_{y0}\tilde{B}_y^*(0,0)P^*\left(p, q + \frac{1}{2}\right) \end{bmatrix} = \sum_{m=-N/2}^{N/2-1} \sum_{n=-N/2}^{N/2-1} \begin{bmatrix} \tilde{G}'_{xx}(m,n) & \tilde{G}'_{xy}(m,n) \\ \tilde{G}'_{yx}(m,n) & \tilde{G}'_{yy}(m,n) \end{bmatrix} \cdot$$

$$\sum_{q'=-N/2}^{N/2-1} \sum_{p'=-N/2}^{N/2-1} e^{-j[(k_{xm}p' - k_{xm}p)\Delta x + (k_{yn'}q' - k_{yn}q)\Delta y]} \begin{bmatrix} C_x(p',q') \\ C_y(p',q') \end{bmatrix} \tag{3.39}$$

in which,

$$P(p,q) = e^{-j(k_x^i p\Delta x + k_y^i q\Delta y)}$$

and

$$\tilde{G}'_{xx}(m,n) = \frac{1}{N^2} \sum_{r=-\infty}^{\infty} \sum_{s=-\infty}^{\infty} \tilde{G}^e_{xx}(m',n')\tilde{B}_x(m',n')\tilde{B}_x^*(m',n')$$

$$\tilde{G}'_{xy}(m,n) = \frac{1}{N^2} \sum_{r=-\infty}^{\infty} \sum_{s=-\infty}^{\infty} \tilde{G}^e_{xy}(m',n')\tilde{B}_y(m',n')\tilde{B}_x^*(m',n')$$

$$\times e^{j\left(k_{xm'}(\Delta x/2) - k_{yn'}(\Delta y/2)\right)}$$

$$\tilde{G}'_{yx}(m,n)=\frac{1}{N^2}\sum_{r=-\infty}^{\infty}\sum_{s=-\infty}^{\infty}\tilde{G}^{e}_{yx}(m',n')\tilde{B}_x(m',n')\tilde{B}^{*}_y(m',n')$$
$$\times e^{-j(k_{xm'}(\Delta x/2)-k_{yn'}(\Delta y/2))}$$
$$\tilde{G}'_{yy}(m,n)=\frac{1}{N^2}\sum_{r=-\infty}^{\infty}\sum_{s=-\infty}^{\infty}\tilde{G}^{e}_{yy}(m',n')\tilde{B}_y(m',n')\tilde{B}^{*}_y(m',n') \tag{3.40}$$

E_{x0} and E_{y0} are the amplitudes of the x and y components of the incident field at $z = 0$, respectively. When $N \times N$ FFT is used, $n' = n + sN$ and $m' = m + rN$; $-N/2 \le m, n \le N/2 - 1$. Following the MoM procedure, the unknown weight coefficients $C_x(p,q)$ and $C_y(p,q)$ can be obtained either by a direct solution of the matrix equation (3.39) or using an iterative procedure like the CGM in case of large matrix size as described in Appendix D. The more efficient way to evaluate the matrix elements is first to compute the double summations in (3.40) and then carry out the double summations in (3.39) via the use of FFT. The Fourier transforms of the testing and basis functions given in (3.37) are

$$\tilde{B}_x(m,n)=\left[\frac{\sin\left(\frac{k_{xm}\Delta x}{2}\right)}{\frac{k_{xm}\Delta x}{2}}\right]^2\frac{\sin\left(\frac{k_{yn}\Delta y}{2}\right)}{\frac{k_{yn}\Delta y}{2}}$$
$$\tilde{B}_y(m,n)=\frac{\sin\left(\frac{k_{xm}\Delta x}{2}\right)}{\frac{k_{xm}\Delta x}{2}}\left[\frac{\sin\left(\frac{k_{yn}\Delta y}{2}\right)}{\frac{k_{yn}\Delta y}{2}}\right]^2 \tag{3.41}$$

The doubly infinite summations in (3.40) need to be truncated to $N \times N$ if FFT procedures are used. In this case, it is seen that only $N \times N$ Floquet harmonics are retained in the doubly infinite summation when an $N \times N$ FFT is employed. This approximation is used and is valid only if the contributions of the remainder of the Floquet harmonics are negligible.

After determining the weighting coefficients of the basis functions, the induced currents can be expressed in terms of Floquet harmonics at $z = 0$:

$$\underline{J}(x,y)=\sum_{m}\sum_{n}\tilde{\underline{J}}_{mn}\Psi_{mn}(x,y) \tag{3.42}$$

in which

$$\Psi_{mn}(x,y)=\frac{1}{\sqrt{a\cdot b}}e^{j(k_{xm}x+k_{yn}y)}$$

$$k_{xm} = k_x^i + \frac{2\pi m}{a} = k \sin\theta^i \cos\phi^i + \frac{2\pi m}{a};$$

$$k_{yn} = k_y^i + \frac{2\pi n}{b} = k \sin\theta^i \sin\phi^i + \frac{2\pi n}{b} \tag{3.43}$$

and

$$\tilde{\underline{J}}_{mn} = \int_{\partial S} \underline{J}(x', y')\Psi_{mn}^{*}(x', y')dx'dy' \tag{3.44}$$

Because the FSS surface is planar, the fields can be divided into those TE and TM to $\hat{z}$—the normal of the surface, such that different z-directed vector potentials can represent the TM and TE cases [6]. These z-directed vector potentials can be represented in terms of superposition of Floquet modes:

$$\underline{A}_{TM}^{\pm}(x, y, z) = \hat{z}A_{TM}^{\pm} = \hat{z}\sum_m \sum_n a_{TM,mn}^{\pm}\Psi_{mn}(x, y)e^{\pm jk_{zmn}z}$$

$$\underline{F}_{TE}^{\pm}(x, y, z) = \hat{z}F_{TE}^{\pm} = \hat{z}\sum_m \sum_n a_{TE,mn}^{\pm}\Psi_{mn}(x, y)e^{\pm jk_{zmn}z} \tag{3.45}$$

In the TM case, the magnetic vector potential can be represented by its z-component $A_{TM}^{\pm}$ and in the TE case, the electric vector potential can be represented by its z-component $F_{TE}^{\pm}$, such that the scattered electric and magnetic fields can be described by

$$\underline{E}^s = \underline{E}_{TM}^s + \underline{E}_{TE}^s = \frac{1}{j\omega\varepsilon}\left(\nabla\nabla + k^2\bar{\bar{I}}\right)\underline{A}_{TM} - \nabla \times \underline{F}_{TE}$$

$$\underline{H}^s = \underline{H}_{TM}^s + \underline{H}_{TE}^s = \nabla \times \underline{A}_{TM} + \frac{1}{j\omega\mu}\left(\nabla\nabla + k^2\bar{\bar{I}}\right)\underline{F}_{TE} \tag{3.46}$$

Since the TM field is uniquely determined by the z-component of the electric field, the TM potential coefficients $a_{TMmn}^{\pm}$ can be found by relating the x- and y-components of the electric field described in (3.46), assuming $\underline{F}_{TE} = 0$ [6]

$$E_x^s = \frac{1}{j\omega\varepsilon}\frac{\partial^2 A_{TM}}{\partial x \partial z} = \frac{1}{j\omega\varepsilon}\sum_m \sum_n a_{TMmn}^{+}k_{xm}k_{zmn}\Psi_{mn}(x, y)e^{-jk_{zmn}z}$$

$$E_y^s = \frac{1}{j\omega\varepsilon}\frac{\partial^2 A_{TM}}{\partial y \partial z} = \frac{1}{j\omega\varepsilon}\sum_m \sum_n a_{TMmn}^{+}k_{yn}k_{zmn}\Psi_{mn}(x, y)e^{-jk_{zmn}z} \tag{3.47}$$

to the x and y components of the TM electric fields described in (3.25):

$$E_x^s = \sum_m \sum_n \left[Z_{mn}^e \cos^2\theta_{mn}\tilde{J}_{xmn} + Z_{mn}^e \sin\theta_{mn}\cos\theta_{mn}\tilde{J}_{ymn}\right]\Psi_{mn}(x, y)e^{-jk_{zmn}z}$$

$$E_y^s = \sum_m \sum_n \left[Z_{mn}^e \sin\theta_{mn}\cos\theta_{mn}\tilde{J}_{xmn} + Z_{mn}^e \sin^2\theta_{mn}\tilde{J}_{ymn}\right]\Psi_{mn}(x, y)e^{-jk_{zmn}z} \tag{3.48}$$

The comparison considers the total field at $z = 0$ and the result is

$$a^+_{TM,mn} = \frac{-1}{2j(k^2_{xm} + k^2_{yn})}(k_{xm}\tilde{J}_{xmn} + k_{yn}\tilde{J}_{ymn})$$
$$a^-_{TM,mn} = -a^+_{TM,mn} \tag{3.49}$$

Similar derivation can be performed for the TE case, assuming $\underline{A}_{TM} = 0$. In this case, the electric field x- and y-components can be written in the form

$$E^s_x = -\frac{\partial F_{TE}}{\partial y} = -\sum_m \sum_n a^+_{TEmn} jk_{yn} \Psi_{mn}(x, y)e^{-jk_{zmn}z}$$
$$E^s_y = \frac{\partial F_{TE}}{\partial x} = \sum_m \sum_n a^+_{TEmn} jk_{xm} \Psi_{mn}(x, y)e^{-jk_{zmn}z} \tag{3.50}$$

and the x- and y-field components based on (3.25) are

$$E^s_x = \sum_m \sum_n \left[Z^h_{mn}\sin^2\theta_{mn}\tilde{J}_{xmn} - Z^h_{mn}\sin\theta_{mn}\cos\theta_{mn}\tilde{J}_{ymn}\right]\Psi_{mn}(x, y)e^{-jk_{zmn}z}$$
$$E^s_y = \sum_m \sum_n \left[-Z^h_{mn}\sin\theta_{mn}\cos\theta_{mn}\tilde{J}_{xmn} + Z^e_{mn}\cos^2\theta_{mn}\tilde{J}_{ymn}\right]\Psi_{mn}(x, y)e^{-jk_{zmn}z} \tag{3.51}$$

Comparison of the total field at $z = 0$ using (3.50) and (3.51) results in

$$a^+_{TE,mn} = \frac{-j\omega\mu}{2k_{zmn}(k^2_{xm} + k^2_{yn})}(k_{yn}\tilde{J}_{xmn} - k_{xm}\tilde{J}_{ymn})$$
$$a^-_{TE,mn} = a^+_{TE,mn} \tag{3.52}$$

With the coefficients found for the z-directed TE and TM vector potentials, the decomposition is complete and the scattered fields may be found from (3.47). The specular reflection and transmission of the co-polarized fields can be computed based on the $m,n = 0,0$ harmonics, i.e.,

$$T^{p,00}_{p,00} = 1 + \frac{a^+_{p,00}}{a^i_{p,00}}; \quad R^{p,00}_{p,00} = \frac{a^-_{p,00}}{a^i_{p,00}} \tag{3.53}$$

in which p denotes the polarization TE or TM, the subscripts indicate the incident field and superscripts the scattered field. Similarly, the cross-polarized coefficients can be computed by

$$T^{TM,00}_{TE,00} = \frac{a^+_{TM,00}}{a^i_{TE,00}}\left(\frac{\mu}{\varepsilon}\right)^{1/2}; \quad T^{TE,00}_{TM,00} = \frac{a^+_{TE,00}}{a^i_{TM,00}}\left(\frac{\varepsilon}{\mu}\right)^{1/2}$$
$$R^{TM,00}_{TE,00} = \frac{a^-_{TM,00}}{a^i_{TE,00}}\left(\frac{\mu}{\varepsilon}\right)^{1/2}; \quad R^{TE,00}_{TM,00} = \frac{a^-_{TE,00}}{a^i_{TM,00}}\left(\frac{\varepsilon}{\mu}\right)^{1/2} \tag{3.54}$$

An alternative approach to the spectral analysis used in the MoM procedure with the basis functions described in (3.36) is the MoM boundary integral (BI) combined with the resonant mode expansion (RME) [13] . In this method the basis functions are more computational efficient since they are derived based on the current distributions at the eigenvalues of the structure.

In the following, all numerical simulations were performed with HFSS commercial software from ANSYS. Fig. 3.8 shows the normal incidence reflection and transmission of a square patch array printed on a dielectric slab with dielectric constant $\varepsilon_r = 3.8$, $\tan\delta = 0.015$ and thickness, h. The patch dimensions are $L_x = L_y = 10$ mm and the unit cell size $a = b = 20$ mm.

The effect of the dielectric layer is seen to decrease the resonant frequency of the FSS if h increases. Fig. 3.9 shows the reflection and transmission of a circular patch array printed on a dielectric slab with thickness $h = 2$ mm and dielectric constant $\varepsilon_r = 3.8$, $\tan\delta = 0.015$ at $f = 10.4$ GHz as a function of the angle of incidence and two orthogonal polarizations TM and TE. The unit cell dimensions $a = b = 20$ mm and the patch radius $R = 6.25$ mm.

At the resonance frequency of the FSS 10.4 GHz, one can observe that for TM polarization, the reflection is maximum and is stable up to 24 deg, while the reflection of the TE polarization is sensitive to the incident angle and can be considered as full reflector only few degrees from normal incidence. In addition, it is seen that higher-order Floquet modes occur when $\theta^i > 26$ deg. Fig. 3.10 shows the performance of an FSS comprised of cross-shaped patches in a square grid. The FSS is illuminated by a uniform plane wave incident from the normal direction. Different values of the dielectric constant ε_r of the substrate were considered in the analysis.

The simulations were performed with HFSS. The equivalent circuit of the FSS shown in Fig. 3.10 includes a serial LC resonant circuit such that at the resonance the FSS performs as a perfect reflector. Increase in the dielectric constant of the slab reduces the full reflection frequency. Fig. 3.11 exemplifies the complementary FSS geometry of the crossed dipoles shown in Fig. 3.10, i.e., an FSS with the Jerusalem cross slot in its unit cell.

The equivalent circuit in this case is a parallel LC resonant circuit such that at resonance the FSS performs as a perfect transparent.

A major drawback of FSSs previously discussed is their narrow frequency bandwidth, which is a direct result from being a resonant

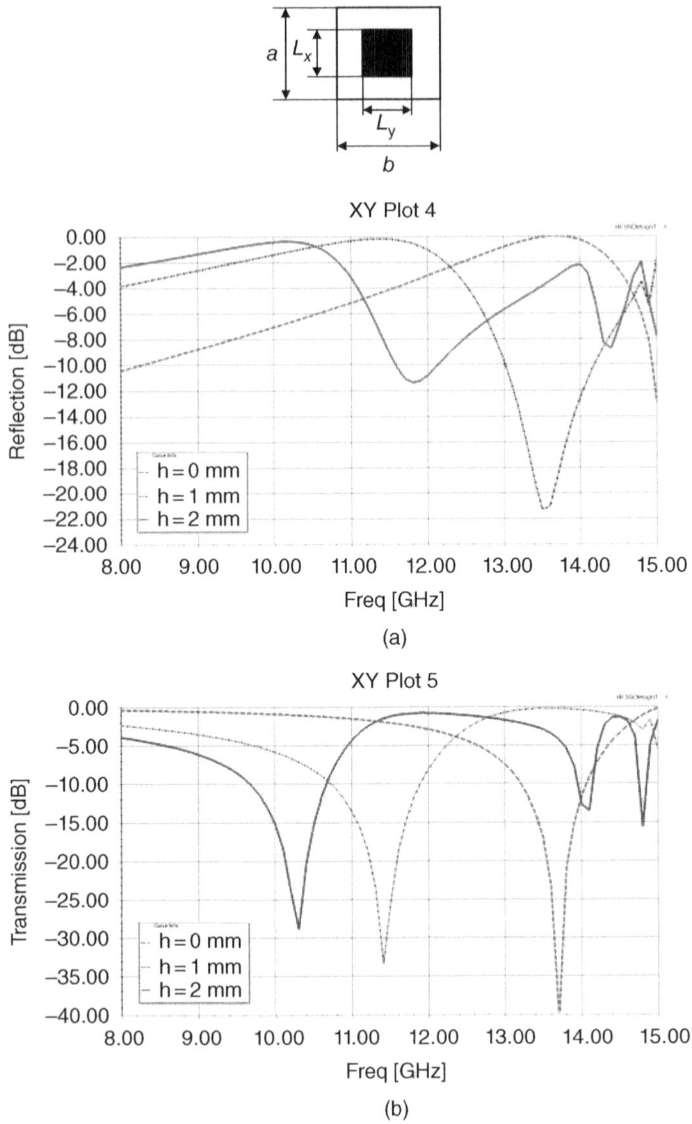

Figure 3.8 Normal incidence: (a) reflection [dB]; and (b) transmission [dB] vs. frequency for an array of square patches printed on dielectric slabs with various thicknesses, h; $a = b = 20$ mm, patch dimensions $L_x = L_y = 10$ mm, $\varepsilon_r = 3.8$, $\tan\delta = 0.015$; all simulations performed with HFSS.

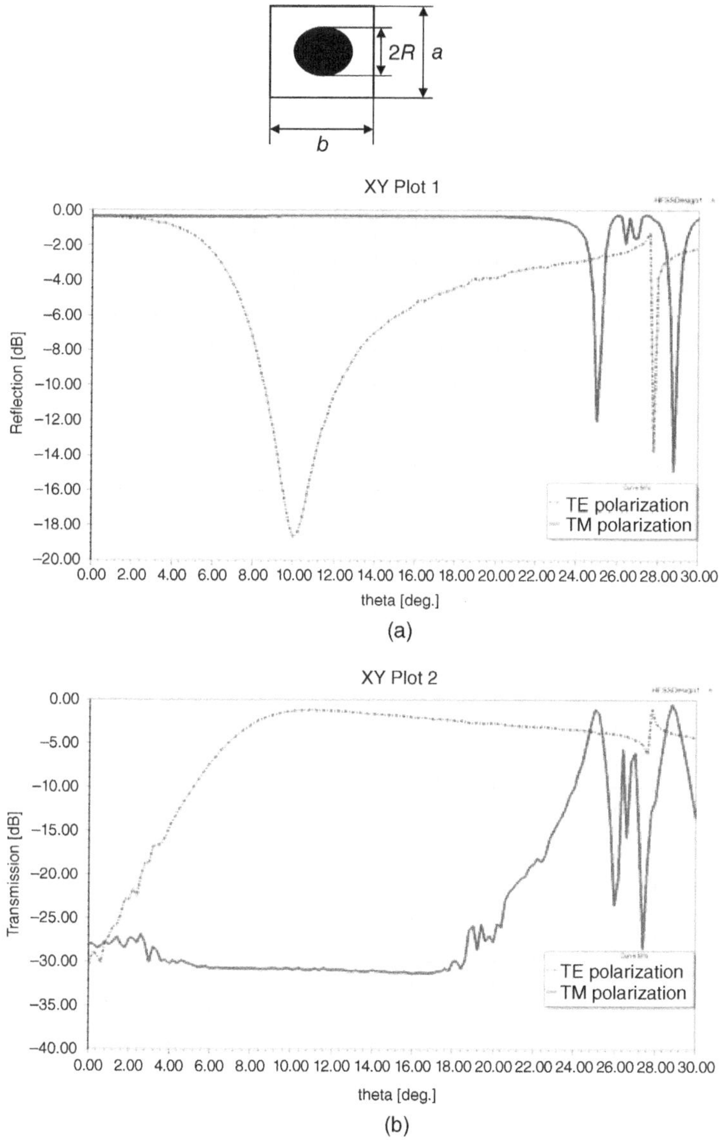

Figure 3.9 (a) reflection and (b) transmission vs. incident angle θ^i for an array of circular patches; unit cell dimensions $a = b = 20$ mm, patch radius $R = 6.25$ mm, printed on a dielectric slab with thickness $h = 2$ mm and dielectric constant $\varepsilon_r = 3.8$, $\tan\delta = 0.015$ at $f = 10.4$ GHz. All simulations performed with HFSS.

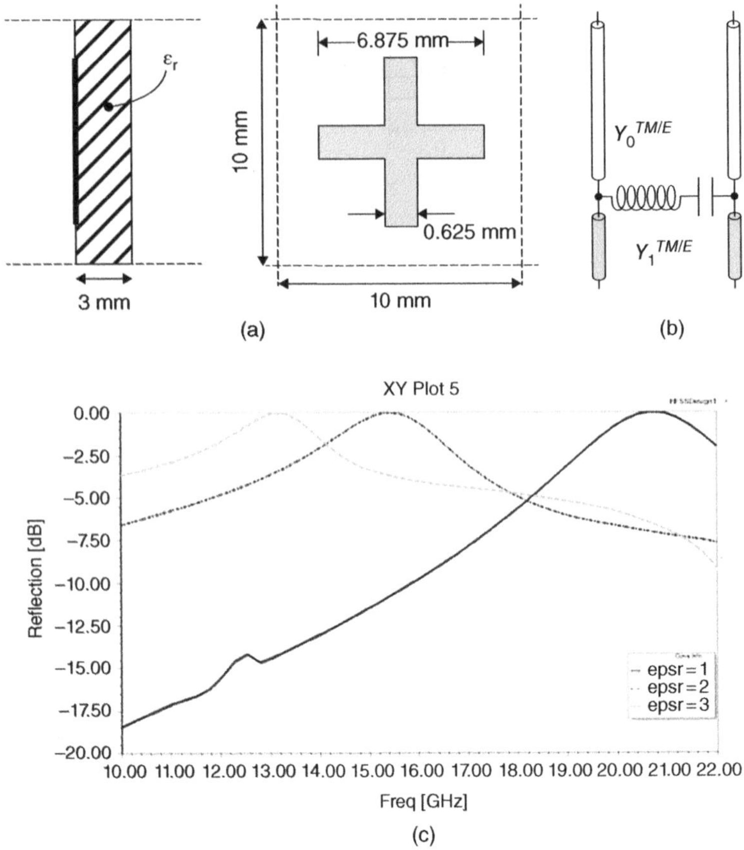

Figure 3.10 FSS made of a crossed dipole on a thin dielectric substrate: (a) side and front view of the unit cell; (b) equivalent circuit of the FSS; and (c) reflection performance for normal incidence and different dielectric constants of the substrate based on HFSS simulations.

structure. A possible way to widen their frequency bandwidth is to couple two resonant circuits, and through their coupling the frequency bandwidth is extended. This concept is implemented in the FSS shown in Fig. 3.12, in which two loops operating at two adjacent resonant frequencies are coupled in the FSS unit cell. The result is an extended frequency bandwidth, as demonstrated in the transmission as a function of frequency plot shown in Fig. 3.12, with the inner loop dimension *d* being the analysis study parameter.

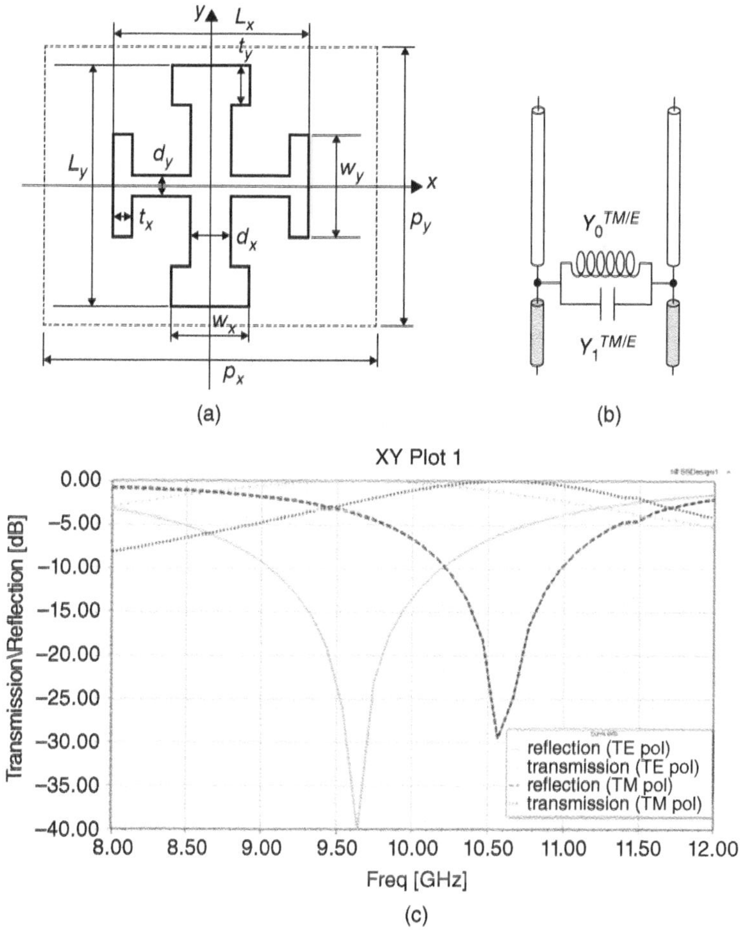

Figure 3.11 FSS made of a Jerusalem cross slot: (a) front view of the unit cell with parameters $L_x = 9.75$ mm, $L_y = 12.28$ mm, $W_x = 3.9$ mm, $W_y = 5.11$ mm, $d_x = 1.95$ mm, $d_y = 1.02$ mm, $t_x = 0.975$ mm, $t_y = 2.04$ mm, $p_x = 16.57$ mm, $p_y = 14.33$ mm; (b) equivalent circuit of the FSS; and (c) transmission/reflection performance as a function of frequency for TM and TE polarizations and normal incidence based on HFSS simulations.

An alternative design to widen the FSS frequency bandwidth is to embed a conductive FSS with square holes 13 × 13 mm and unit cell dimensions 14 × 14 mm in between two dielectric slabs, as demonstrated in Fig. 3.13.

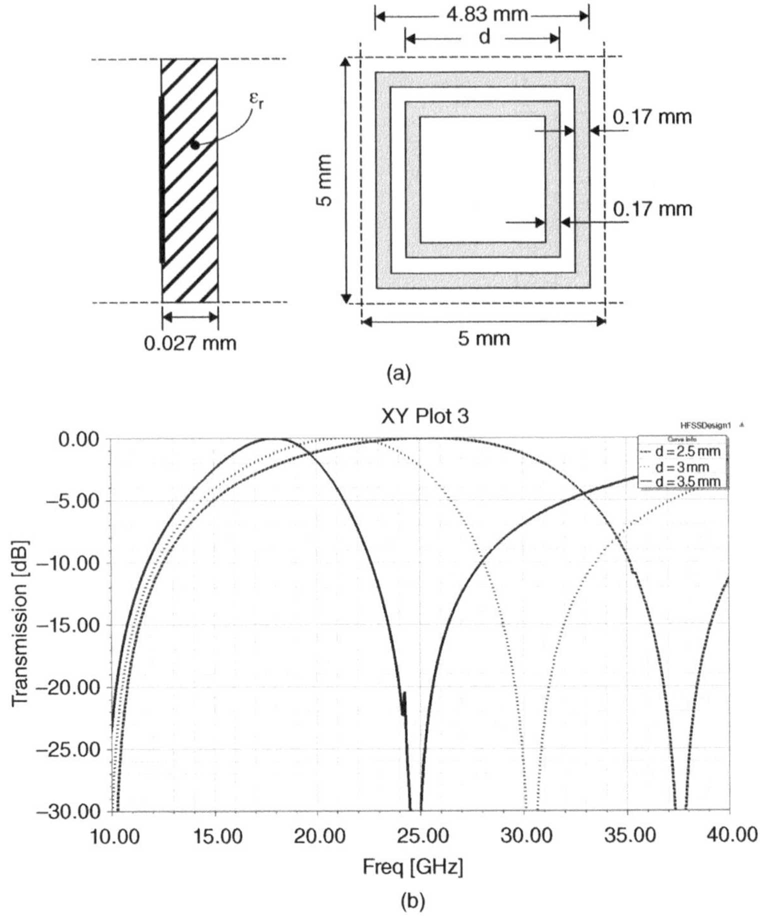

Figure 3.12 FSS made of two square loops on a thin dielectric substrate with ε_r = 3.5: (a) side and front view of the unit cell; and (b) transmission performance at normal incidence for different dimensions *d* of the inner loop based on HFSS simulations.

In Fig. 3.13a, we can see the transmission as a function of frequency of a dielectric slab with thickness d = 7 mm, dielectric constant ε_r = 4.6, tan δ = 0.0015, and different incident angles for TM and TE polarizations. At 10GHz, the slab thickness is $\lambda_g/2$, and accordingly, its transmittance is maximal. One can observe that the optimal frequency increases as the incident angle increases. In Fig. 3.13b, the transmittance of a slotted FSS with optimal frequency at 10 GHz

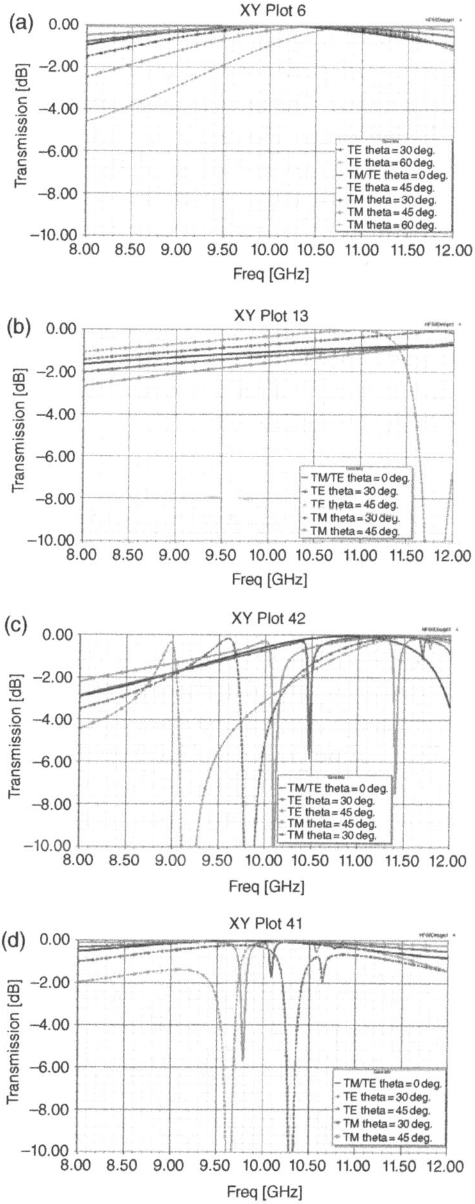

Figure 3.13 Transmission vs. frequency for TM polarization (solid lines) and TE polarization (dashed lines): (a) a dielectric $\lambda_g/2$ radome panel; (b) a slotted FSS surface made of copper with conductivity $\sigma = 5.8 \times 10^7$ S/m; (c) a combination of the FSS on top of the dielectric $\lambda_g/2$ radome panel; (d) the FSS is positioned in the center of the dielectric $\lambda_g/2$ radome panel. All results are based on HFSS simulations.

is plotted. In this scenario, in contrast to the dielectric slab case the optimal frequency decreases as the incident angle increases. In Fig. 3.13c, the FSS and the dielectric slab are combined with the FSS placed on top. The nulls in the transmission plot indicate an excitation of higher-order Floquet modes in the dielectric medium, which absorbs some of the lower modes power. Finally, in Fig. 3.13d, we can notice that if the FSS is placed at the center of the dielectric slab, the optimal transmittance is stable with the angle of incidence, and we can also notice a significant increase in the frequency bandwidth compared to previous cases. In a similar fashion to the previous case, the nulls in the transmission indicate on excitation of higher-order Floquet modes in the dielectric medium. This result indicates that hybridization of a dielectric slab with an FSS dramatically improves the radome performance.

3.2 Scattering Analysis of Multilayer FSS Structures

This section describes the scattering analysis methodology for multiple screens geometry arranged in cascade. The material in this section is based primarily on the papers [4] and [14]. The general configuration to which the cascade analysis is applicable is shown in Fig. 3.14.

The method is also applicable to the problem of FSS screens embedded in a multilayer dielectric medium discussed in section 3.1. Using this method, the scattering parameters of periodic surfaces and dielectric layers may be cascaded, thus reducing the formulation from one

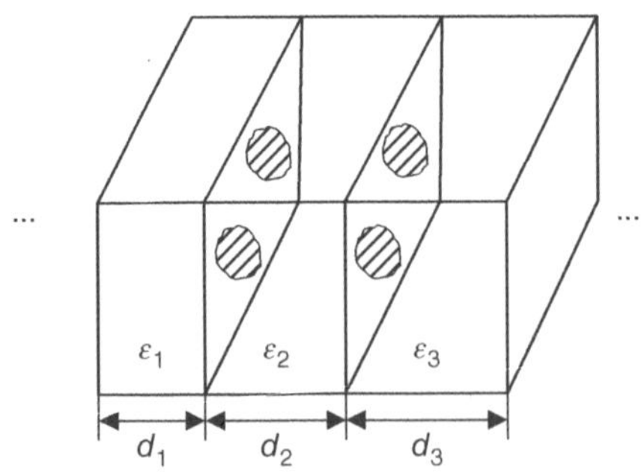

Figure 3.14 The general multilayer periodic geometry.

that treats the entire structure to one that treats each element separately. This allows a simplification in the analysis as long as the number of harmonics calculated and included in the cascade connection is large enough to give an accurate solution. The use of the Floquet modes to form the scattering parameters used in the cascade connection requires the calculation of the fields scattered from the periodic surface for multiple incident modes. The scattering parameters of the surface are then stored to be used with any combination of dielectric layers or other periodic surfaces. This is in contrast to the formulation that analyzes the screen in its entirety, where it is necessary to consider only a single incident mode. Therefore, the cascade formulation gives the flexibility of treating the general case and it does so at the expense of calculating the full set of surface-scattering parameters. The scattering parameters of each periodic structure in the cascade assembly are found from the scattered Floquet modes for a free-standing periodic surface, as described in (3.47) and (3.50) for TE and TM polarizations.

The electric and magnetic fields, exhibiting periodicities equal to that of the periodic surface, are represented by an infinite superposition of vector Floquet modes weighted by the complex amplitude sequence $\{a^{\pm}_{pmn}\}$,

$$\underline{E}(x,y,z) = \sum_{pmn} a^{\pm}_{pmn}\left[\underline{e}_{t_{pmn}}(x,y) + \hat{z}e_{z_{pmn}}(x,y)\right] e^{\pm jk_{zmn}z}$$
$$\underline{H}(x,y,z) = \sum_{pmn} a^{\pm}_{pmn}\left[\underline{h}_{t_{pmn}}(x,y) + \hat{z}h_{z_{pmn}}(x,y)\right] e^{\pm jk_{zmn}z} \tag{3.55}$$

where the fields are propagating in the $\pm\hat{z}$ direction. The summation is over an infinite number of vector modes ($-\infty < m, n < \infty$) and over both polarization states (p = TE and TM). The vector components of (3.55) transverse to boundaries are $\underline{e}_t$ and $\underline{h}_t$ with e_z and h_z being the components normal to boundaries. Taking the transverse component of the scalar Floquet modes $\Psi_{mn}(x,y)$ described by (3.42) as the $\hat{z}$-directed potential, the vector modes are found as

$$\underline{e}_{t_{TE,mn}}(x,y) = -\nabla \times \hat{z}\Psi_{mn}(x,y)$$
$$\underline{h}_{t_{TM,mn}}(x,y) = \nabla \times \hat{z}\Psi_{mn}(x,y) \tag{3.56}$$

where the modes are divided into those TE and TM to $\hat{z}$. Performance of the operation in (3.56) results in

$$\underline{e}_{t_{TE,mn}}(x,y) = -j(k_{yn}\hat{x} - k_{xm}\hat{y})\Psi_{mn}(x,y)$$
$$\underline{h}_{t_{TM,mn}}(x,y) = j(k_{yn}\hat{x} - k_{xm}\hat{y})\Psi_{mn}(x,y) \tag{3.57}$$

Substitution of (3.56) in Maxwell's eqs. (3.14) and (3.16) results in the relations

$$\hat{z} \times \underline{e}_{t_{TE,mn}} = \pm Z^h_{mn} \underline{h}_{t_{TE,mn}}; \quad Z^h_{mn} = \frac{\omega\mu}{k_{zmn}}$$
$$\underline{h}_{t_{TM,mn}} \times \hat{z} = \pm \frac{1}{Z^e_{mn}} \underline{e}_{t_{TM,mn}}; \quad Z^e_{mn} = \frac{k_{zmn}}{\omega\varepsilon} \tag{3.58}$$

in which $Z^{e,h}_{mn}$ are the modes characteristic impedances and the transverse fields are propagating in the $\pm\hat{z}$ direction. The $\hat{z}$ -components are similarly found:

$$h_{z_{TE,mn}}(x,y) = \frac{1}{j\omega\mu}\left(k^2 - k^2_{zmn}\right)\Psi_{mn}(x,y)$$
$$e_{z_{TM,mn}}(x,y) = \frac{1}{j\omega\varepsilon}\left(k^2 - k^2_{zmn}\right)\Psi_{mn}(x,y) \tag{3.59}$$

The vector modes constitute an orthogonal set of functions such that for the TE modes:

$$\int_{\partial S}\left(\underline{e}_{t_{TEmn}} \times \underline{h}^*_{t_{TEm'n'}}\right)\cdot(\pm\hat{z})ds = \frac{1}{Z^{h*}_{mn}}\int_{\partial S}\left(\underline{e}_{t_{TEmn}} \times \left(\hat{z}\times\underline{e}^*_{t_{TEm'n'}}\right)\right)\cdot(\pm\hat{z})ds$$
$$= \begin{cases} \frac{1}{Z^{h*}_{mn}}\left\|\underline{e}_{t_{TEmn}}\right\|^2; & mn = m'n' \\ 0 & ; \ mn \neq m'n' \end{cases} \tag{3.60}$$

It should be noted that because the originating medium is lossless, the transverse components of the vector modes are real throughout the dielectric structure. Similarly, for the TM modes:

$$\int_{\partial S}\left(\underline{e}_{t_{TMmn}} \times \underline{h}^*_{t_{TMm'n'}}\right)\cdot(\pm\hat{z})ds = \begin{cases} Z^{e*}_{mn}\left\|\underline{h}_{t_{TMmn}}\right\|^2; & mn = m'n' \\ 0 & ; \ mn \neq m'n' \end{cases} \tag{3.61}$$

Central to the cascade connection of dielectric layers and periodic surfaces is the calculation of the normalized Floquet voltage waves. As an extension of the Floquet vector modes defined above, these traveling waves are identical to the traveling waves defined for the guided modes used in microwave circuit theory. The normalized voltage waves reduce the formulation from one that considers the vector field quantities (Floquet vector modes) to one that considers only a scalar quantity with known traveling wave dependence and with amplitudes given by an element of the sequence $\{a^{\pm}_{pmn}\}$. The normalized Floquet voltage waves at a plane z are

$$V^{\pm}_{pmn}(z) = a^{\pm}_{pmn}\left[\int_{\partial S}\left(\underline{e}_{t_{pmn}} \times h^*_{t_{pmn}}\right)\cdot(\pm\hat{z})ds\right]^{1/2} e^{\pm jk_{zmn}z} \tag{3.62}$$

for fields propagating in the $\pm\hat{z}$ direction. The complex amplitude coefficient $a^{\pm}_{pmn}$ is not within the square root so as to preserve the phase of the wave associated with each mode. Using the orthogonality relations in (3.60) and (3.61), the normalized voltage wave for each mode simplifies to

$$V^{\pm}_{TEmn}(z) = a^{\pm}_{TEmn}\frac{1}{(Z^{h^*}_{mn})^{1/2}}\left\|\underline{e}_{tTEmn}\right\| e^{\pm jk_{zmn}z}$$

$$V^{\pm}_{TMmn}(z) = a^{\pm}_{TMmn}(Z^{e^*}_{mn})^{1/2}\left\|\underline{h}_{tTMmn}\right\| e^{\pm jk_{zmn}z} \tag{3.63}$$

From (3.57), one can realize that

$$\left\|\underline{e}_{tTEmn}\right\| = \left\|\underline{h}_{tTMmn}\right\| = \left(k^2_{xm} + k^2_{yn}\right)^{1/2} \tag{3.64}$$

Fig. 3.15 shows a sketch of the terminal plane defining the normalized voltage waves.

The scattering parameters that will characterize a terminal plane are defined in the usual manner:

$$\mathbf{V}^s = \mathbf{S}\mathbf{V}^I \tag{3.65}$$

in which $\mathbf{V}^s$ and $\mathbf{V}^I$ denote the unknown scattered (S) and known incident (I) normalized voltage wave vectors, respectively. Each vector contains all the TE and TM waves outlined above as elements with the

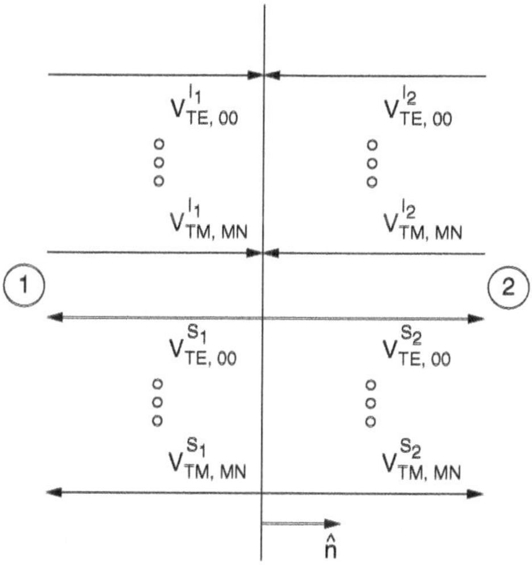

Figure 3.15 A terminal plane defining the normalized incident and reflected voltage waves.

appropriate direction of propagation. The vectors at terminal plane z are

$$\mathbf{V}^I = \begin{bmatrix} V^{I_1} \\ V^{I_2} \end{bmatrix}; \quad \mathbf{V}^S = \begin{bmatrix} V^{S_1} \\ V^{S_2} \end{bmatrix} \tag{3.66}$$

with the direction of propagation relative to $\hat{n}$ indicated by the arrows in Fig. 3.15. Accordingly, eq. (3.65) can be rewritten in the form

$$\begin{bmatrix} V^{S_1} \\ V^{S_2} \end{bmatrix} = \begin{bmatrix} \mathbf{S}_{11} & \mathbf{S}_{12} \\ \mathbf{S}_{21} & \mathbf{S}_{22} \end{bmatrix} \begin{bmatrix} V^{I_1} \\ V^{I_2} \end{bmatrix} \tag{3.67}$$

where **S** has been divided into submatrices. In general, each submatrix may be rectangular, i.e., for I incident modes, there will be J scattered modes resulting in the submatrices having I columns and J rows. This may correspond to a single mode incident upon a periodic surface, scattering ideally into an infinite number (but truncated to a finite number) of modes on each side of the surface. Since, in general, each component in the cascade assembly will be considered as an arbitrary element where the number of incident and scattered modes is not known *a priori*, and since the component may be cascaded with other arbitrary components, the scattering matrices will be calculated for an equal number of incident and scattered modes. The submatrices will be of order $2MN$, (M,N) being the total number of modes in the $(\hat{x}, \hat{y})$ directions, with the coefficient 2 taking into consideration the TE and TM states.

The scattering parameters of a periodic surface are found from the scattered Floquet modes calculated for a free-standing periodic surface, as explained in section 3.1. The coupling of an incident mode into a forward scattered mode, i.e., transmitted mode for a surface at an arbitrary plane z,

$$T_{(pmn)_i}^{(pmn)_s} = \frac{V_{(pmn)_s}^{S_2}(z)}{V_{(pmn)_i}^{I_1}(z)} \tag{3.68}$$

while the coupling into a backscattered mode, i.e., reflected mode, is

$$R_{(pmn)_i}^{(pmn)_s} = \frac{V_{(pmn)_s}^{S_1}(z)}{V_{(pmn)_i}^{I_1}(z)} \tag{3.69}$$

where the subscript indicates incident modes, and the superscript indicates scattered modes. The index p denotes the polarization TE or TM. Since both TE and TM modes may be incident, and since the surface allows coupling of energy from one state into another, there will be four transmission and four reflection coefficients associated with each pair of incident and reflected modes. Substituting (3.63) into (3.68) and (3.69) results in the four sets of coefficients. The co-polarized transmission coefficients of a periodic surface are:

$$T_{TE(mn)_i}^{TE(mn)_s} = \frac{a^+_{TE(mn)_i}\delta_{(mn)_i(mn)_s} + a^+_{TE(mn)_s}}{a^i_{TE(mn)_i}}\left[\frac{Z^{h*}_{(mn)_i}}{Z^{h*}_{(mn)_s}}\right]^{1/2}\frac{\|\underline{e}_{t_{TE(mn)_s}}\|}{\|\underline{e}_{t_{TE(mn)_i}}\|}$$

$$T_{TM(mn)_i}^{TM(mn)_s} = \frac{a^+_{TM(mn)_i}\delta_{(mn)_i(mn)_s} + a^+_{TM(mn)_s}}{a^i_{TM(mn)_i}}\left[\frac{Z^{e*}_{(mn)_s}}{Z^{e*}_{(mn)_i}}\right]^{1/2}\frac{\|\underline{h}_{t_{TM(mn)_s}}\|}{\|\underline{h}_{t_{TM(mn)_i}}\|} \quad (3.70)$$

where $\delta_{(mn)_i(mn)_s}$ is the Kronecker delta function, which enables the addition of the incident field mode to the identical forward-scattered mode. This addition is absent in all other cases due to the orthogonality of the modes. The cross-polarized transmission coefficients are:

$$T_{TE(mn)_i}^{TM(mn)_s} = \frac{a^+_{TM(mn)_s}}{a^i_{TE(mn)_i}}[Z^{e*}_{(mn)_s}Z^{h*}_{(mn)_i}]^{1/2}\frac{\|\underline{h}_{t_{TM(mn)_s}}\|}{\|\underline{e}_{t_{TE(mn)_i}}\|}$$

$$T_{TM(mn)_i}^{TE(mn)_s} = \frac{a^+_{TE(mn)_s}}{a^i_{TM(mn)_i}}\left[\frac{1}{Z^{e*}_{(mn)_i}Z^{h*}_{(mn)_s}}\right]^{1/2}\frac{\|\underline{e}_{t_{TE(mn)_s}}\|}{\|\underline{h}_{t_{TM(mn)_i}}\|} \quad (3.71)$$

Similarly, the reflection scattering parameters of a periodic structure can be computed. The co-polarized terms are:

$$R_{TE(mn)_i}^{TE(mn)_s} = \frac{a^-_{TE(mn)_s}}{a^i_{TE(mn)_i}}\left[\frac{Z^{h*}_{(mn)_i}}{Z^{h*}_{(mn)_s}}\right]^{1/2}\frac{\|\underline{e}_{t_{TE(mn)_s}}\|}{\|\underline{e}_{t_{TE(mn)_i}}\|}$$

$$R_{TM(mn)_i}^{TM(mn)_s} = \frac{a^-_{TM(mn)_s}}{a^i_{TM(mn)_i}}\left[\frac{Z^{e*}_{(mn)_s}}{Z^{e*}_{(mn)_i}}\right]^{1/2}\frac{\|\underline{h}_{t_{TM(mn)_s}}\|}{\|\underline{h}_{t_{TM(mn)_i}}\|} \quad (3.72)$$

and the reflection cross-polarized parameters are:

$$R_{TE(mn)_i}^{TM(mn)_s} = \frac{a^-_{TM(mn)_s}}{a^i_{TE(mn)_i}}\left[Z^{e*}_{(mn)_s}Z^{h*}_{(mn)_i}\right]^{1/2}\frac{\|\underline{h}_{t_{TM(mn)_s}}\|}{\|\underline{e}_{t_{TE(mn)_i}}\|}$$

$$R_{TM(mn)_i}^{TE(mn)_s} = \frac{a^-_{TE(mn)_s}}{a^i_{TM(mn)_i}}\left[\frac{1}{Z^{e*}_{(mn)_i}Z^{h*}_{(mn)_s}}\right]^{1/2}\frac{\|\underline{e}_{t_{TE(mn)_s}}\|}{\|\underline{h}_{t_{TM(mn)_i}}\|} \quad (3.73)$$

Usually, a homogeneous dielectric layer is the second element in a cascade assembly. In this case, each mode, having the form of a plane wave, will have scattering parameters simply related to the transmission and reflection coefficients of the mode, and since there is no coupling between modes at the interface, the scattering

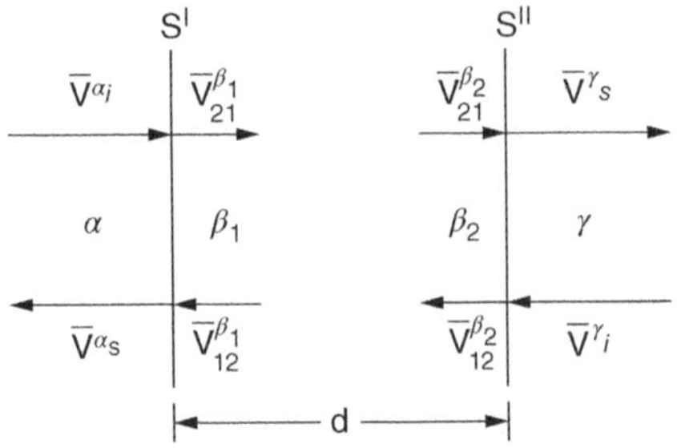

Figure 3.16 Cascade connection of scattering matrices $\mathbf{S}^{\mathrm{I}}$ and $\mathbf{S}^{\mathrm{II}}$.

parameter matrices are diagonal. For a dielectric boundary between media 1 and 2 (see Fig. 3.4), the scattering matrix is antisymmetric and reciprocal: $\mathbf{S}_{11} = -\mathbf{S}_{22} = \mathbf{R}$; $\mathbf{S}_{12} = \mathbf{S}_{21} = \mathbf{T}$, being related to the medium impedances as follows:

$$R^{(pmn)_s}_{(pmn)_i} = \frac{V^{S_1}_{(pmn)_s}(0)}{V^{I_1}_{(pmn)_i}(0)} = \begin{cases} \dfrac{Z^2_{pmn} - Z^1_{pmn}}{Z^2_{pmn} + Z^1_{pmn}}, & i = s \\ 0, & i \neq s \end{cases}$$

$$T^{(pmn)_s}_{(pmn)_i} = \frac{V^{S_2}_{(pmn)_s}(0)}{V^{I_1}_{(pmn)_i}(0)} = \begin{cases} \dfrac{2(Z^1_{pmn} Z^2_{pmn})^{1/2}}{Z^2_{pmn} + Z^1_{pmn}}, & i = s \\ 0, & i \neq s \end{cases} \tag{3.74}$$

With the scattering parameters of the two elements derived, it only remains to outline the procedure to cascade an arbitrary number of elements to find the composite scattering parameters.

Fig. 3.16 illustrates two components separated by a distance d, characterized by scattering matrices $\mathbf{S}^{\mathrm{I}}$ and $\mathbf{S}^{\mathrm{II}}$. These matrices may characterize, e.g., a surface embedded in a medium a distance d away from a dielectric boundary separating a second medium, or, e.g., two dielectric boundaries separating a dielectric layer of thickness d from neighboring media.

The normalized voltage waves to the left of the composite structure are labeled α, to the right γ, and between elements β_1 and β_2, respectively, at the terminal planes of each element. Since the waves between elements are the combination of the incident and scattered waves of each element, the individual incident and scattered waves cannot be isolated; hence, the combination is labeled $V^{\beta_{1,2}}_{12}$ and $V^{\beta_{1,2}}_{21}$, respectively, at each terminal plane.

For the component characterized by $\mathbf{S}^{\mathrm{I}}$, the defining equations for the scattered waves due to incident waves are

$$\begin{bmatrix} V^{\alpha_s} \\ V_{21}^{\beta_1} \end{bmatrix} = \begin{bmatrix} \mathbf{S}_{\alpha\alpha} & \mathbf{S}_{\alpha\beta_1} \\ \mathbf{S}_{\beta_1\alpha} & \mathbf{S}_{\beta_1\beta_1} \end{bmatrix} \begin{bmatrix} V^{\alpha_i} \\ V_{12}^{\beta_1} \end{bmatrix} \tag{3.75}$$

and for the component characterized by $\mathbf{S}^{\mathrm{II}}$,

$$\begin{bmatrix} V_{12}^{\beta_2} \\ V^{\gamma_s} \end{bmatrix} = \begin{bmatrix} \mathbf{S}_{\beta_2\beta_2} & \mathbf{S}_{\beta_2\gamma} \\ \mathbf{S}_{\gamma\beta_2} & \mathbf{S}_{\gamma\gamma} \end{bmatrix} \begin{bmatrix} V_{21}^{\beta_2} \\ V^{\gamma_i} \end{bmatrix} \tag{3.76}$$

Since the normalized voltage waves at terminal planes are traveling waves, the waves at terminal planes are simply related as

$$\mathbf{V}_{21}^{\beta_2} = \mathbf{P}\mathbf{V}_{21}^{\beta_1}; \quad \mathbf{V}_{12}^{\beta_1} = \mathbf{P}\mathbf{V}_{12}^{\beta_2} \tag{3.77}$$

where $\mathbf{P}$ is a diagonal propagator matrix, with a diagonal element being $e^{-jk_{zmn}d}$ for each mn^{th} mode. The relations between the normalized voltage waves of each component may be combined to produce a composite scattering matrix defined as

$$\begin{bmatrix} V^{\alpha_s} \\ V^{\gamma_s} \end{bmatrix} = \begin{bmatrix} \mathbf{S}_{\alpha\alpha}^{\Sigma} & \mathbf{S}_{\alpha\gamma}^{\Sigma} \\ \mathbf{S}_{\gamma\alpha}^{\Sigma} & \mathbf{S}_{\gamma\gamma}^{\Sigma} \end{bmatrix} \begin{bmatrix} V^{\alpha_i} \\ V^{\gamma_i} \end{bmatrix} \tag{3.78}$$

where Σ indicates the scattering matrices characterizing the composite structure. An intermediate step in the matrix combination is the calculation of the internal waves between elements β_1 and β_2. The knowledge of these waves then allows the calculation of fields at any point in between the boundaries. Combining (3.75) and (3.76) and using (3.77) gives

$$\begin{bmatrix} V_{12}^{\beta_1} \\ V_{21}^{\beta_2} \end{bmatrix} = \begin{bmatrix} \mathbf{H}_2\mathbf{S}_{\beta_2\beta_2}\mathbf{P}\mathbf{S}_{\beta_1\alpha} & \mathbf{H}_2\mathbf{S}_{\beta_2\gamma} \\ \mathbf{H}_1\mathbf{S}_{\beta_1\alpha} & \mathbf{H}_1\mathbf{S}_{\beta_1\beta_1}\mathbf{P}\mathbf{S}_{\beta_2\gamma} \end{bmatrix} \begin{bmatrix} V^{\alpha_i} \\ V^{\gamma_i} \end{bmatrix} \tag{3.79}$$

where $\mathbf{H}_1 = (\mathbf{P}^{-1} - \mathbf{S}_{\beta_1\beta_1}\mathbf{P}\mathbf{S}_{\beta_2\beta_2})^{-1}$ and $\mathbf{H}_2 = (\mathbf{P}^{-1} - \mathbf{S}_{\beta_2\beta_2}\mathbf{P}\mathbf{S}_{\beta_1\beta_1})^{-1}$. Further manipulations yield the composites scattering parameters:

$$\begin{aligned}
\mathbf{S}_{\alpha\alpha}^{\Sigma} &= \mathbf{S}_{\alpha\alpha} + \mathbf{S}_{\alpha\beta_1}\mathbf{H}_2\mathbf{S}_{\beta_2\beta_2}\mathbf{P}\mathbf{S}_{\beta_1\alpha} \\
\mathbf{S}_{\alpha\gamma}^{\Sigma} &= \mathbf{S}_{\alpha\beta_1}\mathbf{H}_2\mathbf{S}_{\beta_2\gamma} \\
\mathbf{S}_{\gamma\alpha}^{\Sigma} &= \mathbf{S}_{\alpha\beta_2}\mathbf{H}_1\mathbf{S}_{\beta_2\alpha} \\
\mathbf{S}_{\gamma\gamma}^{\Sigma} &= \mathbf{S}_{\gamma\gamma} + \mathbf{S}_{\gamma\beta_2}\mathbf{H}_1\mathbf{S}_{\beta_1\beta_1}\mathbf{P}\mathbf{S}_{\beta_2\gamma}
\end{aligned} \tag{3.80}$$

A multicomponent screen may be built by repeating the above procedure, using the composite scattering parameters of the initial two components of the screen as the first scattering parameter matrix, with the scattering parameters of the next component being the second matrix. By repeatedly adding components in this fashion, a general screen is constructed from individual components with the layering being arbitrary.

An alternative method to compute the total scattering matrix of the multilayer structure makes use of the attractive matrix multiplication feature of the total transmission matrix of a multilayer (cascaded) structure, as shown in Fig. 3.17.

The transmission and scattering matrices are defined as

$$\begin{bmatrix} a_1 \\ b_1 \end{bmatrix} = \begin{bmatrix} T_{11} & T_{12} \\ T_{21} & T_{22} \end{bmatrix} \begin{bmatrix} b_2 \\ a_2 \end{bmatrix}$$

$$\begin{bmatrix} b_1 \\ b_2 \end{bmatrix} = \begin{bmatrix} S_{11} & S_{12} \\ S_{21} & S_{22} \end{bmatrix} \begin{bmatrix} a_1 \\ a_2 \end{bmatrix} \quad (3.81)$$

The transformation between the [S] and [T] matrices can be derived out of (3.81) after some algebraic manipulations:

$$[T] = \begin{bmatrix} S_{21}^{-1} & -S_{22}S_{21}^{-1} \\ S_{11}S_{21}^{-1} & S_{12} - S_{11}S_{21}^{-1}S_{22} \end{bmatrix}$$

$$[S] = \begin{bmatrix} T_{12}T_{11}^{-1} & T_{11} - T_{12}T_{11}^{-1}T_{21} \\ T_{11}^{-1} & -T_{12}T_{11}^{-1} \end{bmatrix} \quad (3.82)$$

As such, in the first step of the algorithm the scattering parameters of the individual layers are computed using numerical analysis,

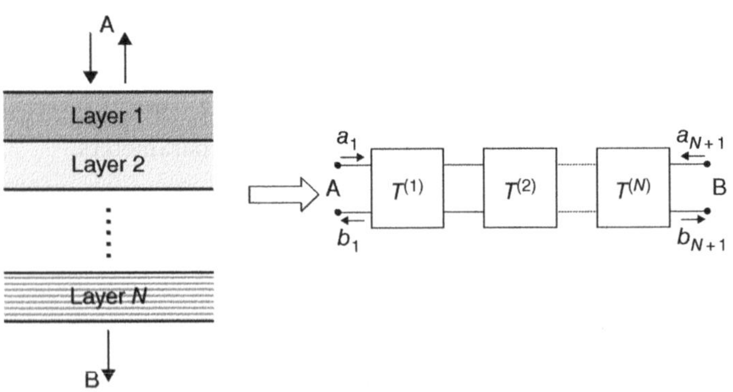

Figure 3.17 A multilayer periodic medium and its equivalent transmission matrices.

as described in section 3.1. In the second step, each scattering matrix is transformed into a transmission matrix using (3.82). The total transmission matrix is calculated using matrix multiplication for all the transmission matrices.

$$T_{total} = T^{(N)} \cdots T^{(2)} T^{(1)} \tag{3.83}$$

In the next step, the total transmission matrix is transformed into a scattering matrix and all the scattering parameters are extracted using (3.82). Given that the total field, $E(\omega,x,y)$ is known as outlined above, the actual magnitude of the different harmonics (Floquet modes) can be calculated by projecting the total field on each harmonics:

$$E^{m,n}(\omega) = \frac{1}{ab} \int_0^a \int_0^b E(\omega, x, y) e^{jk_{xm}x} e^{jk_{yn}y} dx\, dy \tag{3.84}$$

where k_{xm} and k_{yn} are given by (3.8). Eq. (3.84) can be rewritten in the discrete form as

$$E^{m,n}(\omega) = \frac{1}{N_x N_y} \sum_{i=0}^{N_x} \sum_{j=0}^{N_y} E(\omega, i\Delta x, j\Delta y) e^{jk_{xm} i\Delta x} e^{jk_{yn} j\Delta y} \tag{3.85}$$

where N_x and N_y are the total number of cells in the x and y directions, respectively, and Δx and Δy are the cell sizes in the x and y directions, respectively.

The multilayer geometry shown in Fig. 3.18 consists of two identical FSS structures of dipole elements. The dipole length is 12 mm and its width 3 mm. The periodicity is 15 mm in both x and y directions [15]. The substrate has a thickness of 6 mm and a relative permittivity

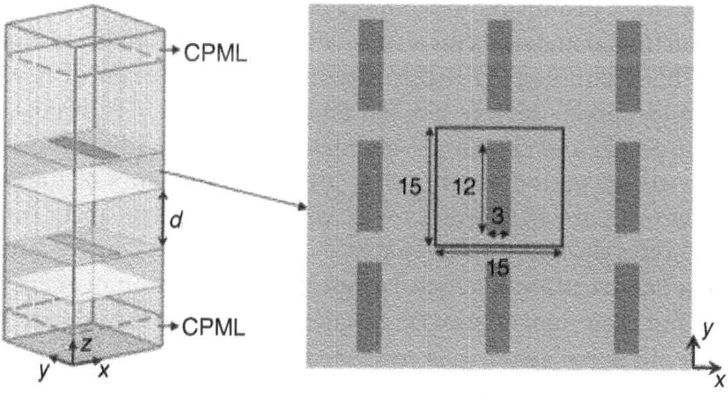

Figure 3.18 Geometry of an FSS unit cell composed of two identical dipoles cascaded (all dimensions in mm) [15].

of $\varepsilon_r = 2.2$. The structure is illuminated by a TE^z normally incident plane wave (with polarization along the *y*-axis). The frequency range is 0–16 GHz. Convolutional perfectly matched layers (CPML) are used as the absorbing boundaries at the top and the bottom of the computational domain. The goal was to determine the distance (*d*) after which all the magnitudes of the higher harmonics reached 40 dB below the magnitude of the incident electric field.

To validate the cascading technique, several air gaps were analyzed and compared with simulations of the entire structure. From Fig. 3.19 and Fig. 3.20, it can be noticed that when the gap size (*d*) is less than 15.5 mm, the cascading technique using only the dominant mode is not accurate (higher harmonics than −40 dB), especially at high frequencies.

To analyze the structure accurately with a gap size (*d*) less than 15.5 mm, the cascading technique should include all the harmonics that have magnitudes greater than −40 dB compared to incidence. Fig. 3.21 compares the results of the cascading technique while using only the dominant mode, and then while using the dominant mode and the first two harmonics ($M_{1,0}$) and ($M_{-1,0}$). Notice that including the two harmonics in the cascaded analysis enhances the accuracy of the results.

3.3 Metamaterial Radomes

Metamaterials technology was introduced a few years ago, and since then it is the most promising technology in the development of new customized artificial materials for a variety of civilian and military applications. This technology is highly attractive in the design of new microwave and optical devices like flat lenses, delay lines with positive speed-up phase, and cloaking. Metamaterials are actually artificial materials based on periodic structures. The basic elements of these periodic structures are of different types: printed, dielectric, or ferromagnetic. Careful design of these materials enables us to control and tailor their geometrical and electrical properties, such as homogeneity, anisotropy, and their dependence on frequency to obtain the required performance. In some cases, it is possible to obtain double negative materials (DNG), with both permeability and permittivity of the metamaterial being negative. This feature results in some interesting effects for a plane wave incident on such a material like negative refraction

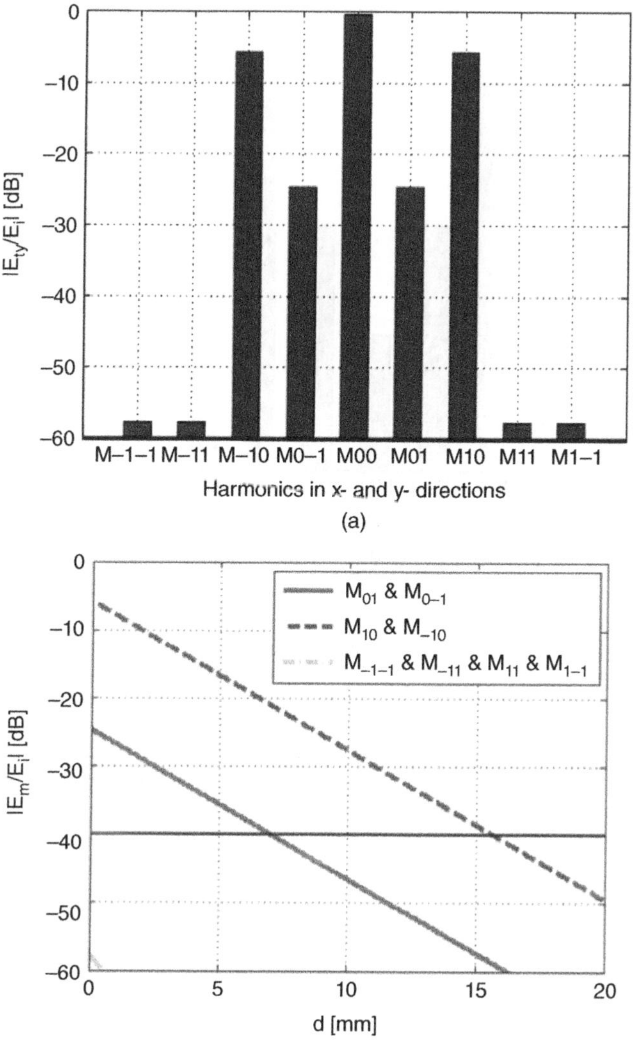

Figure 3.19 The first eight transmitted harmonics at 16 GHz (normalized to the incident field): (a) amplitudes; (b) the decay with distance of the first eight harmonics [15].

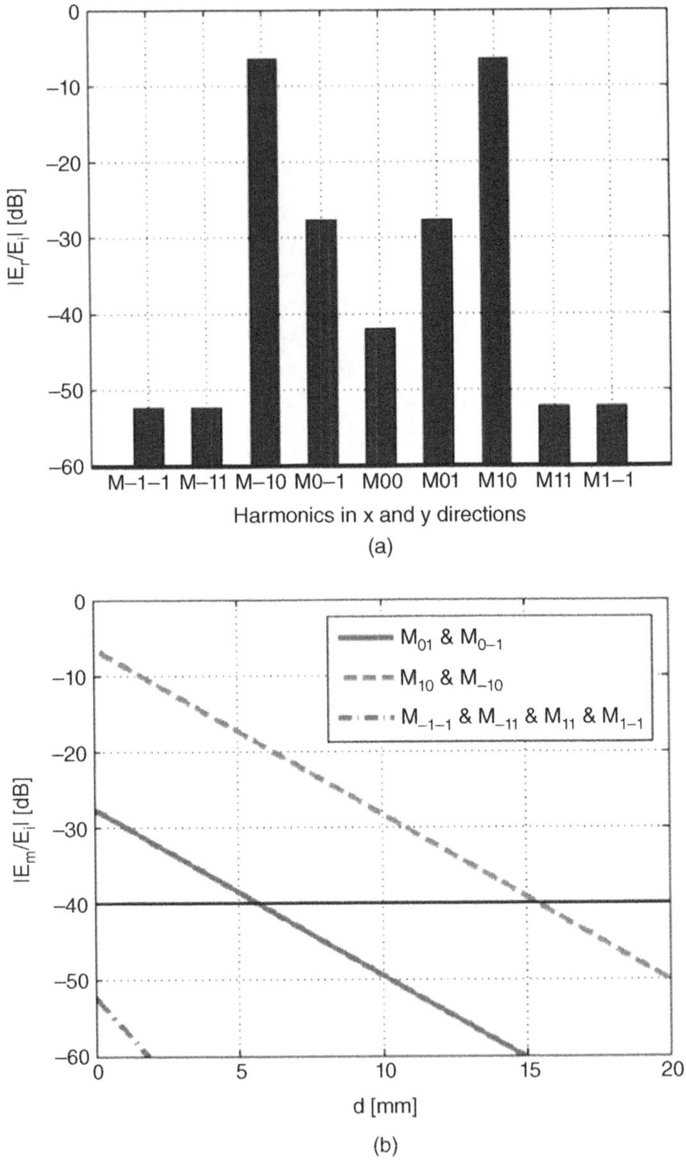

Figure 3.20 The first eight reflected harmonics at 16 GHz (normalized to the incident field): (a) amplitudes; (b) the decay with distance of the first eight harmonics [15].

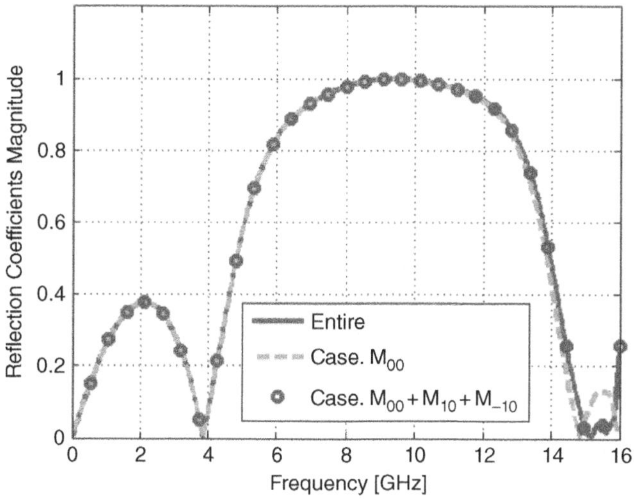

Figure 3.21 The reflection coefficients of two identical dipole FSS for normal incidence TE^z case with $d = 7$ mm [15].

coefficient and negative phase velocity. In addition, metamaterials may have special anisotropic and bi-anisotropic properties [16, 17]. The two-dimensional (2D) version of metamaterials is a straight extension of FSS structures, with the difference being the periodic structure unit cell size. For the FSS structures the operational frequency starts when the unit cell size is on the order of $\lambda/2$, while in case of metamaterials, the unit cell size is smaller than 0.1λ in the operational band. Accordingly, for metamaterials quasi-static analysis is applicable. Therefore, assigning effective constitutive parameters to a metamaterial, within a specific frequency band, is reasonable.

Besides satisfactory performance with respect to high frequency of operation, the metamaterial structures have some other excellent features compared with conventional FSSs:

- Insensitivity of the frequency response (transmission and reflection) to the incoming wave angle of incidence.
- Harmonic free frequency response.
- Ease of achieving broadband/multiband characteristics with superior reflection/transmission efficiency.
- More compact and thin.

An example of a combination of a metamaterial resistive periodic surface and an interdigitated FSS to realize a radome with a narrow pass-band to allow the enclosed antenna transmission and reception and a wide band above the pass-band for maximum absorption is described in [3] and will be discussed in the following. Usually, wide-band absorption is obtained by placing a lossy surface on top of a ground plane. Accordingly, an FSS is designed for a pass-band with a wide reflection band out of the pass-band, such that it can be used as a ground plane for an external absorbing structure. The wide-band absorbing structure is synthesized by placing an additional resistive FSS on top of an air spacer attached to the metallic FSS. Within the frequency range where the FSS behaves as a ground plane, the multilayer absorbs impinging signals as a wide-band high-impedance surface absorber [2, 18]. The geometry of the proposed structure is shown in Fig. 3.22 [3].

The top layer is a resistive FSS backed by an air spacer and a metallic pass-band FSS. In order to obtain low insertion loss of the entire structure within the operating pass-band of the antenna, the resistive FSS has to be almost transparent in this band. Since the operating bands of the absorbing and transmissive parts of the radome are different,

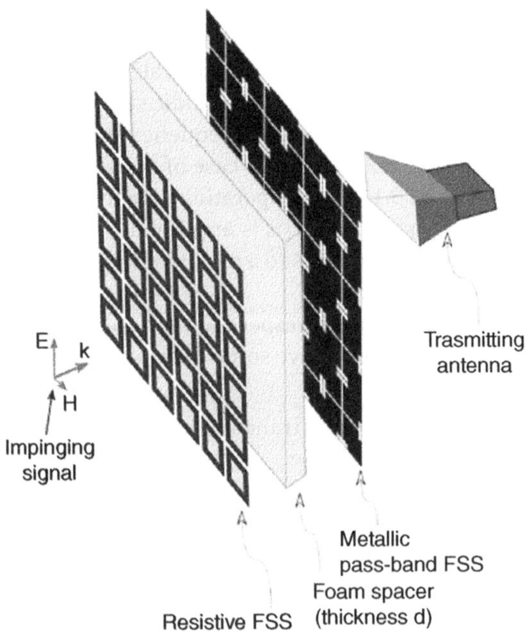

Figure 3.22 Geometry of the transmitter/absorber radome [3].

it is expected that the periodicity of the metallic FSS and the resistive FSS are dissimilar. This aspect leads to the following drawbacks: first, the periodicity of the metallic FSS can't exceed one wavelength at the maximum operating frequency of the absorber to avoid grating lobes in the absorber working band, and second, the ratio of the unit cell size of the metallic FSS and the unit cell size of the resistive FSS should be a rational number.

The equivalent network of the investigated structure is shown in Fig. 3.23 and described in [3].

In the equivalent circuit, the metallic pass-band FSS is represented by a parallel L_pC_p circuit, if the filter is realized with a simple FSS unit cell such as a slotted cross or a slotted square loop. If the unit cell of the FSS is a slotted Jerusalem cross, a series inductance L_s other than the parallel L_pC_p is needed for the equivalent circuit representation [18]. The capacitive resistive FSS is instead approximated by a series RLC circuit [19]. The reflection and transmission coefficients of the composite structure are obtained for both normal and oblique incidence by computing the input impedance of the circuit in Fig. 3.23. The impedance across the dielectric slab is computed by

$$Z_v^{TE,TM} = Z_m^{TE,TM} \frac{[Z_L^{TE,TM} + jZ_m^{TE,TM} \tan(\beta_m d)]}{[Z_m^{TE,TM} + jZ_L^{TE,TM} \tan(\beta_m d)]} \quad (3.86)$$

where Z_L is the load impedance and Z_m, β_m, k_m are:

$$\beta_m = \sqrt{k_m^2 - k_t^2},\ k_t = k_0 \sin(\theta),\ k_m = k_0\sqrt{\varepsilon_r \mu_r}$$
$$Z_m^{TE} = \frac{\omega\mu}{\beta_m};\ Z_m^{TM} = \frac{\beta_m}{\omega\varepsilon} \quad (3.87)$$

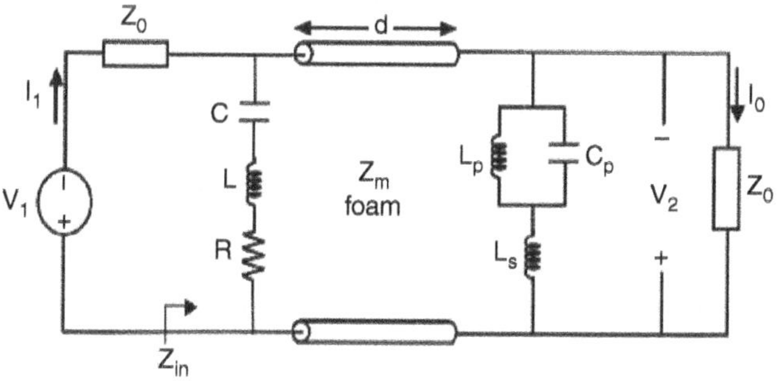

Figure 3.23 Transmission line equivalent circuit of the transmissive/absorbing radome [3].

θ is the incident angle, k_0 is the free space propagation constant, and k_t is the transverse wavevector. Once the input impedance of the structure is derived, the reflection coefficient S_{11}^r in receive mode is obtained by

$$S_{11}^r = \frac{Z_{in}^{TE,TM} - Z_0^{TE,TM}}{Z_{in}^{TE,TM} + Z_0^{TE,TM}};\ Z_0^{TE} = \frac{\eta_0}{\cos\theta};\ Z_0^{TM} = \eta_0 \cos\theta;\ \eta_0 = \sqrt{\frac{\mu_0}{\varepsilon_0}} \tag{3.88}$$

where Z_{in} is the input impedance of the entire structure. Similarly, the transmission coefficient S_{21}^r in receive mode can be evaluated using the *ABCD* representation of the equivalent circuit, as described in Chapter 2:

$$S_{21}^r = \frac{2}{A + B/Z_0^{TE,TM} + CZ_0^{TE,TM} + D} \tag{3.89}$$

The terms *A*, *B*, *C*, *D* are the elements of the transmission line matrix of the entire system, which is evaluated as the product of the three cascaded matrices [20]:

$$\begin{bmatrix} A & B \\ C & D \end{bmatrix} = \begin{bmatrix} 1 & 0 \\ \frac{1}{Z_{FSS}^{res}} & 1 \end{bmatrix} \begin{bmatrix} \cos(\beta_m d) & jZ_m \sin(\beta_m d) \\ j\frac{\sin(\beta_m d)}{Z_m} \cos(\beta_m d) & \cos(\beta_m d) \end{bmatrix} \begin{bmatrix} 1 & 0 \\ \frac{1}{Z_{FSS}^{metal}} & 1 \end{bmatrix} \tag{3.90}$$

where Z_{FSS}^{res} and Z_{FSS}^{metal} are the approximate impedances of the resistive and the metallic FSS, respectively. The inductances and the capacitances of the employed FSSs are computed by matching the normal incidence full-wave response of the elements and the characteristic impedance of the incoming medium. The procedure is described in [19]. The energy absorbed by the structure is obtained through the following relation:

$$A_p^{dB} = 10 \log\left(|S_{11}^r|^2 + |S_{21}^r|^2\right) \tag{3.91}$$

As already explained, the metallic FSS in the radome structure has to operate as a pass-band with a wide-reflection stop-band such that it can be used as a ground plane for the absorbing structure. The simplest FSS element that operates as a pass-band filter is a slotted cross-periodic structure in a metallic ground plane. The center frequency of the pass-band occurs when the unit cell size is approximately $\lambda/2$. The second pass-band occurs when the cell size is approximately λ. This feature may limit the stop-band in which the FSS is required to perform as a perfect ground. A possible way

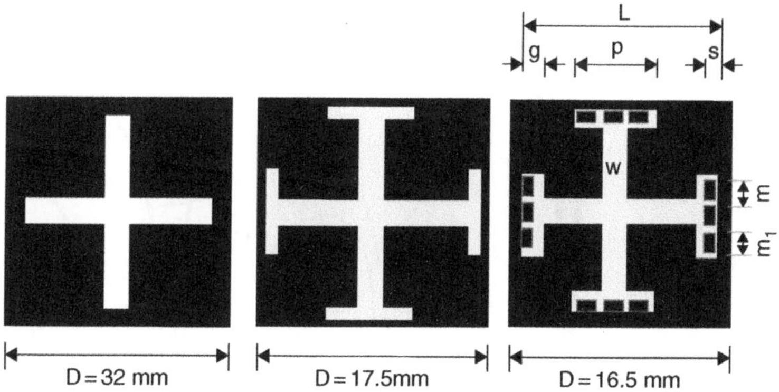

Figure 3.24 Geometries for the metallic FSS: (a) slotted cross; (b) slotted Jerusalem cross; (c) slotted interdigitated Jerusalem cross.

to increase the stop-band of the FSS without affecting its pass-band center frequency is to reduce its unit cell size and change the geometry of the periodic structure element. This goal can be achieved if the simple slotted cross is replaced by a slotted Jerusalem cross. To reduce further the unit cell size the slotted Jerusalem cross is replaced by an interdigitated slotted Jerusalem cross, as shown in Fig. 3.24.

In [3] the required pass-band is 4.2–4.9 GHz and its required absorbing band is 8–18 GHz, with insertion loss higher than 15 dB. The replacement of the standard cross with the Jerusalem cross reduces the unit cell size from 32 mm to 17.5 mm, and further replacement with the interdigitated Jerusalem cross reduces the unit cell size to 16.5 mm without affecting the center frequency of the pass-band, as shown in the simulation results in Fig. 3.25.

According to the previous finding, the designed bottom metallic FSS is employed as a ground plane of the outer absorber. On the other hand, the absorbing structure has to fulfill the essential requirement to be transparent in the antenna-working band (4.2–4.9 GHz). The wide-band radar absorber based on the top resistive FSS is shown in Fig. 3.22. The wide-band absorber is formed by a suitably designed square loop shaped resistive FSS on top of a thin grounded air substrate. To meet the required radome performance, the free space resonance of the resistive FSS should be located around 15 GHz, as described in [18]. All the elements are characterized by the same periodicity D, i.e., 11 mm (equal to $\lambda/2$ at 13.7 GHz). The side lengths

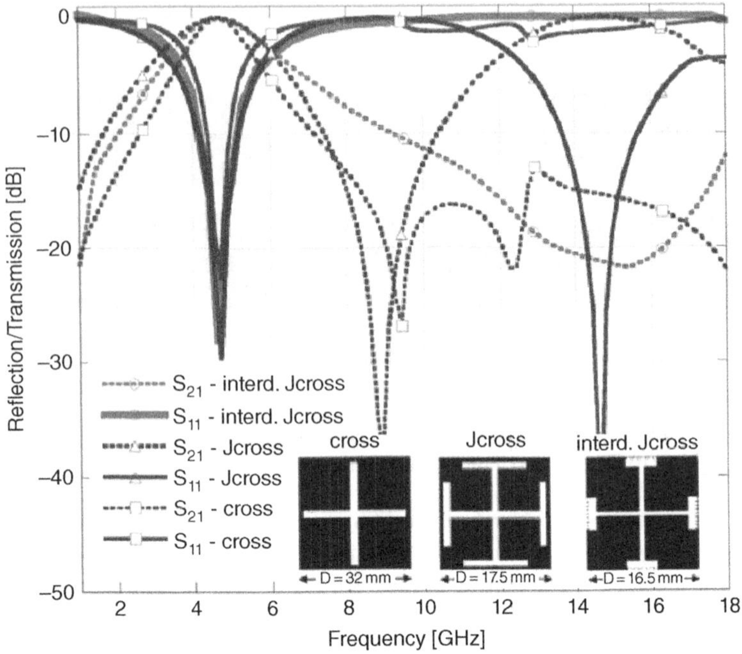

Figure 3.25 Normal incidence reflection/transmission of FSS with slotted cross, slotted Jerusalem cross, and interdigitated slotted Jerusalem cross elements [3].

of the loops, from the narrowest to the widest one, are 10/16*D*, 12/16*D*, and 14/16*D*, while the thicknesses of the loop are 1/16*D*, 2/16*D*, and 3/16*D*, respectively. The transmission coefficient in all these cases is shown in Fig. 3.26.

One can observe that in the absorption band (8–18 GHz), the required insertion loss of more than 15 dB is not met. The conclusion is that the FSS loops cannot be just metallic for obtaining the required absorption, and a resistive FSS needs to be used. Moreover, the wide-printed square loop is characterized by high insertion loss in the pass-band, while the narrower square loop yields a much lower insertion loss and therefore is the preferable choice for the pass-band performance. The surface resistance of the loop is 15 [Ω/sq] and the equivalent parameters of the resistive FSS, as shown in Fig. 3.23, are $L = 4.1$ nH, $C = 0.024$ pF, and $R = 260\ \Omega$ based on the parameter study described in [3]. The air substrate thickness is 5 mm ($\lambda/4$ at the center frequency of the absorbing band). The reflection coefficient of the

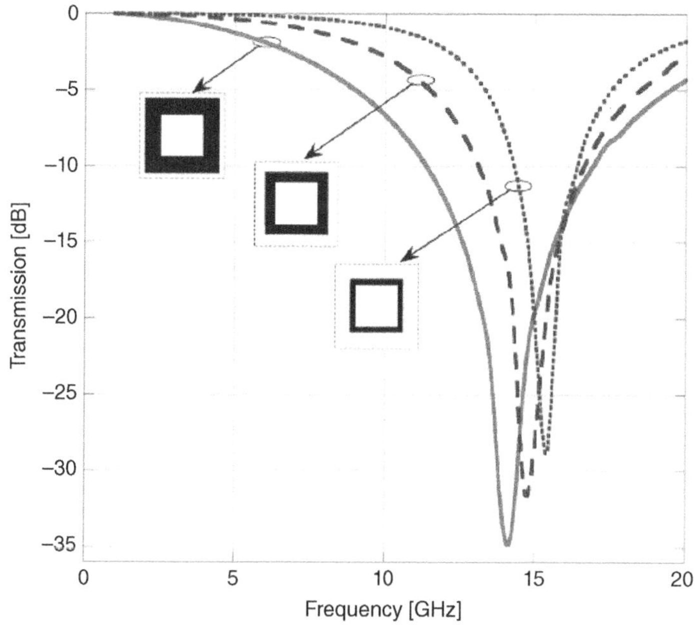

Figure 3.26 Transmission coefficient of three different square loop shaped metallic FSSs. The periodicity of the FSS is 11 mm in all cases [3].

absorbing structure with metallic ground plane, loops with different thicknesses (1/16*D*, 2/16*D*, and 3/16*D*), and surface resistance equal to 15 [Ω/sq] is shown in Fig. 3.27.

One can observe that the widest loop gives the best performance in terms of wide-band absorption, but gives rise to high losses within the antenna pass-band. On the other hand, the narrowest square loop can be classified as the best compromise if the insertion loss in the pass-band is considered the key parameter. Thus, in the following, we will consider the resistive narrowest loop for the analysis of the radome performance.

The entire structure is analyzed with periodic boundary conditions; therefore, if the unit cell size of the metallic FSS is 16.5 mm and the unit cell size of the resistive FSS is 11 mm, the macro unit cell size of the entire structure would be 33 mm and include 4 unit cells of the metallic FSS and 9 unit cells of the resistive FSS. The equivalent circuit parameters of the metallic FSS based on [3] are L_p = 3.74 nH, C_p = 0.313 pF, and L_s = 0.372nH, and the geometrical dimensions

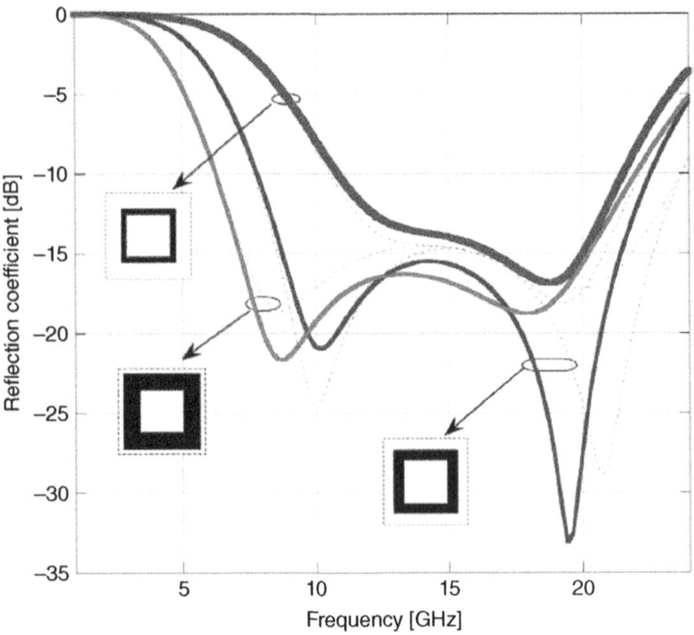

Figure 3.27 Reflection coefficient of the absorbing structure with metallic ground and loops with surface resistance 15 [Ω/sq]. The thickness of the structure is 5 mm and the unit cell size is 11 mm [3].

of the interdigitated slotted Jerusalem cross, as shown in Fig. 3.24, are $L = 13.95$ mm, $p = 4.65$ mm, $w = 0.46$ mm, $m = 0.82$ mm, $m_1 = 0.23$ mm, $s = 0.35$ mm, and $g = 0.93$ mm. In Fig. 3.28, we can see the radome performance in the transmitting mode, and in Fig. 3.29, the radome performance in the receiving mode. By observing the curves, we can notice that the structure is reducing the reflected power by more than 15 dB in the frequency band 10–18 GHz and introduces only 0.3 dB losses within the antenna frequency band 4.2–4.9 GHz.

The results obtained by the equivalent circuit approach compare satisfactorily with the results obtained by simulations with HFSS and CST. In addition, one of the main requirements for a practical radome is the preservation of the transmissive properties as a function of the incident angle. In Fig. 3.30 the transmission coefficient of the radome are shown for 0, 30, and 45 deg incident angles and for TM and TE polarizations [3].

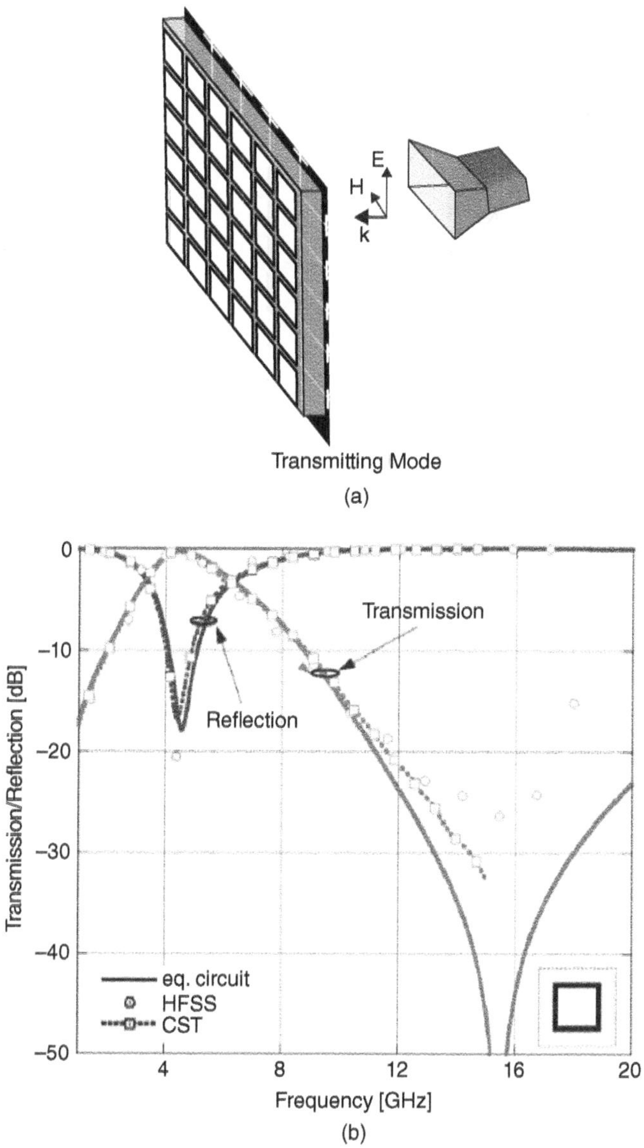

Figure 3.28 The radome in transmission mode: (a) the geometrical configuration in transmit mode; (b) the transmission/reflection performance in transmit mode [3].

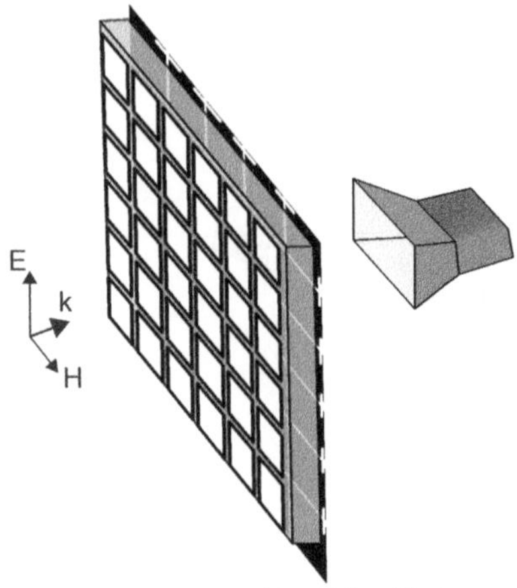

(a)

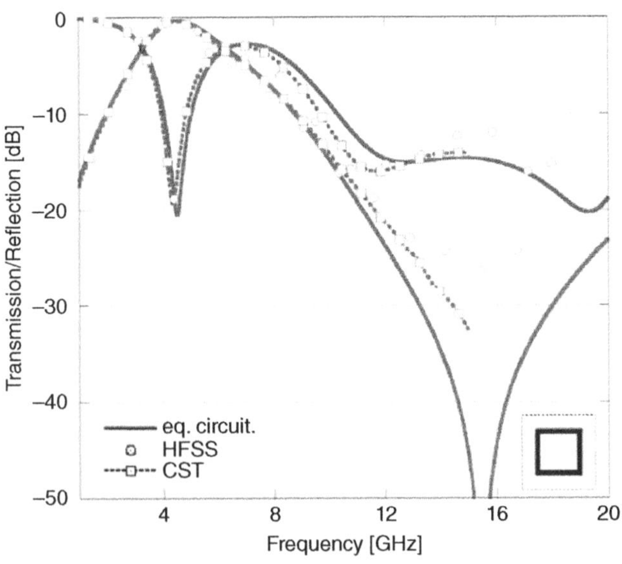

(b)

Figure 3.29 The radome in the receiving mode: (a) geometry in the receiving mode; (b) transmission/reflection in receiving mode; (c) absorption in the receiving mode [3].

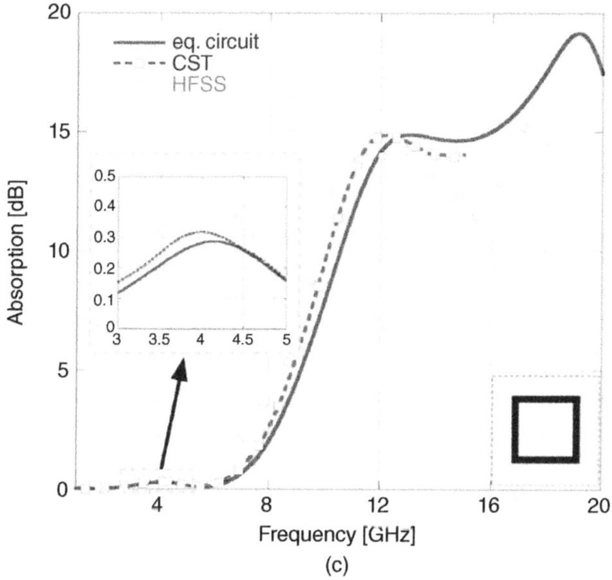

Figure 3.29 *(Continued)*

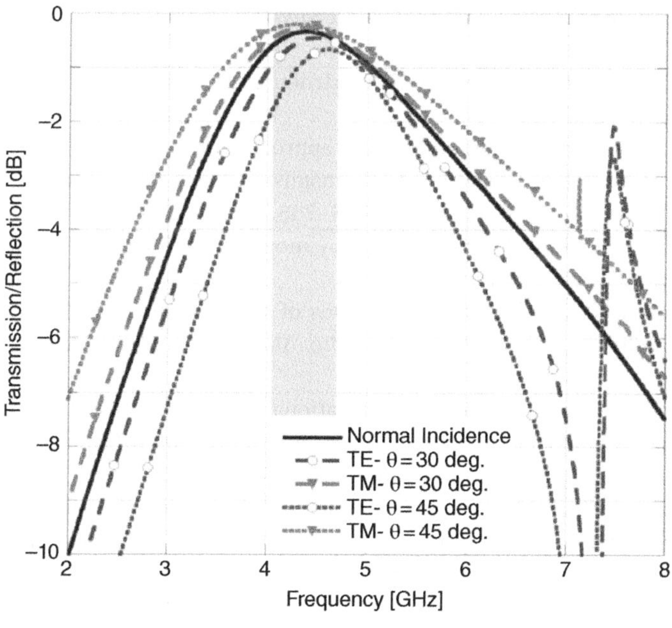

Figure 3.30 Transmission coefficient of the radome for 0, 30, and 45 deg incident angles and for TE and TM polarizations in receive mode [3].

One can observe the stability of the pass-band for 0, 30, and 45 deg incident angles for both polarizations, TE and TM. This feature is typical to metamaterial characteristics.

References

1 Rittenhouse, D. An optical problem, proposed by Mr. Hopkinson, and solved by Mr. Rittenhouse. *Trans. Amer. Phil. Soc.*, 2, 201–206, 1786.

2 Munk, B. Frequency selective surfaces-theory and design. New York: John Wiley, 2000.

3 Costa F, and Monorchio, A. A frequency selective radome with wideband absorbing properties. *IEEE Trans. on Antennas and Propagat.*, 60(6), 2740–2748, 2012.

4 Mittra, R, Chan, CH, and Cwik, TA. Techniques for Analyzing Frequency Selective Surfaces—A Review. *Proceedings of IEEE*, 76(12), 1593–1615, 1988.

5 Cwik, TA, and Mittra, R. Scattering from a periodic array free-standing arbitrarily shaped perfectly conducting or resistive patches. *IEEE Trans. Antennas and Propagat.*, 35(11), 1226–1234, 1987.

6 Harrington, RF. Time-harmonic electromagnetic fields. New York: McGraw-Hill, 1961.

7 T. Itoh. Spectral domain immitance approach for dispersion characteristics of generalized printed transmission lines. *IEEE Trans. Microwave Theory Tech.*, 28, (7), 733–736, 1980.

8 Harrington, RF. Field computation by moment methods. New York: IEEE Press, 1968.

9 Hestenes, MR, and Stiefel, E. Methods of conjugate gradient for solving linear systems, *J. Res. Nat. Bur. Stand.*, 49(6), 409–436, 1952.

10 van den Berg, PM. Iterative computational techniques in scattering based upon the integrated square error criterion. *IEEE Trans. Antennas Propagat.*, 32(10), 1063–1070, 1984.

11 Glisson, AW, and Wilton, DR. Simple and efficient numerical methods for problems of electromagnetic radiation and scattering from surfaces. *IEEE Trans. Antennas Propagat.*, 28(5), 593–603, 1980.

12 Rubin, B.J, and Bertoni, HL. Reflection from a periodically perforated plane using a subsectional current approximation. *IEEE Trans. Antennas Propagat.*, 31(6), 829–836, 1983.
13 Bozzi, M, and Perregrini, L. Analysis of multilayered printed frequency selective surfaces by the MoM/BI-RME method. *IEEE Trans. Antennas Propagat.*, 51(10), 2830–2836, 2003.
14 Cwik, T, and Mittra, R. The cascade connection of planar periodic surfaces and lossy dielectric layers to form an arbitrary periodic screen. *IEEE Trans. Antennas Propagat.*, 35(12), 1397–1405, 1987.
15 ElMahgoub, K, Yang, F, Elsherbeni, AZ, Demir, V, and Chen, J. Analysis of a multilayered periodic structures using a hybrid FDTD/GSM method. *IEEE Antennas Propagat. Magazine,* 54(2), 57–73, 2012.
16 Capolino, F. Applications of metamaterials. New York: CRC Press, 2009.
17 Ziolkowski, RW, and Engheta, N. Electromagnetic metamaterials: Physics and engineering exploration. New York: John Wiley-IEEE Press, 2006.
18 Costa, F, Monorchio, A, and Manara, G. Analyisis and design of ultra thin electromagnetic absorbers comprising resistively loaded high impedance surfaces. *IEEE Trans. Antennas Propag.*, 58(5), 1551–1558, 2010.
19 Costa, F, Monorchio, A, and Manara, G. Efficient analysis of frequency selective surfaces by a simple equivalent circuit approach. *IEEE Antennas Propag. Mag.*, 54(4), 35–48, 2012.
20 Pozar, DM. Microwave Engineering, 2nd edition. New York: John Wiley & Sons, 1998.

Problems

P3.1 Prove eq. (3.6).

P3.2 Prove eqs. (3.30) and (3.31).

P3.3 Prove eq. (3.82).

P3.4 Design a periodic FSS with pass-band 4.2–4.9 GHz and insertion loss smaller than 0.5 dB. Use for the design three types of unit cell elements: slottted cross, slottted Jerusalem cross, and interdigitated slotted cross in an infinite PEC ground plane, as shown in Fig. 3.24. The unit cell sizes are 32 mm, 17.5 mm, and 16.5 mm, respectively. Find the geometry of the unit cell elements and plot the transmission coefficient of the three cases as a function of frequency in the frequency band 2–18 GHz for normal incidence. Identify for each design the grating lobes of the structure.

P3.5 Design an absorptive periodic FSS with rectangular ring elements with thickness 1 mm and with surface resistance of 15 [Ω/sq]. The FSS is backed by the interdigitated FSS evaluated in P3.4 at a distance $d = 5$ mm and unit cell size 11 mm, as shown in Fig. 3.22. The required frequency band of the absorptive FSS is 10–18 GHz, the minimum absorption required in the band is 15 dB, and its pass-band is 4.2–4.9 GHz. Find the geometry of the absorptive FSS and plot the transmission coefficient as a function of frequency for normal incidence.

P3.6 Compute the transmission coefficient of the absorptive periodic FSS designed in P3.5 for TE and TM polarizations and for incident angles 30 deg and 45 deg.

4

Airborne Radomes

Analysis of airborne radome effects on antenna performance has been carried out in the literature for both transmitting and receiving antennas. The radome shape is usually conformal and is determined by aero dynamics considerations like minimum weight, minimum drag, strength of the mechanical structure and the radome sustainability for a wide range temperature. There are several methods to analyze the conformal radome electrical performance as outlined in the following. Each method has its own advantages and deficiencies, and the proper method for the analysis depends on parameters such as the size of the radome in terms of wavelength, the smoothness of its conformal shape (it has tips or not), and its structural texture (multilayer, inhomogeneous, includes FSS, etc.).

Paris [1] has presented in the early 1970s a technique for determining the transmitting antenna pattern of a horn antenna covered by an airborne radome. His procedure is based on calculating the near fields incident on the radome by an aperture integration technique, treating the incident fields as local plane waves with a certain spectrum, applying plane-wave transmission coefficients through the radome using the ABCD transmission matrix formulation [2] as discussed in Chapter 2, and then treating the fields exterior to the radome as equivalent sources of the radiated fields and using physical optics to compute the far field parameters, such as radiation pattern, insertion loss (IL), the insertion phase delay (IPD) and the boresight error (BSE). The IPD is defined as the phase delay introduced by the presence of the wall relative to free-space conditions. The IPD is thus the negative of the phase angle of the insertion transmission coefficient.

Radome Electromagnetic Theory and Design, First Edition. Reuven Shavit.

The Paris method [1] simplifies to the *ray-tracing* or *high-frequency (HF) technique* if the near field of the enclosed antenna incident upon the radome can be approximated by a single plane wave propagating normal to the antenna aperture. A similar approach can be found in [3] to obtain the boresight error (BSE) for a circular aperture in an ogive radome. The ray-tracing method is described in section 4.1 and is especially effective for electrically large radomes with spherical and cylindrical shapes because of its computational efficiency. Higher-order effects generated by tip scatter and surface waves are neglected by the ray-tracing technique. The computation of the radome enclosed antenna radiation pattern is time and memory consuming because the integration on the radome surface is repeated for each observation angle in space. An alternative and more effective procedure is described in section 4.1.1 and is based on a multilevel algorithm [4] using an hierarchical decomposition of the radiating aperture (radome surface).

The plane wave method described in section 4.1 is especially efficient for large (in wavelength) antennas and radomes. In case of a small radome (on the order of a few wavelengths in length) of arbitrary shape and small antennas enclosed, the surface equivalence principle can be used to model the effect of the radome on the transmitted field, in terms of equivalent electric and magnetic surface currents, which radiate in an unbounded medium, as described in section 4.2. This method is called the surface integral equation (SIE) method. The analysis presented is based on [5]. The boundary conditions on the tangential components of the total fields give a coupled set of integral equations involving these currents. The method of moments (MoM) is used to solve numerically these integral equations using triangular basis functions.

An alternate method to the SIE method to analyze antenna radiation in the presence of dielectric radomes is the volume integral equation (VIE) method, as described in section 4.3. In solving the VIE, the radome is modeled by small volume cells of tetrahedron or hexahedron shape so that 3D complex radomes can be modeled accurately. Its significant advantage is that it can analyze inhomogeneous dielectrics. In general, the SIE produces fewer unknowns for homogeneous dielectrics. Improvement of the computational efficiency can be obtained by combining the VIE with the multilevel fast multipole algorithm (MLFMA), as described in section 4.3.1.

The SIE and VIE numerical techniques based on MoM produce highly populated matrices with high condition number, which affect negatively the accuracy of the solution. In case of radomes with arbitrary inhomogeneities, we can use the finite element method (FEM) for the numerical analysis. This method offers three additional advantages:

1. Fill time of the matrix equation is substantially smaller than that with the MoM approach.
2. The resulting matrix is highly sparse and can be efficiently solved
3. The radome shape can be conveniently described using an automatic mesh generation algorithm. This method is described in section 4.4.

In the analysis, the reciprocity theorem is used in conjunction with an *axisymmetric* finite element approach that employs an absorbing boundary condition for mesh truncation. A two-step procedure is used, which initially employs the FEM to determine the near fields within the radome when it is illuminated by a distant source. In the second step, the reciprocity theorem is used to find the far-field pattern of an antenna with a given current distribution enclosed in the radome, from the near-field information determined earlier in the first step. The method presented assumes that the presence of the radome does not have a significant effect on the antenna currents.

4.1 Plane Wave Spectrum Combined with Surface Integration Technique

Plane wave surface integration technique or the high-frequency (HF) method, assumes locally planar surfaces and has been used successfully for the analysis of smooth spherical and cylindrical radomes. Being fast and easy to implement, HF method permits the analysis of smooth tip radomes [1]. This method is also called ray-tracing analysis combined with physical optics (PO) and has been implemented in the commercial software called SAVANT from ANSYS. However, HF method is not accurate in the analysis of sharp radome tip, since the local flat surface assumption is not satisfied at the tip.

A typical airborne radome can be described by a two-dimensional super-quadratic equation revolved about its main axis to obtain a body of revolution family of geometric shapes called super spheroids [6].

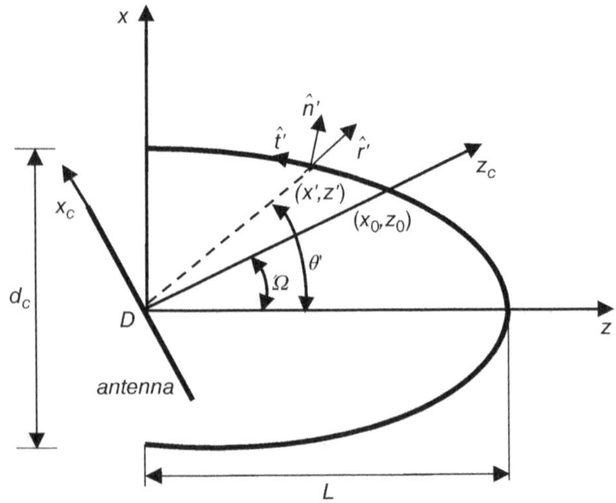

Figure 4.1 The cross section of an airborne radome based on super-spheroids geometry profile.

The two-dimensional super-quadratic equation is shown in Fig. 4.1 and can be described mathematically by:

$$\left(\frac{x'}{d_c/2}\right)^{v}+\left(\frac{z'}{L}\right)^{v}=1 \tag{4.1}$$

in which L is the radome length and d_c is the radome base diameter ($L > d_c\,/2$), v is a positive real number and $D = 2a$ is the aperture antenna diameter.

By revolving this equation about the z-axis, we obtain a body of revolution (BOR) given by

$$x'^2+y'^2=\frac{(d_c/2)^2}{L^2}(L^v-z'^v)^{2/v} \tag{4.2}$$

The radome's surface $f(\rho',\phi',z')$ in cylindrical coordinates is not dependent on ϕ' due to symmetry and can be described by

$$\begin{aligned} f(\rho',\phi',z') &= x'^2+y'^2-\frac{(d_c/2)^2}{L^2}(L^v-z'^v)^{2/v} \\ &= \rho'^2-\frac{(d_c/2)^2}{L^2}(L^v-z'^v)^{2/v}=0 \end{aligned} \tag{4.3}$$

Some interesting shapes can be identified as a function of the parameter v. Thus, if $v = 2$, we obtain prolate spheroid, if $v = 1.381$ yields almost exactly the traditional von Karman radome geometry and

$\nu = 1$ yields the conical shape. The range $1 < \nu < 2$ yields various types of radome cross sections for high-speed aircraft and missiles. The real benefit of this analytical cross section is the fact that all geometry parameters can be obtained analytically. Setting the antenna gimbal point at the coordinate origin, the antenna aperture central ray (normal to the aperture) intersects the radome at (x_0, z_0). One can observe from Fig. 4.1 that the coordinates x_0 and z_0 are interrrelated through

$$x_0 = z_0 \tan \Omega \tag{4.4}$$

where $'\Omega$ is the antenna gimbal angle. Substitution of (4.4) into (4.2) for $y' = 0$ yields the z-coordinate of the point at which the central ray intersects the super-quadratic radome's function

$$z_0 = \frac{L \cdot (d_c/2)}{[L^\nu (\tan \Omega)^\nu + (d_c/2)^\nu]^{1/\nu}}. \tag{4.5}$$

Using (4.3), we can compute the unit normal vector to the radome's surface by

$$\hat{n}' = \frac{\nabla f}{|\nabla f|} = \frac{\partial f}{\partial \rho'}\hat{\rho} + \frac{1}{\rho'}\frac{\partial f}{\partial \phi'}\hat{\phi} + \frac{\partial f}{\partial z'}\hat{z} = \frac{L(L^\nu - z'^\nu)^{1-(1/\nu)}\hat{\rho} + (d_c/2)z'^{\nu-1}\hat{z}}{\{L^2(L^\nu - z'^\nu)^{2-(2/\nu)} + (d_c/2)^2 z'^{2(\nu-1)}\}^{1/2}} \tag{4.6}$$

Similarly, we can compute the unit tangent vector to the radome's surface using (4.6):

$$\hat{t}' = \hat{n}' \times \hat{\phi} = \frac{L(L^\nu - z'^\nu)^{1-(1/\nu)}\hat{z} - (d_c/2)z'^{\nu-1}\hat{\rho}}{\{L^2(L^\nu - z'^\nu)^{2-(2/\nu)} + (d_c/2)^2 z'^{2(\nu-1)}\}^{1/2}} \tag{4.7}$$

Moreover, the incidence angle θ_i is given by

$$\begin{aligned} \cos \theta_i &= (\hat{r}' \cdot \hat{n}') \\ \sin \theta_i &= (\hat{r}' \cdot \hat{t}') \end{aligned} \tag{4.8}$$

In case that the antenna aperture is large in terms of wavelength and the radome surface is in its near field region, we can make the assumption that the field radiated from the aperture can be described by a locally single plane wave propagating in $'\Omega$ direction. In this case $\hat{r}'$ can be written in the form

$$\hat{r}' = \frac{\hat{z} + \tan \Omega \hat{x}}{\sqrt{1 + \tan^2 \Omega}} \tag{4.9}$$

Division of (4.8)b by (4.8)a and substituting of (4.5) into (4.7) and (4.6), yields a convenient form for the incident angles of super-quadratic

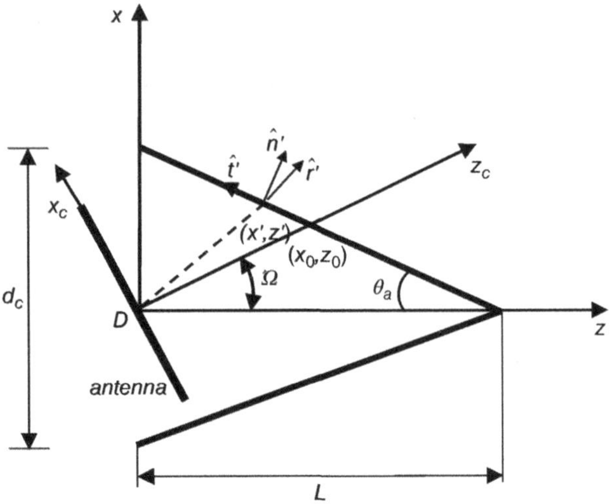

Figure 4.2 Antenna and conical radome geometry.

shapes in the x-z plane as a function of the antenna gimbal angle, i.e.,

$$\tan\theta_i = \frac{[L^{2\nu}(\tan\Omega)^\nu]^{1-(1/\nu)} - (d_c/2)^\nu a^{\nu-2}\tan\Omega}{\tan\Omega[L^{2\nu}(\tan\Omega)^\nu]^{1-(1/\nu)} + (d_c/2)^\nu L^{\nu-2}} \tag{4.10}$$

In the following, we consider the analysis of a conical radome surface ($\nu = 1$) as a typical example of the super-spheroids family radomes. Fig. 4.2 shows an antenna and a conical radome geometry with a length L. The enclosed antenna is a uniformly illuminated circular aperture with diameter $D = 2a$ and linearly y polarized. Its Cartesian coordinate system is (x_c,y_c,z_c) or (ρ_c,ϕ_c,z_c) in cylindrical coordinates. Similarly, the Cartesian radome coordinate system is (x,y,z) and its cylindrical coordinates is (ρ,ϕ,z). The gimbal angle of the aperture is denoted by 'Ω.

The radome geometry as shown in Fig. 4.2 can be defined by eq. (4.11).

$$f(\rho', z') = \sqrt{x'^2 + y'^2}\cos\theta_a + z'\sin\theta_a - L\sin\theta_a = 0 \tag{4.11}$$

Accordingly, the unit normal vector (on the interior surface) of the radome taking in consideration the radome symmetry with respect to ϕ' is defined by

$$\hat{n}' = \frac{\nabla f}{|\nabla f|} = \frac{x'\cos\theta_a}{\sqrt{x'^2 + y'^2}}\hat{x} + \frac{y'\cos\theta_a}{\sqrt{x'^2 + y'^2}}\hat{y} + \sin\theta_a\hat{z} \tag{4.12}$$

The coordinate transformation from the (x,y,z) to the (x_c,y_c,z_c) system can be expressed by the matrix eq.

$$\begin{pmatrix} x_c \\ y_c \\ z_c \end{pmatrix} = \begin{pmatrix} \cos\Omega & 0 & -\sin\Omega \\ 0 & 1 & 0 \\ \sin\Omega & 0 & \cos\Omega \end{pmatrix} \begin{pmatrix} x \\ y \\ z \end{pmatrix} \tag{4.13}$$

Similarly, we can express the coordinate system (x,y,z) in terms of the coordinate system (x_c,y_c,z_c) by matrix inversion of (4.13) to obtain

$$\begin{pmatrix} x \\ y \\ z \end{pmatrix} = \begin{pmatrix} \cos\Omega & 0 & \sin\Omega \\ 0 & 1 & 0 \\ -\sin\Omega & 0 & \cos\Omega \end{pmatrix} \begin{pmatrix} x_c \\ y_c \\ z_c \end{pmatrix} \tag{4.14}$$

We can also express the unit vectors of (x_c,y_c,z_c) system in terms of the unit vectors of the (x,y,z) system, such that

$$\begin{aligned} \hat{x}_c &= \cos\Omega\hat{x} - \sin\Omega\hat{z} \\ \hat{y}_c &= \hat{y} \\ \hat{z}_c &= \sin\Omega\hat{x} + \cos\Omega\hat{z} \end{aligned} \tag{4.15}$$

The incident angle, θ_i at an intercept point of a ray orthogonal to thc antenna aperture and the normal to the internal conical radome surface can be computed by using (4.12) and (4.15)c through

$$\begin{aligned} \theta_i &= \cos^{-1}(\hat{n}' \cdot \hat{z}_c) \\ &= \cos^{-1}\left[\left[\frac{x'\cos\theta_a}{\sqrt{x'^2+y'^2}}\hat{x} + \frac{y'\cos\theta_a}{\sqrt{x'^2+y'^2}}\hat{y} + \sin\theta_a\hat{z}\right] \cdot (\sin\Omega\hat{x} + \cos\Omega\hat{z})\right] \\ &= \cos^{-1}\left[\frac{x'\cos\theta_a\sin\Omega}{\sqrt{x'^2+y'^2}} + \sin\theta_a\cos\Omega\right] \end{aligned} \tag{4.16}$$

If we denote by $d(\theta_i)$ the distance of a ray normal to the antenna aperture traveling to the inner surface of the radome, it can be computed by substituting eq. (4.14) into (4.11) to obtain

$$\begin{aligned} &\sqrt{[x_c\cos\Omega + d(\theta_i)\sin\Omega]^2 + y_c^2}\cos\theta_a + [-x_c\sin\Omega + d(\theta_i)\cos\Omega]\sin\theta_a \\ &- L\sin\theta_a = 0 \end{aligned} \tag{4.17}$$

Eq. (4.17) is a quadratic equation with $d(\theta_i)$ being the unknown

$$\alpha \cdot d(\theta_i)^2 + \beta \cdot d(\theta_i) + \gamma = 0 \tag{4.18}$$

in which

$$\begin{aligned} \alpha &= \sin^2\Omega\cos^2\theta_a - \cos^2\Omega\sin^2\theta_a \\ \beta &= 2x_c\sin\Omega\cos\Omega + 2L\cos\Omega\sin^2\theta_a \\ \gamma &= x_c^2(\cos^2\Omega\cos^2\theta_a - \sin^2\Omega\sin^2\theta_a) - 2Lx_c\sin^2\theta_a\sin\Omega - L^2\sin^2\theta_a + y_c^2\cos^2\theta_a \end{aligned} \tag{4.19}$$

The solution of the quadratic equation (4.18) results in $d(\theta_i)$:

$$d(\theta_i) = \frac{-\beta \pm \sqrt{\beta^2 - 4\alpha\gamma}}{2\alpha} \tag{4.20}$$

The physical solution is a positive $d(\theta_i)$.

The near fields of the circular aperture antenna with y polarization can be calculated in terms of the aperture antenna plane wave spectrum (PWS) $\underline{\Im}(k_{\rho c}, \alpha_c, z_c) = \underline{F}(k_{\rho c}, \alpha_c)e^{-jk_{zc}z_c}$, such that its radiated time harmonic $\{e^{j\omega t}\}$ electric field can be expressed by [3]

$$\begin{aligned} \underline{E}(\rho_c, \phi_c, z_c) &= \frac{1}{4\pi^2}\int_0^{\infty}\int_0^{2\pi} \underline{F}(k_{\rho c}, \alpha_c)e^{-jk_{\rho c}\rho_c\cos(\phi_c-\alpha_c)}e^{-jk_{zc}z_c}k_{\rho c}dk_{\rho c}d\alpha_c \\ \underline{\Im}(k_{\rho c}, \alpha_c, z_c) &= \int_0^{\infty}\int_0^{2\pi} \underline{E}(\rho_c, \phi_c, z_c)e^{jk_{\rho c}\rho_c\cos(\phi_c-\alpha_c)}\rho_c d\rho_c d\phi_c \end{aligned} \tag{4.21}$$

in which $k_{zc}^2 = k^2 - k_{\rho c}^2; k = \omega\sqrt{\mu\varepsilon}$. Evaluation of (4.21)b on the antenna aperture ($z_c = 0$) and taking in consideration that the field on the aperture is y_c polarized results in $F_x(k_{\rho c}, \alpha_c) = 0$. Therefore, $\underline{F}(k_{\rho c}, \alpha_c) = F_y(k_{\rho c}, \alpha_c)\hat{y}_c + F_z(k_{\rho c}, \alpha_c)\hat{z}_c$. Moreover,

$$F_y(k_{\rho c}, \alpha_c) = \int_0^{a}\int_0^{2\pi} E_y(\rho_c, \phi_c, 0)e^{jk_{\rho c}\rho_c\cos(\phi_c-\alpha_c)}\rho_c d\rho_c d\phi_c \tag{4.22}$$

Moreover, for a circularly symmetric aperture distribution the spectrum is independent of α_c, such that $F_y(k_{\rho c}, \alpha_c) = F_y(k_{\rho c})$. This simplifies (4.21)a to

$$\begin{aligned} E_y(\rho_c, \phi_c, z_c) &= \frac{1}{4\pi^2}\int_0^{\infty}\int_0^{2\pi} F_y(k_{\rho c})e^{-jk_{\rho c}\rho_c\cos(\phi_c-\alpha_c)}e^{-jk_{zc}z_c}k_{\rho c}dk_{\rho c}d\alpha_c \\ &= \frac{1}{2\pi}\int_0^{\infty} F_y(k_{\rho c})J_0(k_{\rho c}\rho_c)e^{-jk_{zc}z_c}k_{\rho c}dk_{\rho c} \end{aligned} \tag{4.23}$$

Substitution of (4.21)a into the third Maxwell eq., i.e., $\nabla \cdot \underline{E} = 0$ results into

$$\begin{aligned} \nabla \cdot \underline{E} &= \nabla \cdot \left\{ \frac{1}{4\pi^2}\int_0^{\infty}\int_0^{2\pi} \underline{F}(k_{\rho c}, \alpha_c)e^{-j\underline{k}_c\cdot\underline{r}_c}k_{\rho c}dk_{\rho c}d\alpha_c \right\} \\ &= \frac{1}{4\pi^2}\int_0^{\infty}\int_0^{2\pi} \underline{F}(k_{\rho c}, \alpha_c)\cdot\nabla(e^{-j\underline{k}_c\cdot\underline{r}_c})k_{\rho c}dk_{\rho c}d\alpha_c = 0 \end{aligned} \tag{4.24}$$

in which

$$\begin{aligned} &\underline{r}_c = x_c\hat{x}_c + y_c\hat{y}_c + z_c\hat{z}_c; \\ &\underline{k} = k_{xc}\hat{x}_c + k_{yc}\hat{y}_c + k_{zc}\hat{z}_c = k_{\rho c}\cos\alpha_c\hat{x}_c + k_{\rho c}\sin\alpha_c\hat{y}_c + k_{zc}\hat{z}_c \end{aligned}$$

and since $\nabla(e^{-j\underline{k}\cdot\underline{r}_c}) = -j\underline{k}e^{-j\underline{k}\cdot\underline{r}_c}$, eq. (4.24) simplifies to

$$\begin{aligned} &\underline{F}\cdot\nabla(e^{-j\underline{k}\cdot\underline{r}_c}) = -j\underline{F}\cdot\underline{k}e^{-j\underline{k}\cdot\underline{r}_c} = 0 \\ &\underline{F}\cdot\underline{k} = F_y k_{yc} + F_z k_{zc} = 0 \\ &F_z = -\frac{k_{yc}}{k_{zc}}F_y = -\frac{k_{\rho c}\sin\alpha_c}{k_{zc}}F_y \end{aligned} \tag{4.25}$$

Substitution of (4.25)c into (4.21)a yields

$$\begin{aligned} \underline{E}(\rho_c, \phi_c, z_c) = \frac{1}{4\pi^2}\int_0^\infty \int_0^{2\pi} \left(\hat{y}_c - \hat{z}_c\frac{k_{\rho c}\sin\alpha_c}{k_{zc}}\right) \\ \times F_y(k_{\rho c})e^{-jk_{\rho c}\rho_c\cos(\phi_c-\alpha_c)}e^{-jk_{zc}z_c}k_{\rho c}dk_{\rho c}d\alpha_c \end{aligned} \tag{4.26}$$

Further simplification can be obtained if we perform analytically the integration over α_c and use the Bessel function property [7].

$$\int_0^{2\pi} \begin{Bmatrix} \sin(n\alpha_c) \\ \cos(n\alpha_c) \end{Bmatrix} e^{jk_{\rho c}\rho_c\cos(\phi_c-\alpha_c)}d\alpha_c = 2\pi j^n \begin{Bmatrix} \sin(n\phi_c) \\ \cos(n\phi_c) \end{Bmatrix} J_n(k_{\rho c}\rho_c) \tag{4.27}$$

We then obtain the aperture antenna radiated electric field

$$\begin{aligned} &E_y(\rho_c, \phi_c, z_c) = \frac{1}{2\pi}\int_0^\infty F_y(k_{\rho c})J_0(k_{\rho c}\rho_c)e^{-jk_{zc}z_c}k_{\rho c}dk_{\rho c} \\ &E_z(\rho_c, \phi_c, z_c) = \frac{j\sin\phi_c}{2\pi}\int_0^\infty F_y(k_{\rho c})J_1(k_{\rho c}\rho_c)e^{-jk_{zc}z_c}\frac{k_{\rho c}^2}{k_{zc}}dk_{\rho c} \end{aligned} \tag{4.28}$$

in which $J_0(k_{\rho c}\rho_c)$ and $J_1(k_{\rho c}\rho_c)$ are Bessel functions of zero and first orders, respectively.

To complete the formulation and evaluate the magnetic fields $\underline{H}(x_c, y_c, z_c)$, we use the first Maxwell equation

$$\underline{H} = -\frac{1}{j\omega\mu}\nabla\times\underline{E} = -\frac{1}{j\omega\mu}\nabla\times\left[\frac{1}{4\pi^2}\int_0^\infty\int_0^{2\pi}\underline{F}(k_{\rho c}, \alpha_c)e^{-j\underline{k}\cdot\underline{r}_c}k_{\rho c}dk_{\rho c}d\alpha_c\right] \tag{4.29}$$

Interchanging integration with differentiation and using the vector identity

$$\nabla\times(\alpha\underline{A}) = \alpha\nabla\times\underline{A} + (\nabla\alpha)\times\underline{A} \tag{4.30}$$

and the identity $\nabla(e^{-j\underline{k}\cdot\underline{r}_c}) = -j\underline{k}e^{-j\underline{k}\cdot\underline{r}_c}$ reduces (4.29) to

$$\underline{H} = -\frac{1}{4\pi^2 k\eta}\int_0^\infty\int_0^{2\pi}(\underline{F}(k_{\rho c})\times\underline{k})e^{-j\underline{k}\cdot\underline{r}_c}k_{\rho c}dk_{\rho c}d\alpha_c \tag{4.31}$$

If we substitute (4.25) into (4.31) and perform analytically the integration over α_c, one can show that the radiated magnetic field components by the antenna aperture can be expressed in terms of the plane wave expansion function $F_y(k_{\rho c})$:

$$H_x(\rho_c, \phi_c, z_c) = -\frac{1}{2\pi k\eta}\int_0^{\infty} F_y(k_{\rho c})\left[\left(k^2 - \frac{1}{2}k_{\rho c}^2\right)J_0(k_{\rho c}\rho_c) + \frac{1}{2}k_{\rho c}^2 \cos 2\phi_c J_2(k_{\rho c}\rho_c)\right]$$

$$\times e^{-jk_{zc}z_c}\frac{k_{\rho c}}{k_{zc}}dk_{\rho c}$$

$$H_y(\rho_c, \phi_c, z_c) = -\frac{\sin 2\phi_c}{4\pi k\eta}\int_0^{\infty} F_y(k_{\rho c})J_2(k_{\rho c}\rho_c)e^{-jk_{zc}z_c}\frac{k_{\rho c}^2}{k_{zc}}dk_{\rho c}$$

$$H_z(\rho_c, \phi_c, z_c) = -\frac{j\cos\phi_c}{2\pi k\eta}\int_0^{\infty} F_y(k_{\rho c})J_1(k_{\rho c}\rho_c)e^{-jk_{zc}z_c}k_{\rho c}^2 dk_{\rho c} \tag{4.32}$$

In the special case of a y-polarized uniform circular aperture antenna the spectrum F_y can be computed through (4.22), using (4.27) and the identity 6.561(5) from [8], which results into

$$F_y(k_{\rho c}, \alpha_c) = \int_0^a \int_0^{2\pi} e^{jk_{\rho c}\rho_c \cos(\phi_c - \alpha_c)}\rho_c d\rho_c d\phi_c$$

$$= 2\pi\int_0^a J_0(k_{\rho c}\rho_c)\rho_c d\rho_c = \frac{2\pi a}{k_{\rho c}}J_1(k_{\rho c}a) \tag{4.33}$$

The plane wave spectrum analysis treats the transmission through a point on the radome as the transmission of a local plane wave through a planar sheet tangential to the point on the curved radome surface. The transmission through a given point on the radome can be treated approximately as that of a single plane wave incident in the direction of the Poynting vector of the antenna near field. The unit Poynting vector $\hat{p}$ is obtained from the knowledge of the near fields on the radome inner surface and is given by

$$\hat{p} = \frac{\mathrm{Re}(\underline{E}_i \times \underline{H}_i^*)}{|\mathrm{Re}(\underline{E}_i \times \underline{H}_i^*)|} \tag{4.34}$$

in which $\underline{E}_i$ and $\underline{H}_i$ are the incident electric and magnetic fields, which can be computed through (4.28) and (4.32) at internal points of the radome. The angle of incidence can be computed through

$$\theta_i = \cos^{-1}(\hat{n}' \cdot \hat{p}) \tag{4.35}$$

where $\hat{n}'$ is the normal of the internal surface of the radome and for the conical radome is given by (4.12). The transmission through a given point on the radome can be treated approximately as that of a single plane wave incident in the direction of the Poynting vector of the antenna near field.

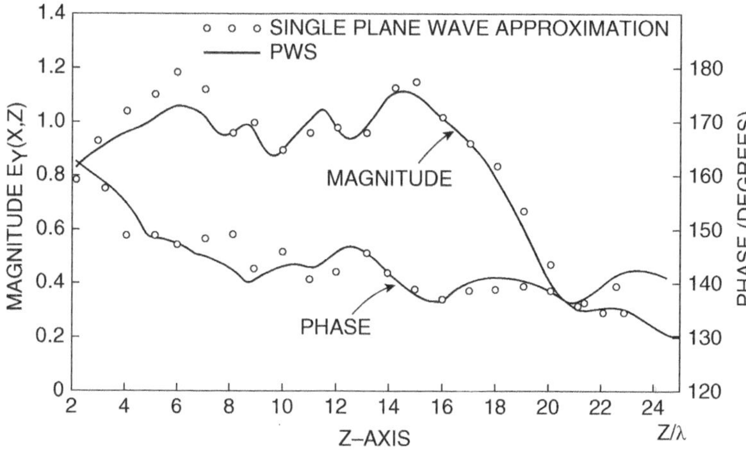

Figure 4.3 Transmitted field on outer radome surface using PWS and single plane wave approximation [3].

Fig. 4.3 shows a comparison between the transmitted field on the outer radome surface using PWS and single plane wave approximation for a radome with gimbal angle '$\Omega = 20$ deg and apex angle $\theta_a = 20$ deg as shown in Fig. 4.2 and the radome is 24.5λ long with a base width of 12λ. A linear polarized 10λ diameter uniform antenna is used in these calculations.

It is seen that the single plane wave approximation for both the magnitude and phase of $E_y(x_c, z_c)$ check well with the full PWS analysis for $z_c/\lambda > 12$. Hence the single plane wave approximation is accurate in that region. However, for $z_c/\lambda < 12$, small deviations are observed between the two techniques. It is concluded that the single plane wave approximation is accurate except for some error in the region within 12λ of the antenna aperture.

Because of the complexity to modify the spectrum of plane waves by the appropriate weighting function at each point over a 3-D radome surface, the single plane wave approximation (ray-tracing technique) is used in the following treatment. Thus, in the transmitted field analysis, the near fields incident at a point on the radome inner surface are decomposed into perpendicular and parallel components in the plane of incidence. The set of vectors $\hat{t}'$, $\hat{n}'$, $\hat{b}'$ commonly known as tangential, normal, and binormal unit vectors are used to decompose the incident fields $(\underline{E}_i, \underline{H}_i)$ into the perpendicular and parallel components. These vectors are related through

$$\hat{b}' = \frac{\hat{n}' \times \hat{p}}{|\hat{n}' \times \hat{p}|} \tag{4.36}$$

and

$$\hat{t}' = \hat{n}' \times \hat{b}' \tag{4.37}$$

with $\hat{n}'$ and $\hat{p}$ given by (4.12) and (4.34), respectively. Thus, the tangential components of the transmitted electric and magnetic fields through the radome are given by

$$\underline{E}_t = [(\hat{b}' \cdot \underline{E}_i)\hat{b}']T_\perp + [(\hat{t}' \cdot \underline{E}_i)\hat{t}']T_\parallel \tag{4.38}$$

and

$$\underline{H}_t = [(\hat{b}' \cdot \underline{H}_i)\hat{b}']T_\parallel + [(\hat{t}' \cdot \underline{H}_i)\hat{t}']T_\perp \tag{4.39}$$

in which $T_\perp$ and $T_\parallel$ are the plane wave transmission coefficients of the radome wall for perpendicular and parallel polarizations. The corresponding reflection coefficients are designated by $\Gamma_\parallel$ and $\Gamma_\perp$, respectively. The method used to find the transmission coefficients is based on the ABCD transmission matrix formulation [2] discussed in Chapter 2. This formulation is well suited to the analysis of multiple layered sandwiched walls and thin etched metal layers, which act as lumped circuit elements and FSS layers.

Next, the surface integration (SI) method is applied in order to determine the pattern distortion and the corresponding boresight error due to the radome. Since the surface integration is performed in the radome coordinates, it is necessary to transform the antenna coordinates into the radome coordinates. The transformation from the antenna coordinates to the radome coordinates is given by (4.14). Accordingly, the enclosed antenna distorted radiation pattern (far field approximation) is given by [9]

$$\underline{E}_p(r,\theta,\phi) = jk\frac{e^{-jkr}}{4\pi r}\left\{\int_{S_a}[\eta(\hat{n}' \times \underline{H}_t) \times \hat{r} + (\hat{n}' \times \underline{E}_t)] \times \hat{r}e^{jk\underline{r}'\cdot\hat{r}}ds'\right\} \tag{4.40}$$

in which $\eta = \sqrt{\mu/\varepsilon}$, $\underline{E}_t, \underline{H}_t$ are calculated in (4.38) and (4.39), $\hat{r} = \sin\theta\cos\phi\hat{x} + \sin\theta\sin\phi\hat{y} + \cos\theta\hat{z}$ and $\hat{n}'$ is given by (4.12).

So now the entire procedure for the radiation pattern computation in the presence of the radome is in place. Given the aperture distribution on the antenna and making the assumption of a single plane illumination of the radome for each antenna/radome relative position configuration the radiation pattern of the enclosed antenna in the radome can be computed. Initially, the plane wave spectrum $F_y(k_\rho)$ is computed using (4.22) and (4.25). Next, for each ray traveling from the antenna aperture to the radome inner surface, the incident angle θ_i

and the traveling distance $d(\theta_i)$ are calculated using (4.16) and (4.20), respectively. The incident field on the inner surface of the radome is computed using (4.28) and (4.32). Next, for each intersecting point on the radome, the normal, tangential, and binormal unit vectors are computed using (4.12), (4.36), and (4.37), respectively. Then, the parallel and perpendicular transmission coefficients $T_{\|}$, $T_{\perp}$ are computed using the methods outlined in Chapter 2 according to the type of the radome. Next, the transmitted electric field through the radome is computed using (4.38) and (4.37). Finally, the distorted radiation pattern of the antenna in the presence of the radome is computed using (4.40) performing the integration in the radome coordinate system.

One problem of interest is to analyze the distortion introduced by the radome as a function of antenna gimbal angle ′Ω. The deviation of the electrical boresight (defined as the beam maximum) from the geometrical boresight axis at ′Ω is defined as the boresight error (BSE). A typical example is shown in Fig. 4.4 [10]. If the antenna is scanning in a particular direction and the boresight error is in the same direction, it is defined to be positive error. In calculating the boresight error

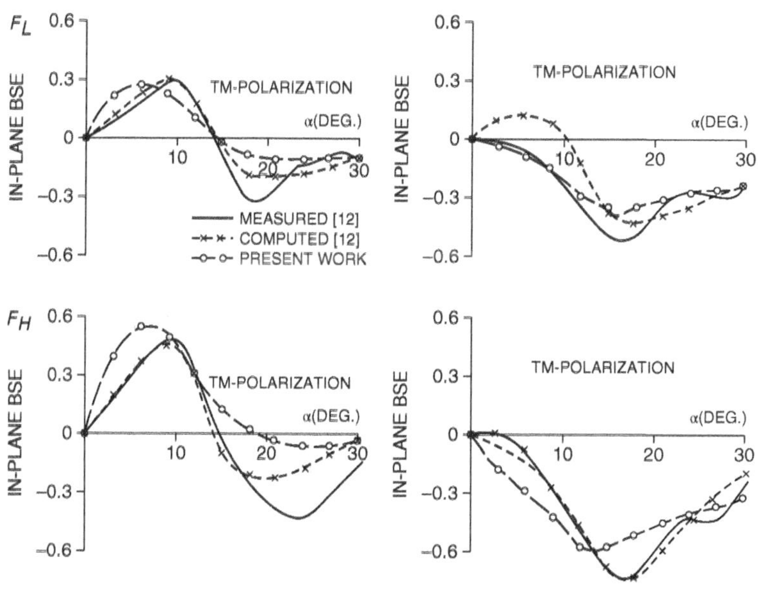

Figure 4.4 Boresight errors for constant wall thickness ogive radome 6.6λ in diameter and 14.4λ in length made of material with dielectric constant, $\varepsilon_r = 5.7$ and operating in the X frequency band with F_L and F_H constituting 2.5% band [10].

several considerations simplify the task. The beam maximum is generally within a fraction of a degree from the gimbal angle 'Ω. Hence, only the radiation pattern at angles near the boresight axis need to be calculated to determine the boresight error. The pattern in the presence of the radome is calculated from (4.40). In addition, the pattern over a small interval enclosing the beam maximum is monotonically decreasing on both sides of the beam maximum and is approximately symmetrical.

The BSE is an intrinsic error that needs to be taken in consideration or calibrated. One possible way to take it in consideration is to measure its value in a testing range for different elevation and azimuth angles on a spatial grid, different polarizations, and discrete frequencies and use these values for calibration for direction findings applications. In case of maximum reception requirement, usually a control system is set to maximize the receiving signal-to-noise ratio, which overrides the antenna pointing coordinates based on the antenna positioner's readings.

A typical example of the conical radome effect on the antenna insertion loss (IL) and radiation pattern is shown in Fig. 4.5. It shows a

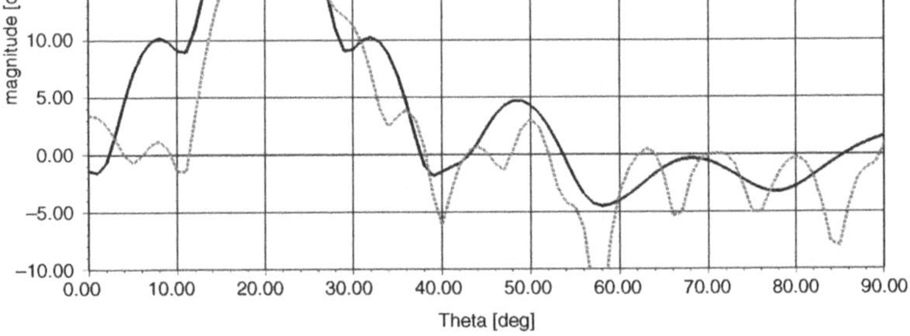

Figure 4.5 Comparison between the radiation patterns of a reflector antenna with diameter $D = 7.5\lambda_0$ and $f/D = 0.28$ inclined $\Omega = 20$ deg from the z-axis (solid line) and that of the reflector covered with a conical radome base diameter $d_c = 11.3\lambda_0$ in, length $L = 15.5\lambda_0$ and $\theta_a = 20$ deg. The radome thickness is $0.15\lambda_0$ with $\varepsilon_r = 5.7$. All simulations are performed with HFSS.

comparison between the radiation patterns of a reflector antenna with diameter $D = 7.5\lambda_0$ and $f/D = 0.28$ inclined $\Omega = 20$ deg from the z-axis and that of the reflector covered with a conical radome with base diameter $d_c = 11.3\lambda_0$, length $L = 15.5\lambda_0$ and $\theta_a = 20$ deg (see Fig. 4.2). The radome is made of a single dielectric layer with $\varepsilon_r = 5.7$ and thickness $0.15\lambda_0$. All simulations are performed with HFSS.

One can observe that the IL of the radome is approximately 2 dB and its boresight error approximately 1 deg. The effect of the radome is also deformation of the sidelobe topology symmetry and level.

Fig. 4.6 shows an additional effect (flash lobe) of the radome presence on the enclosed antenna radiation pattern, and it is due to the specular reflection of the antenna main beam from the radome. The plot shows the location of the flash lobe for different tilt angles of the antenna in the radome. For the radome under consideration the flash lobe is approximately 40 dB down, which in most cases is an acceptable level of degradation.

4.1.1 Multilevel Algorithm for Radiation Pattern Computation

Computation of the radome enclosed antenna radiation pattern using (4.40) is time and memory consuming because the integration is repeated for each observation angle in space. An alternative and more effective procedure is described in [12, 4] based on a multilevel algorithm using a hierarchical decomposition of the radiating aperture (radome surface). In the first phase of the algorithm, are computed the radiation patterns of all subapertures of the finest level over a coarse angular grid. In the next phase, multilevel aggregation of the radiation patterns of neighboring subapertures are used through a phase compensated interpolation to compute the final radiation pattern of the whole aperture. The multilevel algorithm attains the computational complexity of the fast Fourier transform (FFT) algorithm. If we consider $N = kR$, where R is the radius of the smallest sphere circumscribing the aperture and k is the wave number, as a number representing the size of the aperture, the computational complexity can be evaluated as a function of this parameter. The number of points on the aperture surface employed in the physical optics (PO) integration as outlined in (4.40) is proportional to its area, viz. $O(N^2)$. Moreover, based on Nyquist sampling theorem, the minimum number of points required to restore the far field radiation pattern, which is equal to the aperture Fourier transform [13], is

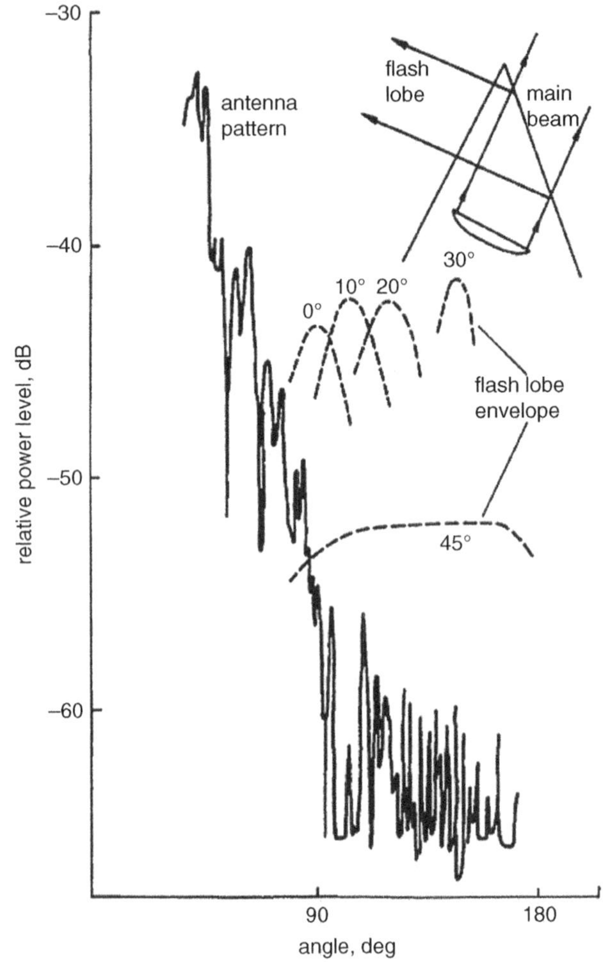

Figure 4.6 Radiation pattern of the antenna in the conical radome with the flash lobe positions for various pointing angles within the radome [11].

also equal to $O(N^2)$. Therefore, the computational complexity of the far-field radiation pattern computed based on (4.40) is $O(N^4)$. The aperture domain decomposition in the proposed method is shown in Fig. 4.7.

In Fig. 4.7 $\underline{r}_n^L$, R_n^L are the center and radius of the sphere circumscribing the subaperture $\underline{S}_n^L$ on level L. The algorithm is based on

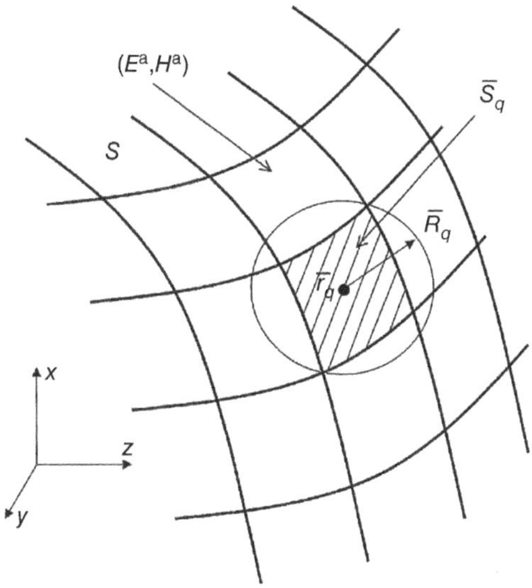

Figure 4.7 Aperture radiation geometry and its domain decomposition.

the observation that the radiation pattern of a finite aperture is a band-limited function of the angles with the angular bandwidth proportional to the linear size of the aperture surface domain. Consequently, the radiation pattern can be evaluated at an arbitrary point by interpolation from its samples, whose number is proportional to the area of the aperture. With this in mind, the aperture (the radome exterior surface) is decomposed into a multilevel hierarchy of subapertures, each of which is considered as a smaller radiating aperture. At every level except for the smallest one, every subaperture is further decomposed into "children" subapertures, whose linear sizes are roughly half of those of their parent. The finest level of subapertures comprises $O(N^2)$ sub-apertures approximately one wavelength in size. The only radiation patterns computed via direct PO integration described in (4.40) are the subapertures on the finest level of decomposition over a very coarse grid of directions. Later, these patterns undergo a multilevel aggregation into the final pattern of the aperture. Each aggregation step involves phase compensation, interpolation, and phase restoration in order to remove rapid phase variations. With $O(\log N)$ levels, the multilevel algorithm attains the $O(N^2 \log N)$ asymptotic complexity in comparison to the $O(N^4)$ computational complexity of the direct evaluation of (4.40). Interestingly, this overall complexity resembles that of the FFT.

The radiation pattern $\underline{U}(\theta,\phi)$ can be defined via the far field $\underline{E}_p(r,\theta,\phi)$ described in (4.40) as

$$\underline{U}(\theta,\phi) = 4\pi r e^{jkr}\underline{E}_p(r,\theta,\phi) \tag{4.41}$$

in the angular sector $\phi \in [\phi_{\min}, \phi_{\max}]$, $[\theta_{\min}, \theta_{\max}]$. $\underline{U}(\theta,\phi)$ can be written in the short form

$$\underline{U}(\theta,\phi) = \int_{S_a} \underline{A}(\hat{r},\underline{r}')e^{jk\underline{r}'\cdot\hat{r}}ds' \tag{4.42}$$

in which $\underline{A}(\hat{r},\underline{r}') = -jk\hat{r}\times[\hat{n}'\times\underline{E}_t(\underline{r}') - \eta\hat{r}\times(\hat{n}'\times\underline{H}_t(\underline{r}'))]$. The formulation is based on a multilevel aperture decomposition. The direct radiation pattern computation using (4.42) is performed only for the smallest subaperture at the highest level of decomposition. As a preprocessing step of the multilevel computational sequence, a hierarchical aperture decomposition is performed. Let $\underline{S}_n^L$ with L = 0,…, M and n=1,…,N^L denote the nth subaperture at level L, where M + 1 is the total number of levels and N^L is the number of subapertures at level L. The zeroth level comprises the total surface of the radome $\underline{S}_1^0$. In general, at each level each "parent" subaperture is subdivided into "children" subapertures, i.e.

$$\underline{S}_m^{L-1} = \bigcup_{n:P^L(n)=m} \underline{S}_n^L \tag{4.43}$$

where $P^L(n)$=m means that the nth subaperture on level L is a child of the parent subaperture m on level L – 1. By each decomposition, the linear dimensions of the subapertures are reduced by a factor of two, such that $R_m^{L-1} \le 2R_n^L$, $P^L(n) = m$. This condition implies that each "parent" will have four "children." The number of divisions depends on the geometry of the radome surface and may be different than two. The radiation pattern of subaperture $\underline{S}_n^L$ is

$$\underline{U}_n^L(\theta,\phi) = e^{-jk\hat{r}\cdot\underline{r}_n^L}\int_{S_n^L} \underline{A}(\hat{r},\underline{r}')e^{jk\underline{r}'\cdot\hat{r}}ds' \tag{4.44}$$

in which the phase factor $e^{-jk\hat{r}\cdot\underline{r}_n^L}$ represents the phase shift of the center of the sphere circumscribing the nth subaperture at level L. The multiplication by this term cancels the pattern rapidly phase varying as a function of $\hat{r}$ and makes the pattern of the subaperture amenable to sampling and interpolation. Accordingly, the full radiation pattern is

$$\underline{U}(\theta,\phi) = \sum_{n=1}^{N^L} e^{jk\hat{r}\cdot\underline{r}_n^L}\underline{U}_n^L(\theta,\phi) \tag{4.45}$$

First, the radiation patterns $\underline{U}_n^M(\theta,\phi)$ at the finest level L=M, comprising the smallest subapertures, are computed through (4.44). Then, based on the parent–child relationship, the parent subaperture pattern can be computed by

$$\underline{U}_m^{L-1}(\theta,\phi) = \sum_{n=1}^{N^L} e^{-jk\hat{r}\cdot(\underline{r}_n^{L-1}-\underline{r}_n^L)}\underline{U}_n^L(\theta,\phi) \tag{4.46}$$

Repeated application of (4.46), first for $L = M$, followed lower and lower level until $L = 1$ is reached, leads to the multilevel evaluation of the full pattern $\underline{U}_1^0(\theta,\phi) = \underline{U}(\theta,\phi)$. Direct application of (4.46), however, produces no computational savings. Economical sampling and interpolation of the radiation patterns on each level are the key to a formulation of an efficient numerical algorithm. Optimal sampling strategy of the radiation pattern depends on the specific shape of the radome surface and are discussed in [14]. For reasons of simplicity and interpolation efficiency, the Cartesian *(θ,ϕ)* grid is chosen in the following. For a source distribution circumscribed by a sphere of radius R centered at origin, it has been shown that the radiation pattern is an essentially band-limited function of ϕ and θ [15]. The number of sample points needed over a 2π angular range should exceed $2kR$. Thus, for each subaperture $\underline{S}_n^L$, the minimum number of angular directions is determined by R_n^L [15]. It is assumed that all subapertures on the same level of decomposition are similar in size. Therefore, to simplify the algorithm the radiation patterns at each level are computed on a 2D grid of directions. In addition, each 2D grid is obtained as a Cartesian product of θ and ϕ grids, thus reducing the interpolation complexity. Let $R_{\max}^L = \max_n R_n^L$ be the maximal subaperture size of level L. Therefore, we require that the number of observation angle samples at level L satisfies:

$$N_\alpha^L = \frac{\Omega_\alpha^L k R_{\max}^L(\alpha_{\max} - \alpha_{\min})}{\pi} + C_\alpha; \alpha = \theta, \phi \tag{4.47}$$

where $\Omega_\alpha^L > 1$ is the oversampling parameter on level L. The empirical constant C_α is needed because of the angular dependence of $\underline{A}(\hat{r},\underline{r}')$, usually is very small. Upon transition between the levels, the grid density for both angles is doubled. Once the number of grid points are determined by (4.47), we can define the 2D grid of observation angles of level L denoted $\{(\theta_i^L,\phi_j^L)\}, i = 1,\ldots,N_\theta^L, j = 1,\ldots,N_\phi^L$. Following (4.44) the radiation patterns at the finest level, $L = M$, are computed by numerical quadrature for directions of the coarsest grid:

$$\underline{U}_n^M(\theta_i^M,\phi_j^M) = \int_{S_n^M} \underline{A}(\hat{r}_{i,j}^M,\underline{r}')e^{jk\hat{r}_{ij}^M\cdot(\underline{r}'-\underline{r}_n^M)}ds' \tag{4.48}$$

for $n=1,\ldots,N^M$, where $\hat{r}_{ij}^M = \hat{r}(\theta_i^M,\phi_j^M)$. The radiation pattern of subaperture m on level $L-1$, $L-1 < M$ can now be computed based on

(4.42) as a sum of interpolated patterns of its children subapertures of level L

$$\underline{U}_m^{L-1}(\theta_i^{L-1}, \phi_j^{L-1}) = \sum_{n:P^L(n)=m} e^{-jk\hat{r}_{ij}^{L-1}\cdot(\underline{r}_m^{L-1}-\underline{r}_n^L)} \underline{U}_n^L(\theta_i^{L-1}, \phi_j^{L-1}) \tag{4.49}$$

Where

$$\underline{U}_n^L(\theta_i^{L-1}, \phi_j^{L-1}) = \sum_{i'} a_{ii'}^L \underline{U}_n^L(\theta_{i'}^L, \phi_j^{L-1}) \tag{4.50}$$

and

$$\underline{U}_n^L(\theta_{i'}^L, \phi_j^{L-1}) = \sum_{j'} b_{jj'}^L \underline{U}_n^L(\theta_{i'}^L, \phi_{j'}^L) \tag{4.51}$$

Eqs. (4.50) and (4.51) represent θ and ϕ interpolations with $a_{ii'}^L$ and $b_{jj'}^L$ representing their respective interpolation coefficients. Moreover, in the interpolation, p points from the neighborhood are used. Usually, $p = 2$ is used for linear interpolation and $p = 4$ for the cubic one. In case of $p = 4$, the polynomial coefficients are $a_{\mathrm{ii}} = b_{\mathrm{jj}} = (-1/16, 9/16, 9/16, -1/16)$. Central interpolation (where each interpolation target point is located in the middle between two groups of $p/2$ grid points) is used to ensure uniform accuracy and exponential convergence of the interpolation error versus p. Near the boundaries of the angular range of interest, central interpolation requires additional points laying beyond the range boundaries. An example of the grid points and interpolation coefficients for cubic interpolation is shown in Fig. 4.8.

Note that the coarse grid points coincide with roughly half of the fine grid points, as shown by the vertical arrows in Fig. 4.8. Thus, interpolation is performed only for the "new" midpoints.

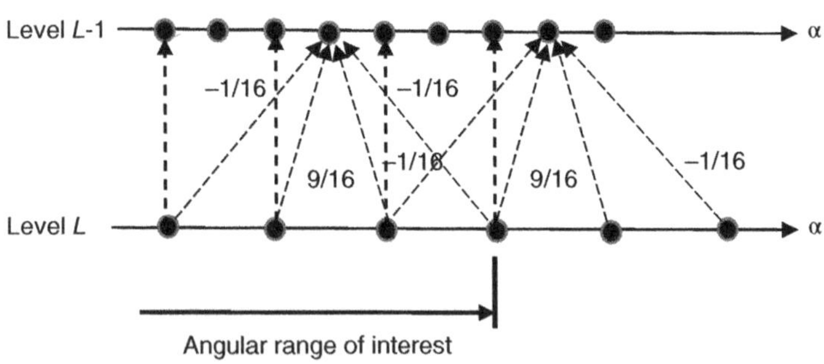

Figure 4.8 Cubic interpolation between grids of levels L and $L - 1$ in the range and near range boundary. Two additional points beyond range of interest are used. The central cubic interpolation weights are shown.

4.2 Surface Integration Technique Based on Equivalence Principle

The plane wave method described in section 4.1 is especially efficient for large (in wavelength) antennas and radomes. In case of a small radome of arbitrary shape and small antennas enclosed, the surface equivalence principle can be used to model the effect of the radome on the transmitted field. The radiated field is expressed in terms of equivalent electric and magnetic surface currents radiating in an unbounded medium. The analysis presented in the following is based on [5]. The boundary conditions on the tangential components of the total fields give a coupled integral equations set involving these currents. The MoM [16] is used to solve these integral equations numerically.

Fig. 4.9 shows a radome of arbitrary shape, which is excited by a transmitting antenna, S_T and its radiation is received by an external antenna, S_R.

Without loss of generality and for simplicity, we assume that the radome is a one-layer structure. In case of a multilayer radome, the procedure outlined below should be repeated given the number of layers of the radome. The outer surface of the radome is denoted by S_0 and the inner surface is denoted by S_1. The radome electrical parameters are (ε_2, μ_2) and is placed in a homogeneous medium with parameters (ε_1, μ_1). The incident fields $(\underline{E}_R^{inc}, \underline{H}_R^{inc})$ and $(\underline{E}_t^{inc}, \underline{H}_t^{inc})$, denote the field produced by external and internal sources to the radome radiating into medium (ε_1, μ_1). We are interested in finding the total field $(\underline{E}, \underline{H})$ at

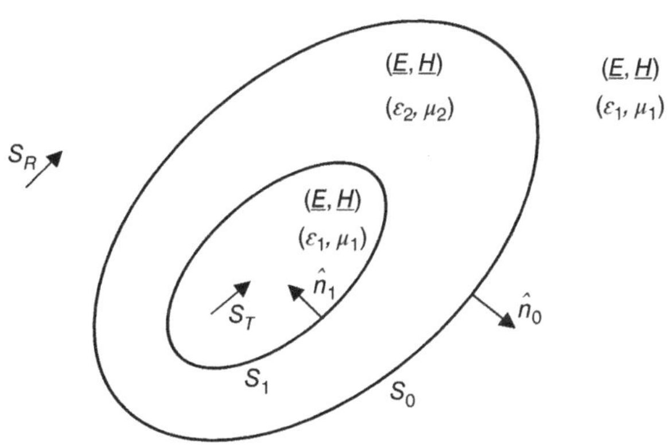

Figure 4.9 The geometry of an arbitrary shaped radome excited by an embedded antenna and an external receiving antenna.

an arbitrary external point of the radome. This field is the sum of the total incident field and the field scattered by the radome.

According to the equivalence principle [17], the problem described in Fig. 4.9 can be solved by considering three simpler but equivalent problems shown in Fig. 4.10. Fig. 4.10a represents the external equivalence scenario. Here, the radome and the transmit-mode source S_T are replaced by equivalent electric and magnetic surface currents $\underline{J}_0$ and $\underline{J}_{m0}$ residing on S_0. The entire space is now characterized by the parameters (ε_1, μ_1) and the source S_R as shown in Fig. 4.10a. It is assumed that the total field at any point external to S_0 is the same as the total field at the same point of Fig. 4.9 and that the total field at any internal point to S_0 of Fig. 4.10a is zero. In other words, the incident field $(\underline{E}_R^{inc}, \underline{H}_R^{inc})$ plus the field produced by the surface currents $\underline{J}_0$ and $\underline{J}_{m0}$ radiating in the unbounded medium (ε_1, μ_1) of Fig. 4.10a give the correct total field $(\underline{E}, \underline{H})$ at external points of S_0.

In the external problem, at any internal point (to S_0) the scattered field cancels the incident field. Hence,

$$\begin{aligned} \underline{E}_1^-(\underline{J}_0, \underline{J}_{mo})_{\tan} &= -\underline{E}_{R,\tan}^{inc} \quad on \ S_0^- \\ \underline{H}_1^-(\underline{J}_0, \underline{J}_{mo})_{\tan} &= -\underline{H}_{R,\tan}^{inc} \quad on \ S_0^- \end{aligned} \tag{4.52}$$

where the subscript "tan" denotes the tangential component, S_0^- represents the surface just inside S_0 and $\underline{E}_1(\underline{J}_0, \underline{J}_{m0})$ and $\underline{H}_1(\underline{J}_0, \underline{J}_{m0})$ represent the electric and magnetic fields produced by the surface currents $\underline{J}_0$ and $\underline{J}_{m0}$ (at a point on S_0) when they radiate into the unbounded medium (ε_1, μ_1). Using the equivalence theorem requires

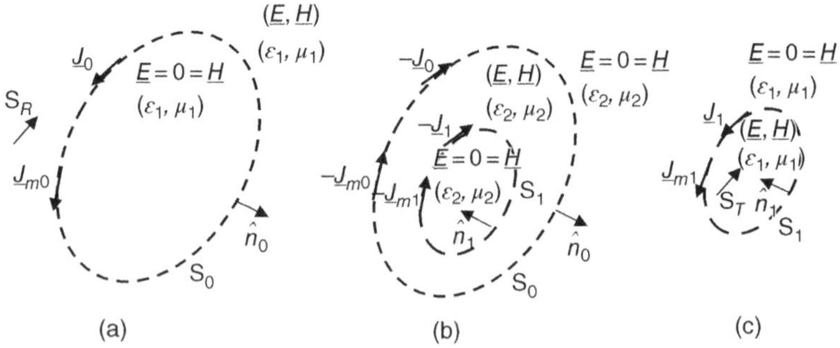

Figure 4.10 The equivalence regions for the radome analysis: (a) equivalence for the external problem; (b) equivalence for the region bounded by S_0 and S_1; (c) equivalence for the internal problem.

that $\underline{J}_0 = \hat{n}_0 \times \underline{H}_0^+; \underline{J}_{m0} = -\hat{n}_0 \times \underline{E}_0^+$, in which $\hat{n}_0$ represents the outward normal vector on S_0, and $(\underline{E}_0^+, \underline{H}_0^+)$ denote the total field just outside S_0 in Fig. 4.10a. To compute the fields in the region bounded by S_0 and S_1, the equivalence theorem is used again. Here the whole space is characterized by the parameters (ε_2, μ_2) and the impressed sources S_R and S_T are now replaced by equivalent surface currents $-\underline{J}_0$ and $-\underline{J}_{m0}$ residing on S_0 and equivalent surface currents $-\underline{J}_1$ and $-\underline{J}_{m1}$ residing on S_1. These four currents, radiating into the bounded medium (ε_2, μ_2) produce the total field $(\underline{E}, \underline{H})$ at any point in the region bounded by the surfaces S_0 and S_1 and they produce null field at any point outside this region:

$$\begin{aligned}
&\underline{E}^+_{2,\tan}(-\underline{J}_0, -\underline{J}_{m0}) + \underline{E}^-_{2,\tan}(-\underline{J}_1, -\underline{J}_{m1}) = 0 \ \ on \ \ S_0^+ \\
&\underline{E}^-_{2,\tan}(-\underline{J}_0, -\underline{J}_{m0}) + \underline{E}^+_{2,\tan}(-\underline{J}_1, -\underline{J}_{m1}) = 0 \ \ on \ \ S_1^+ \\
&\underline{H}^+_{2,\tan}(-\underline{J}_0, -\underline{J}_{m0}) + \underline{H}^-_{2,\tan}(-\underline{J}_1, -\underline{J}_{m1}) = 0 \ \ on \ \ S_0^+ \\
&\underline{H}^-_{2,\tan}(-\underline{J}_0, -\underline{J}_{m0}) + \underline{H}^+_{2,\tan}(-\underline{J}_1, -\underline{J}_{m1}) = 0 \ \ on \ \ S_1^+
\end{aligned} \tag{4.53}$$

where S_0^+ and S_1^+, respectively, denote the surface just outside S_0 and S_1, the subscript 2 means that the currents are radiating into the bounded medium (ε_2, μ_2) and the superscript + means that the fields are computed just outside (in $\hat{n}_0$ and $\hat{n}_1$ directions) the region bounded by S_0 and S_1. The discontinuity of the fields across S_0 and S_1 requires that $\underline{J}_1 = \hat{n}_1 \times \underline{H}_{1n}^+; \underline{J}_{m1} = -\hat{n}_1 \times \underline{E}_{1n}^+$ where $(\underline{E}_{1n}^+, \underline{H}_{1n}^+)$ denote the total field just outside S_1 (in $\hat{n}_1$ direction) in Fig. 4.10.

Finally, to realize the field internal to S_1 of Fig. 4.10, the radome and the source S_R are replaced by equivalent electric and magnetic surface currents $\underline{J}_1$ and $\underline{J}_{m1}$ residing on S_1 and radiating into the internal medium (ε_1, μ_1) as shown in Fig. 4.10c. Note that the original transmit-mode source S_T is also in Fig. 4.10c. The field produced by S_T and the surface currents $\underline{J}_1$ and $\underline{J}_{m1}$ in Fig. 4.10c, give the correct total field at any point inside S_1. However, these two fields sum to zero at any point outside S_1. Hence,

$$\begin{aligned}
&\underline{E}_1^-(\underline{J}_1, \underline{J}_{m1})_{\tan} = -\underline{E}^{inc}_{T,\tan} \ \ on \ \ S_1^- \\
&\underline{H}_1^-(\underline{J}_1, \underline{J}_{m1})_{\tan} = -\underline{H}^{inc}_{T,\tan} \ \ on \ \ S_1^-
\end{aligned} \tag{4.54}$$

Here S_1^- denotes the surface just outside S_1 in the $-\hat{n}_1$ direction. Eqs. (4.52) to (4.54) represent eight coupled integral equations for the four unknown surface currents $\underline{J}_0, \underline{J}_{m0}, \underline{J}_1$, and $\underline{J}_{m1}$. Various combinations of these equations lead to different formulations. The E-field formulation consists of eqs. (4.52)a, (4.53)a,b, and (4.54)a, while the H-field formulation consists of eqs. (4.52)b, (4.53)c,d, and (4.54)b. To reduce the eight coupled integral equations to four integral equations as the number of unknowns, we sum eq. (4.52)a and (4.53)a, sum eq. (4.53)b and (4.54)a, sum eq. (4.52)b and (4.53)c, and sum eq. (4.53)d and (4.54)b to obtain the following:

$$\underline{E}^-_{1,\tan}(\underline{J}_0,\underline{J}_{m0})+\underline{E}^+_{2,\tan}(-\underline{J}_0,-\underline{J}_{m0})+\underline{E}^-_{2,\tan}(-\underline{J}_1,-\underline{J}_{m1})=-\underline{E}^{inc}_{R,\tan}\quad on\ S_0 \tag{4.55}$$

$$\underline{E}^-_{1,\tan}(\underline{J}_1,\underline{J}_{m1})+\underline{E}^+_{2,\tan}(-\underline{J}_1,-\underline{J}_{m1})+\underline{E}^-_{2,\tan}(-\underline{J}_0,-\underline{J}_{m0})=-\underline{E}^{inc}_{T,\tan}\quad on\ S_1 \tag{4.56}$$

$$\underline{H}^-_{1,\tan}(\underline{J}_0,\underline{J}_{m0})+\underline{H}^+_{2,\tan}(-\underline{J}_0,-\underline{J}_{m0})+\underline{H}^-_{2,\tan}(-\underline{J}_1,-\underline{J}_{m1})=-\underline{H}^{inc}_{R,\tan}\quad on\ S_0 \tag{4.57}$$

$$\underline{H}^-_{1,\tan}(\underline{J}_1,\underline{J}_{m1})+\underline{H}^+_{2,\tan}(-\underline{J}_1,-\underline{J}_{m1})+\underline{H}^-_{2,\tan}(-\underline{J}_0,-\underline{J}_{m0})=-\underline{H}^{inc}_{T,\tan}\quad on\ S_1 \tag{4.58}$$

The superscripts are used on the fields that are not continuous across S_0 or S_1. These summations also avoid the excitations of interior resonances in the numerical solution. Equations (4.55) to (4.58) can be rewritten in the following forms:

$$\begin{aligned}&[\underline{E}^-_{1,\tan}(\underline{J}_0)+\underline{E}^+_{2,\tan}(-\underline{J}_0)]+\underline{E}^-_{2,\tan}(-\underline{J}_1)+[\underline{E}^-_{1,\tan}(\underline{J}_{m0})+\underline{E}^+_{2,\tan}(-\underline{J}_{m0})]\\&\quad+\underline{E}^-_{2,\tan}(-\underline{J}_{m1})=-\underline{E}^{inc}_{R,\tan}\qquad on\ S_0\end{aligned} \tag{4.59}$$

$$\begin{aligned}&\underline{E}^-_{2,\tan}(-\underline{J}_0)+[\underline{E}^-_{1,\tan}(\underline{J}_1)+\underline{E}^+_{2,\tan}(-\underline{J}_1)]+\underline{E}^-_{2,\tan}(-\underline{J}_{m0})\\&\quad+[\underline{E}^-_{1,\tan}(\underline{J}_{m1})+\underline{E}^+_{2,\tan}(-\underline{J}_{m1})]=-\underline{E}^{inc}_{T,\tan}\qquad on\ S_1\end{aligned} \tag{4.60}$$

$$\begin{aligned}&[\underline{H}^-_{1,\tan}(\underline{J}_0)+\underline{H}^+_{2,\tan}(-\underline{J}_0)]+\underline{H}^-_{2,\tan}(-\underline{J}_1)+[\underline{H}^-_{1,\tan}(\underline{J}_{m0})+\underline{H}^+_{2,\tan}(-\underline{J}_{m0})]\\&\quad+\underline{H}^-_{2,\tan}(-\underline{J}_{m1})=-\underline{H}^{inc}_{R,\tan}\qquad on\ S_0\end{aligned} \tag{4.61}$$

$$\begin{aligned}&\underline{H}^-_{2,\tan}(-\underline{J}_0)+[\underline{H}^-_{1,\tan}(\underline{J}_1)+\underline{H}^+_{2,\tan}(-\underline{J}_1)]+\underline{H}^-_{2,\tan}(-\underline{J}_{m0})\\&\quad+[\underline{H}^-_{1,\tan}(\underline{J}_{m1})+\underline{H}^+_{2,\tan}(-\underline{J}_{m1})]=-\underline{H}^{inc}_{T,\tan}\qquad on\ S_1\end{aligned} \tag{4.62}$$

For convenience, we express the electric and magnetic fields in terms of magnetic and electric vector potentials $\underline{A}$ and $\underline{F}$ and scalar electric and magnetic potentials Φ and Ψ, respectively. The magnetic vector potential $\underline{A}$ and scalar electric potential Φ are related directly to the conductive currents, $\underline{J}_0$, $\underline{J}_1$ and the electric vector potential $\underline{F}$ and scalar magnetic potential Ψ are related to the magnetic currents $\underline{J}_{m0}$, $\underline{J}_{m1}$ as shown below:

$$\begin{aligned}&\underline{E}(\underline{J})=-jk_i\eta_i\underline{A}-\nabla\Phi=\frac{\eta_i}{jk_i}[\nabla(\nabla\cdot\underline{A})+k_i^2\underline{A}];i=1,2\\&\underline{E}(\underline{J}_m)=-\nabla\times\underline{F}\\&\underline{H}(\underline{J})=\nabla\times\underline{A}\\&\underline{H}(\underline{J}_m)=-j\frac{k_i}{\eta_i}\underline{F}-\nabla\Psi=\frac{1}{jk_i\eta_i}[\nabla(\nabla\cdot\underline{F})+k_i^2\underline{F}]\end{aligned} \tag{4.63}$$

in which $\eta_i = \sqrt{\frac{\mu_i}{\varepsilon_i}}$ and $k_i = \omega\sqrt{\mu_i\varepsilon_i}$ with (i = 1,2). The electric and magnetic vector potentials are related to the conductive and magnetic currents through the Green's function of an elementary electric or magnetic dipole radiating in an unbounded medium [17] such that

$$\begin{aligned}
\underline{A} &= \int_S \underline{J}(\underline{r}')\frac{e^{-jk_i|\underline{r}-\underline{r}'|}}{4\pi|\underline{r}-\underline{r}'|}ds \\
\underline{F} &= \int_S \underline{J}_m(\underline{r}')\frac{e^{-jk_i|\underline{r}-\underline{r}'|}}{4\pi|\underline{r}-\underline{r}'|}ds \\
\Phi &= -\frac{\eta_i}{jk_i}\nabla\cdot\underline{A} = -\frac{\eta_i}{jk_i}\int_S \underline{J}(\underline{r}')\cdot\nabla'\left[\frac{e^{-jk_i|\underline{r}-\underline{r}'|}}{4\pi|\underline{r}-\underline{r}'|}\right]ds \\
\Psi &= -\frac{1}{jk_i\eta_i}\nabla\cdot\underline{F} = -\frac{1}{jk_i\eta_i}\int_S \underline{J}_m(\underline{r}')\cdot\nabla'\left[\frac{e^{-jk_i|\underline{r}-\underline{r}'|}}{4\pi|\underline{r}-\underline{r}'|}\right]ds
\end{aligned} \tag{4.64}$$

where $|\underline{r}-\underline{r}'| = \sqrt{(x-x')^2+(y-y')^2+(z-z')^2}$. Evaluation of $(\underline{E}_R^{inc}, \underline{H}_R^{inc})$ and $(\underline{E}_t^{inc}, \underline{H}_t^{inc})$ on the surfaces S_0 and S_1 is possible using eq. (4.64) using the known current distributions. Substitution of (4.64) into (4.59)–(4.62) results in four coupled integral equations with the unknowns $\underline{J}_0$, $\underline{J}_1$, $\underline{J}_{m0}$, $\underline{J}_{m1}$ on the surfaces S_0 and S_1.

This set of integral equations can be solved numerically using the MoM [16]. In applying MoM, the surfaces S_0 and S_1 are approximated by triangular patches, basis functions for the electric and magnetic currents associated with each edge of each triangle approximating the surfaces. Thus,

$$\begin{aligned}
\underline{J}_0 &= \sum_{m=1}^{N_0} I_{0m}\underline{f}_{0m} \\
\underline{J}_{m0} &= \sum_{m=1}^{N_0} K_{0m}\underline{f}_{0m} \\
\underline{J}_1 &= \sum_{m=1}^{N_i} I_{1m}\underline{f}_{1m} \\
\underline{J}_{m1} &= \sum_{m=1}^{N_i} K_{1m}\underline{f}_{1m}
\end{aligned} \tag{4.65}$$

Here $\underline{f}_{0m}$, (m=1,...N_0) denote the basis functions associated with the mth edge on S_0, and $\underline{f}_{1m}$, (m=1,...N_1) denote the basis functions associated with the mth edge on S_1. The vector basis functions $\underline{f}$ used above are identical to the functions used in [18] and [19].

It is convenient to realize that each basis function is associated with an interior of the patch model and it vanishes everywhere on S except

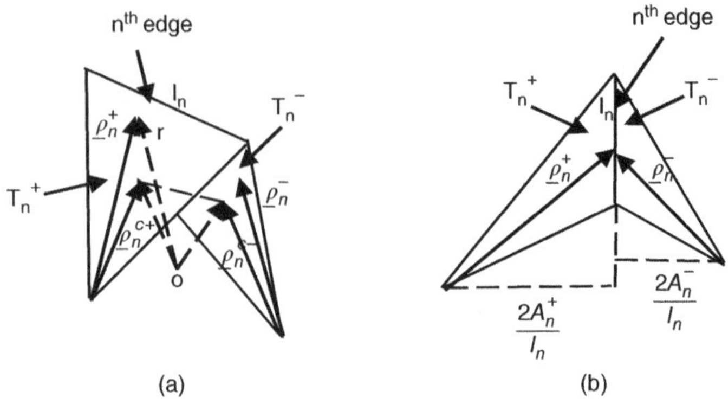

Figure 4.11 (a) coordinates of common edge associated with two triangles; (b) geometry for normal component of basis function at common edge.

in the two triangles attached to that edge. Fig. 4.11 shows two such triangles, T_n^+ and T_n^-, corresponding to the nth edge of a triangulated surface modeling a scatterer. Points in T_n^+ may be designated either by the position vector $\underline{r}$ defined with respect to global origin O, or by the position vector $\underline{\rho}_n^+$ defined with respect to the free vertex of T_n^+. Similar remarks apply to the position vector $\underline{\rho}_n^-$ except that it is directed toward the free vertex of T_n^-. The plus or minus designation of the triangles is determined by the choice of a positive current reference direction for the nth edge, the reference for which is assumed to be from T_n^+ to T_n^-.

The triangular vector basis function associated with the nth edge is defined by

$$\underline{f}_n(\underline{r}) = \begin{cases} \dfrac{l_n}{2A_n^+}\underline{\rho}_n^+, & \underline{r} \ in \ T_n^+ \\ \dfrac{l_n}{2A_n^-}\underline{\rho}_n^-, & \underline{r} \ in \ T_n^- \\ 0, & otherwise \end{cases} \tag{4.66}$$

where l_n is the length of the edge and $A_n^{\pm}$ is the area of triangle $T_n^{\pm}$ as shown in Fig. 4.11. Here are five properties of the basis function $\underline{f}_n(\underline{r})$:

1. The current has no component normal to the boundary (which excludes the common edge) of the surface formed by the triangle pair T_n^+ and T_n^-, and hence no line charges exist along this boundary.
2. The component of current normal to the nth edge is constant and continuous across the edge, as may be seen in Fig. 4.11, which shows

that the normal component of $\underline{\rho}_n^+$ along edge n as the base and the height expressed as $\frac{2A_n^\pm}{l_n}$. This factor normalizes $\underline{f}_n(\underline{r})$ such that its flux density normal to edge n is unity, ensuring continuity of current normal to the edge. This result, together with the first property, implies that all edges of T_n^+ and T_n^- are free of line charges.

3. The surface divergence of $\underline{f}_n(\underline{r})$, which is proportional to the surface charge density associated with the basis element, is

$$\nabla \cdot \underline{f}_n = \begin{cases} \frac{l_n}{A_n^+}, & \underline{r} \ in \ T_n^+ \\ -\frac{l_n}{A_n^-}, & \underline{r} \ in \ T_n^- \\ 0 & , \ otherwise \end{cases} \tag{4.67}$$

since the surface divergence (cylindrical coordinates) in $T_n^\pm$ is $\pm\frac{1}{\rho_n^\pm}\frac{\partial(\rho_n^\pm f_n)}{\partial \rho_n^\pm}$. The charge density is thus constant in each triangle, the total charge associated with the triangle pair T_n^+ and T_n^- is zero, and the basis functions for the charge evidently have the form of pulse doublets.

4. The moment of $\underline{f}_n(\underline{r})$ is given by $(A_n^+ + A_n^-)\underline{f}_n^{avg}$ and defined by

$$\int_{T_n^+ + T_n^-} \underline{f}_n \, ds \equiv (A_n^+ + A_n^-)\underline{f}_n^{avg} = \frac{l_n}{2}(\underline{\rho}_n^{c+} + \underline{\rho}_n^{c-}) = \frac{l_n}{2}(\underline{r}_n^{c+} - \underline{r}_n^{c-}) \tag{4.68}$$

in which $\underline{\rho}_n^{c\pm}$ is the vector between the free vertex and the centroid of $T_n^\pm$ with $\underline{\rho}_n^{c-}$ directed toward the vertex and $\underline{\rho}_n^{c+}$ directed away from the vertex, as shown in Fig. 4.11, and $\underline{r}_n^{c\pm}$ is the vector from global origin O to the centroid of $T_n^\pm$.

5. Furthermore, since the normal component of $\underline{f}_n(\underline{r})$ at the nth edge is unity, each coefficient in (4.65) may be interpreted as the normal component of current density flowing past the nth edge. Also, the basis functions are independent in each triangle since the current normal to the nth edge is an independent quantity.

In order to find the electric and magnetic current coefficients in the integral eqs. (4.59)–(4.62), these combined integral equations are scalar multiplied with N testing functions taking advantage of their orthogonality properties. An efficient way to choose these testing functions is to pick testing functions equal to the basis functions (Galerkin method). Evaluation of the electric and magnetic current coefficients is possible by scalar multiplication of eqs. (4.59)–(4.62) with the testing functions $\underline{f}_m(\underline{r})$ and integration. This operation reduces the functional form of the eqs. (4.59)–(4.62) to a corresponding partitioned matrix equation.

$$\begin{bmatrix} ETJ0 & E2J1 & ETM0 & E2M1 \\ E2J0 & ETJ1 & E2M0 & ETM1 \\ HTJ0 & H2J1 & HTM0 & H2M1 \\ H2J0 & HTJ1 & H2M0 & HTM1 \end{bmatrix} \begin{bmatrix} I_0 \\ I_1 \\ K_0 \\ K_i \end{bmatrix} = \begin{bmatrix} V_0^E \\ V_1^E \\ V_0^H \\ V_1^H \end{bmatrix} \tag{4.69}$$

The nth element of the vector I_0 in (4.65) is I_{0n} and the mth element of I_1 is I_{1m}. Similarly, the mth elements of K_0 and K_1 are K_{0m} and K_{1m}, respectively. The nth elements of the vectors V_0^E and V_0^H are, respectively, equal to the symmetric products of $\underline{f}_{0n}$ with $-\underline{E}^{inc}_{R,\tan}$ and $-\underline{H}^{inc}_{R,\tan}$. Similarly, the nth elements of the vectors V_1^E and V_1^H are, respectively, equal to the symmetric product of the testing function $\underline{f}_{1n}$ with $-\underline{E}^{inc}_{T,\tan}$, and $-\underline{H}^{inc}_{T,\tan}$.

The submatrices of the square moment matrix in (4.69) are named according to the following convention. The first letter denotes the field (E for electric and H for magnetic). The third letter (with subscript) denotes the source of this field. The second character shows the media in which the source radiates. If this is 2, the source is assumed to be radiating in (ε_2, μ_2). If it is the letter T, then the source is assumed to radiate in both media, and the fields are added to give the total field.

Using (4.54) and evaluating the potentials Φ and Ψ at the centroid of the respective triangle results in the following scalar products yields the following terms:

$$\begin{aligned} \left\langle \begin{Bmatrix} \nabla\Phi \\ \nabla\Psi \end{Bmatrix}, \underline{f}_m \right\rangle &= -\left\langle \begin{Bmatrix} \Phi \\ \Psi \end{Bmatrix}, \nabla\cdot\underline{f}_m \right\rangle \\ &= l_m \left[\frac{1}{A_m^-}\int_{T_m^-} \begin{Bmatrix} \Phi \\ \Psi \end{Bmatrix} ds - \frac{1}{A_m^+}\int_{T_m^+} \begin{Bmatrix} \Phi \\ \Psi \end{Bmatrix} ds \right] \\ &\simeq l_m \left[\begin{Bmatrix} \Phi(\underline{r}_m^{c-}) \\ \Psi(\underline{r}_m^{c-}) \end{Bmatrix} - \begin{Bmatrix} \Phi(\underline{r}_m^{c+}) \\ \Psi(\underline{r}_m^{c+}) \end{Bmatrix} \right] \end{aligned} \tag{4.70}$$

in which the scalar multiplication is defined by $\langle \underline{f}, \underline{g} \rangle = \int_S \underline{f}\cdot\underline{g}ds$. Similarly, using (4.66), we obtain

$$\begin{aligned} \left\langle \begin{Bmatrix} \nabla\times\underline{A} \\ \nabla\times\underline{F} \end{Bmatrix}, \underline{f}_m \right\rangle &= \frac{l_m}{2A_m^+}\int_{T_m^+} \underline{\rho}_m^+ \cdot \left[\begin{Bmatrix} \nabla\times\underline{A} \\ \nabla\times\underline{F} \end{Bmatrix} \right]^+ ds \\ &+ \frac{l_m}{2A_m^-}\int_{T_m^-} \underline{\rho}_m^- \cdot \left[\begin{Bmatrix} \nabla\times\underline{A} \\ \nabla\times\underline{F} \end{Bmatrix} \right]^- ds \end{aligned} \tag{4.71}$$

An additional term to be considered is

$$\left\langle \begin{Bmatrix} \underline{E}^{inc} \\ \underline{H}^{inc} \\ \underline{A} \\ \underline{F} \end{Bmatrix}, \underline{f}_m \right\rangle = l_m \left[\frac{1}{2A_m^+} \int_{T_m^+} \begin{Bmatrix} \underline{E}^{inc} \\ \underline{H}^{inc} \\ \underline{A} \\ \underline{F} \end{Bmatrix} \cdot \underline{\rho}_m^+ ds + \frac{1}{2A_m^-} \int_{T_m^-} \begin{Bmatrix} \underline{E}^{inc} \\ \underline{H}^{inc} \\ \underline{A} \\ \underline{F} \end{Bmatrix} \cdot \underline{\rho}_m^- ds \right]$$

$$\simeq \frac{l_m}{2} \left[\begin{Bmatrix} \underline{E}^{inc}(\underline{r}_m^{c+}) \\ \underline{H}^{inc}(\underline{r}_m^{c+}) \\ \underline{A}(\underline{r}_m^{c+}) \\ \underline{F}(\underline{r}_m^{c+}) \end{Bmatrix} \cdot \underline{\rho}_m^{c+} + \begin{Bmatrix} \underline{E}^{inc}(\underline{r}_m^{c-}) \\ \underline{H}^{inc}(\underline{r}_m^{c-}) \\ \underline{A}(\underline{r}_m^{c-}) \\ \underline{F}(\underline{r}_m^{c-}) \end{Bmatrix} \cdot \underline{\rho}_m^{c-} \right] \quad (4.72)$$

Moreover,

$$\begin{aligned} \underline{E}^{inc+}(\underline{r}_m^{c+}) &= \underline{E}^{inc}(\underline{r}_m^{c+}) \\ \underline{E}^{inc-}(\underline{r}_m^{c-}) &= \underline{E}^{inc}(\underline{r}_m^{c-}) \\ \underline{H}^{inc+}(\underline{r}_m^{c+}) &= \underline{H}^{inc}(\underline{r}_m^{c+}) \\ \underline{H}^{inc-}(\underline{r}_m^{c-}) &= \underline{H}^{inc}(\underline{r}_m^{c-}) \end{aligned} \quad (4.73)$$

In the above equations, the vector and scalar potentials $\underline{A}$, $\underline{F}$, Φ, Ψ are given by (4.64). Eq. (4.71) containing the curl operations can be further simplified if we recognize that

$$\begin{aligned} \nabla \times \underline{F} &= \int_S \underline{J}_m(\underline{r}') \times \nabla' \frac{e^{-jk_i|\underline{r}-\underline{r}'|}}{4\pi|\underline{r}-\underline{r}'|} ds \\ \nabla \times \underline{A} &= \int_S \underline{J}(\underline{r}') \times \nabla' \frac{e^{-jk_i|\underline{r}-\underline{r}'|}}{4\pi|\underline{r}-\underline{r}'|} ds \end{aligned} \quad (4.74)$$

in which

$$\nabla' \frac{e^{-jk_i|\underline{r}-\underline{r}'|}}{|\underline{r}-\underline{r}'|} = (\underline{r}-\underline{r}')(1 - jk_i|\underline{r}-\underline{r}'|) \frac{e^{-jk_i|\underline{r}-\underline{r}'|}}{|\underline{r}-\underline{r}'|^3} \quad (4.75)$$

with $i = 1,2$. Using eqs. (4.70), (4.72), and (4.74), we can evaluate the submatrix elements in (4.69) to be

$$\begin{aligned} ETJ0_{mn} = l_m \Bigg[& \frac{\underline{\rho}_m^{c+}}{2} \cdot \sum_{i=1}^{2} (jk_i\eta_i)\underline{A}_{i_{mn}}^+ + \frac{\underline{\rho}_m^{c-}}{2} \cdot \sum_{i=1}^{2} (jk_i\eta_i)\underline{A}_{i_{mn}}^- \\ & + \sum_{i=1}^{2} \left(\frac{-\eta_i}{jk_i} \right) [\Phi_{i_{mn}}^- - \Phi_{i_{mn}}^+] \Bigg] \end{aligned} \quad (4.76)$$

in which

$$\underline{A}_{i_{mn}}^{\pm} = \int_{T_n^+ + T_n^-} \underline{f}_n(\underline{r}') G_i(\underline{r}_m^{c\pm}, \underline{r}') ds(\underline{r}') \quad ; G_i(\underline{r}_m^{c\pm}, \underline{r}') = \frac{e^{-jk_i|\underline{r}_m^{c\pm}-\underline{r}'|}}{4\pi|\underline{r}_m^{c\pm}-\underline{r}'|} \quad (4.77)$$

and

$$\Phi^{\pm}_{i_{mn}} = \int_{T_n^+ + T_n^-} \left[\nabla' \cdot \underline{f}_n(\underline{r}')\right] G_i(\underline{r}^{c\pm}_m, \underline{r}') ds(\underline{r}') \tag{4.78}$$

Moreover,

$$E2J1_{mn} = jk_2\eta_2 \frac{l_m}{2}(\underline{\rho}^{c+}_m \cdot \underline{A}^+_{2_{mn}} + \underline{\rho}^{c-}_m \cdot \underline{A}^-_{2_{mn}}) + j\frac{\eta_2 l_m}{k_2}(\Phi^-_{2_{mn}} - \Phi^+_{2_{mn}}) \tag{4.79}$$

Next,

$$ETM0_{mn} = \sum_{i=1}^{2} P^+_{i_{mn}} + \sum_{i=1}^{2} P^-_{i_{mn}} \tag{4.80}$$

where

$$P^{\pm}_{i_{mn}} = \frac{l_m}{2A^{\pm}_m} \int_{T^{\pm}_m} \underline{\rho}^{\pm}_m \cdot \left[\int_{T_n^+ + T_n^-} \underline{f}_n \times \nabla' G_i(\underline{r}^{\pm}_m, \underline{r}') ds(\underline{r}')\right] ds(\underline{r}) \tag{4.81}$$

and

$$E2M1_{mn} = P^+_{2_{mn}} + P^-_{2_{mn}} \tag{4.82}$$

Similarly, the submatrices *E2J0*, *ETJ1*, *E2M0*, and *ETM1* can be computed. Using eqs. (4.69), (4.72), and (4.74) enables us to evaluate the magnetic field submatrices:

$$\begin{aligned} HTJ0_{mn} = l_m \Bigg[&\frac{\underline{\rho}^{c+}_m}{2} \cdot \sum_{i=1}^{2} \left(\frac{jk_i}{\eta_i}\right) \underline{F}^+_{i_{mn}} + \frac{\underline{\rho}^{c-}_m}{2} \cdot \sum_{i=1}^{2} \left(\frac{jk_i}{\eta_i}\right) \underline{F}^-_{i_{mn}} \\ &+ \sum_{i=1}^{2} \left(\frac{-1}{jk_i\eta_i}\right) [\Psi^-_{i_{mn}} - \Psi^+_{i_{mn}}] \Bigg] \end{aligned} \tag{4.83}$$

in which $\underline{F}^{\pm}_{i_{mn}} = \underline{A}^{\pm}_{i_{mn}}$ and $\Psi^{\pm}_{i_{mn}} = \Phi^{\pm}_{i_{mn}}$. Moreover,

$$H2J1_{mn} = \frac{jk_2}{\eta_2}\frac{l_m}{2}(\underline{\rho}^{c+}_m \cdot \underline{F}^+_{2_{mn}} + \underline{\rho}^{c-}_m \cdot \underline{F}^-_{2_{mn}}) + \frac{-l_m}{jk_2\eta_2}(\Psi^-_{2_{mn}} - \Psi^+_{2_{mn}}) \tag{4.84}$$

and

$$HTM0_{mn} = \sum_{i=1}^{2} Q^+_{i_{mn}} + \sum_{i=1}^{2} Q^-_{i_{mn}} \tag{4.85}$$

where $Q^{\pm}_{i_{mn}} = P^{\pm}_{i_{mn}}$ and

$$H2M1_{mn} = Q^+_{2_{mn}} + Q^-_{2_{mn}} \tag{4.86}$$

Similarly, the submatrices *H2J0, HTJ1, H2M0,* and *HTM1* can be evaluated. Using (4.72), the elements of electric and magnetic field excitation can be evaluated:

$$V_{0_m}^E = \frac{l_m}{2}[\underline{\rho}_m^{c+} \cdot \underline{E}_R^{inc}(\underline{r}_m^{c+}) + \underline{\rho}_m^{c-} \cdot \underline{E}_R^{inc}(\underline{r}_m^{c-})] \tag{4.87}$$

$$V_{1_m}^E = \frac{l_m}{2}[\underline{\rho}_m^{c+} \cdot \underline{E}_t^{inc}(\underline{r}_m^{c+}) + \underline{\rho}_m^{c-} \cdot \underline{E}_t^{inc}(\underline{r}_m^{c-})] \tag{4.88}$$

$$V_{0_m}^H = \frac{l_m}{2}[\underline{\rho}_m^{c+} \cdot \underline{H}_R^{inc}(\underline{r}_m^{c+}) + \underline{\rho}_m^{c-} \cdot \underline{H}_R^{inc}(\underline{r}_m^{c-})] \tag{4.89}$$

$$V_{1_m}^H = \frac{l_m}{2}[\underline{\rho}_m^{c+} \cdot \underline{H}_t^{inc}(\underline{r}_m^{c+}) + \underline{\rho}_m^{c-} \cdot \underline{H}_t^{inc}(\underline{r}_m^{c-})] \tag{4.90}$$

As a follow-up of the previous discussion, we need to evaluate the vector and scalar potential integrals in eqs. (4.77), (4.78) relevant to the MoM solution. Fig. 4.12 illustrates a face pair with an observation point on face p and with source currents residing on face q. Each of the three basis functions that may exist simultaneously in T^q is proportional to one of the vectors $\underline{\rho}_1$, $\underline{\rho}_2$, or $\underline{\rho}_3$ defined in the figure. Each vector $\underline{\rho}_i$, $i = 1,2,3$, is shown directed away from its associated vertex in the figure, but would be directed toward the vertex if the current reference direction for the associated edge were into the triangle.

Consequently,

$$\underline{\rho}_i = \pm(\underline{r}' - \underline{r}_i); \quad i = 1,2,3 \tag{4.91}$$

where the positive sign is used if the positive current reference direction is out of T^q and the negative sign is used otherwise. The integrals

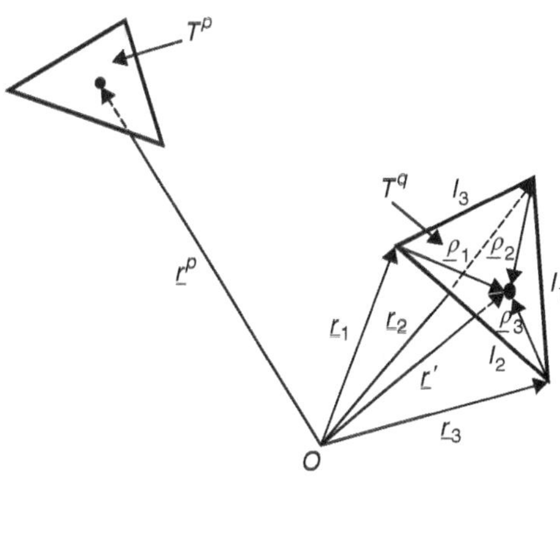

Figure 4.12 Local coordinates and edges for source triangle T^q with observation point in triangle T^p.

under consideration are the magnetic vector potential

$$\underline{A}_{i_{pq}} = \int_{T^q} \left(\frac{l_i}{2A_q}\right) \underline{\rho}_i \frac{e^{-jkR^p}}{4\pi R^p} ds' \tag{4.92}$$

and the electric scalar potential,

$$\Phi_{i_{pq}} = -\frac{\eta_i}{jk_i} \int_{T^q} \left(\frac{l_i}{A_q}\right) \frac{e^{-jkR^p}}{4\pi R^p} ds' \tag{4.93}$$

associated with the ith basis function on face q observed at the centroid of face p. Similar derivation can be applied for the electric vector potential $\underline{F}$, and scalar magnetic potential, Ψ. In (4.92), (4.93),

$$R^p = |\underline{r}^{cp} - \underline{r}'|; \quad i = 1,2,3 \tag{4.94}$$

where $\underline{r}^{cp}$ is the position vector of the centroid of face p. Integrals (4.77), (4.78) are most conveniently evaluated by transforming from the global coordinate system to a local system of coordinates defined within T^q. To define these coordinates, note that the vectors $\underline{\rho}_i$ in Fig. 4.12 divide T^q into three subtriangles of areas A_1, A_2, and A_3, with l_1, l_2, and l_3, respectively, as one of their sides. The areas are not independent; however, since they must satisfy $A_1 + A_2 + A_3 = A_q$. We now introduce the so-called normalized area coordinates:

$$\xi = \frac{A_1}{A_q}, \quad \eta = \frac{A_2}{A_q}, \quad \varsigma = \frac{A_3}{A_q} \tag{4.95}$$

which, because of the area constraint, must satisfy $\xi + \eta + \zeta = 1$. Note that all three coordinates vary between zero and unity in T^q and that at the triangle corners $\underline{r}_1$, $\underline{r}_2$, $\underline{r}_3$ and the triplet (ξ, η, ζ) take on the values (1,0,0), (0,1,0), and (0, 0, 1), respectively. The transformation from Cartesian to normalized area coordinates may be written in vector form:

$$\underline{r}' = \xi \underline{r}_1 + \eta \underline{r}_2 + \varsigma \underline{r}_3 \tag{4.96}$$

where ξ, η, and ζ are subject to the constraint that their summation is equal to one. Thus, surface integrals over T^q transform to

$$\int_{T^q} g(\underline{r}) ds = 2A_q \int_0^1 \int_0^{1-\eta} g[\xi \underline{r}_1 + \eta \underline{r}_2 + (1 - \xi - \eta)\underline{r}_3] d\xi d\eta \tag{4.97}$$

Using (4.91), (4.94), (4.95), and (4.96) enables to evaluate (4.92) and (4.93) to be

$$\underline{A}_{i_{pq}} = \pm \frac{l_i}{4\pi} (\underline{r}_1 I_{\xi_{pq}} + \underline{r}_2 I_{\eta_{pq}} + \underline{r}_3 I_{\varsigma_{pq}}) \tag{4.98}$$

and

$$\Phi_{i_{pq}} = \mp \frac{l_i}{j2\pi\omega\varepsilon_i} I_{pq} \tag{4.99}$$

where

$$\begin{aligned}
I_{pq} &= \int_0^1 \int_0^{1-\eta} \frac{e^{-jkR^P}}{R^P} d\xi d\eta \\
I_{\xi_{pq}} &= \int_0^1 \int_0^{1-\eta} \xi \frac{e^{-jkR^P}}{R^P} d\xi d\eta \\
I_{\eta_{pq}} &= \int_0^1 \int_0^{1-\eta} \eta \frac{e^{-jkR^P}}{R^P} d\xi d\eta \\
I_{\varsigma_{pq}} &= I_{pq} - I_{\xi_{pq}} - I_{\eta_{pq}}
\end{aligned} \tag{4.100}$$

Thus only three independent integrals, (4.100)a to (4.100)c, must be numerically evaluated for each combination of face pairs p and q. Numerical evaluation of the integrals (4.100)a to (4.100)c may be accomplished by using numerical quadrature techniques specially developed for triangular domains. However, for the terms in which $p = q$ the integrands are singular, and for these cases the singular portion of each integrand must be removed and integrated analytically.

If we follow the MoM procedure [16], the matrix equation (4.69) is solved by matrix inversion. In case of large matrices with high condition number the efficient way to solve the matrix equation (4.69) is by an iterative method such as the conjugate gradient method [20] as described in Appendix D. For an N-unknown problem, the memory requirement is $O(N^2)$ and the computational complexity is $O(N^2)$ per iteration. Such a memory requirement and computational complexity is too high for analyzing electrically large radomes and must be reduced to make the solution more efficient. The adaptive integral method (AIM) [21] can be used as an alternative to reduce the computational complexity and memory requirements of MoM. In the AIM algorithm, the impedance matrix (4.69) is represented as a sum of near-field and far-field zone components. The near-field component of the impedance matrix is computed by using the conventional MoM with Galerkin discretization scheme, utilizing arbitrary local-support functions. Since by definition a given current element possesses only a limited number of near neighbors, it results in a sparse matrix. The far-field matrix are computed using fast Fourier transforms (FFT), which provides storage and operation count reduction for problems with sufficiently large number of unknowns. For surface scatterers, the computational complexity and memory requirements of AIM are $O(N^{1.5}\log N)$ and $O(N^{1.5})$, respectively. An additional effective option for the matrix equation solution is the multilevel fast-multipole algorithm (MLFMA) described in [22], which exhibits $O(N\log N)$ computational complexity, when sufficiently high levels of implementation are involved: Hence, the complexity of MLFMA is favorable to that of AIM. Nonetheless, the AIM competes with the MLFMA because it involves an FFT algorithm that avoids the complex interpolation and interpolation operators in the MLFMA. Also, the complexity of the AIM can be further reduced by proper integration with other specific techniques, such as the domain decomposition strategy.

Fig. 4.13 shows a comparison of radiation patterns of MoM solution, hybrid PO-MoM, and antenna without radome for an antenna aperture tilted 10 deg. The length of the ogive-shaped radome is 10λ and its

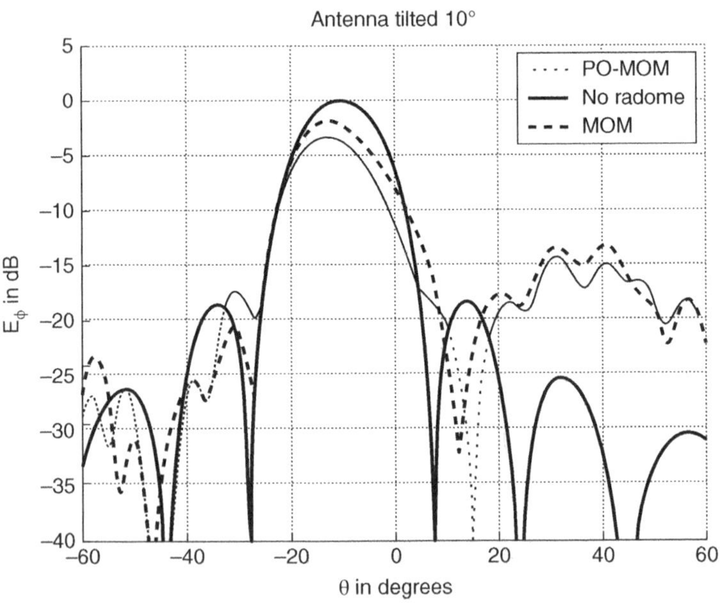

Figure 4.13 Comparison of radiation patterns of MoM solution, hybrid PO-MoM, and antenna without the ogive radome for an antenna aperture tilted 10 deg. The radome's length is 10λ and its diameter is 5λ. The antenna is circular with 4λ in diameter. The radome thickness is 0.2λ and its dielectric constant $\varepsilon_r = 4$ [23].

diameter is 5λ. The radome thickness is 0.2λ and its dielectric constant $\varepsilon_r = 4$ [23].

A nice agreement can be observed between the numerical solutions of the radiation patterns of the antenna with the radome.

4.3 Volume Integration Formulation Methods

The volume integral equation (VIE) combined with the multilevel fast multipole algorithm (MLFMA) is an alternate method to the surface integral equation (SIE) to analyze antenna radiation in the presence of dielectric radomes. In solving the VIE, the radome is modeled by small-volume cells of tetrahedron or hexahedron shape so that 3D complex radomes can be modeled accurately. Its significant advantage is that it can analyze inhomogeneous dielectrics. In general, the SIE produces less number of unknowns for homogeneous dielectrics.

In this section, we present a general-purpose MoM method for effective solution of the VIE based on [24]. The efficiency of the numerical solution is increased due to the application of the MLFMA [22]. Consider a dielectric shell of arbitrary shape residing in a homogeneous background medium of infinite extent with (ε_b, μ_b) as permittivity and permeability parameters as shown in Fig. 4.14.

For simplicity of the formulation it is assumed that the permittivity $\varepsilon(\underline{r}')$ is position dependent and the permeability of the material is constant. The source of excitation is located at a fixed position in the background medium (inside or outside the radome). By equivalence principle, the total radiation is the sum of the antenna's primary radiation $\underline{E}^i(\underline{r})$ in the absence of the radome and the radiation of the induced

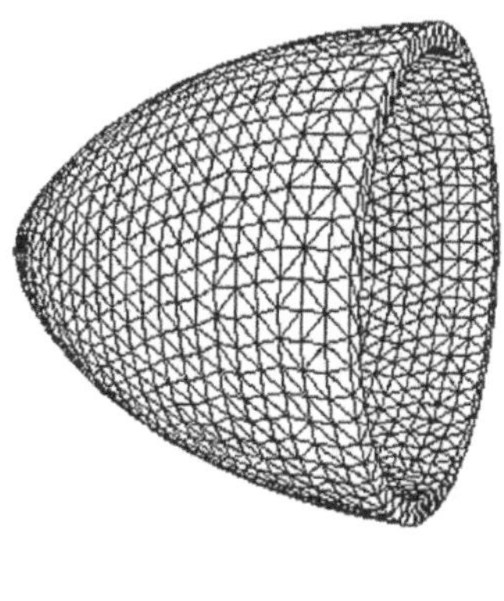

Figure 4.14 The tetrahedron mesh of an ogive-shaped radome.

volume polarization currents of the dielectric, $\underline{J}_V$ in the background medium [17], i.e.,

$$\underline{E}(\underline{r}) = \underline{E}^i(\underline{r}) - j\omega\mu_b \int_V G(\underline{r},\underline{r}')\underline{J}_V(\underline{r}')dv' + \frac{1}{j\omega\varepsilon_b}\nabla \int_V G(\underline{r},\underline{r}')\nabla'\cdot\underline{J}_V(\underline{r}')dv' \tag{4.101}$$

where V is the radome region, $\underline{E}(\underline{r})$ is the total electric field and $G(\underline{r},\underline{r}') = \exp(-jk_bR)/(4\pi R)$ is the 3D scalar Green's function for the background media. $R = |\underline{r} - \underline{r}'|$ is the distance from a source point $\underline{r}'$ to a field point $\underline{r}$ and $k_b = \omega\sqrt{\varepsilon_b\mu_b}$ is the wavenumber of the background medium. By writing $\varepsilon(\underline{r}) = \varepsilon_b + [\varepsilon(\underline{r}) - \varepsilon_b]$ in Maxwell's equations one identifies that the volume current is related to the total electric field by

$$\underline{J}_V(\underline{r}) = j\omega[\varepsilon(\underline{r}) - \varepsilon_b]\underline{E}(\underline{r}) = j\omega\varepsilon(\underline{r})\chi(\underline{r})\underline{E}(\underline{r}) \tag{4.102}$$

in which $\chi(\underline{r}) = 1 - \varepsilon_b/\varepsilon(\underline{r})$. Accordingly, eq. (4.101) can be rewritten in the form

$$\frac{1}{j\omega\varepsilon(\underline{r})\chi(\underline{r})}\underline{J}_V(\underline{r}) + j\omega\mu_b \int_V G(\underline{r},\underline{r}')\underline{J}_V(\underline{r}')dv' - \frac{1}{j\omega\varepsilon_b}\nabla \int_V G(\underline{r},\underline{r}')\nabla'\cdot\underline{J}_V(\underline{r}')dv' = \underline{E}^i(\underline{r}) \tag{4.103}$$

Eq. (4.103) is the volume integral equation (VIE) with the unknown function $\underline{J}_V(\underline{r})$. The MoM can be applied to solve the VIE. In applying a subdomain MoMs, one needs to subdivide the dielectric region into small volume cells. Cubic cells are used in [25, 26, 27] and tetrahedron cells in [28] and [29]. The matrix equation in the MoM can be solved using the conjugate gradient fast Fourier transform (CGFFT) [26] algorithm using the structured cubic model. However, it is not accurate in modeling structures with curved boundaries and if the dielectric contrast is high. On the other hand, the tetrahedron shape and hexahedron shape are suitable for modeling arbitrarily shaped dielectric structures. Hence, both of the two shapes are considered in the implementation of the algorithm presented here. An example of the tetrahedron mesh for an ogive-shaped radome is shown in Fig. 4.14.

To solve the integral eq. (4.103), the unknown vector $j\omega\varepsilon(\underline{r})\underline{E}(\underline{r})$ is represented by a set of N basis functions $\underline{f}_j$, j=1,2,…,N, such that

$$\underline{J}_V(\underline{r}) = [1 - \varepsilon_b/\varepsilon(\underline{r})]j\omega\varepsilon(\underline{r})\underline{E}(\underline{r}) = \chi(\underline{r})\sum_{j=1}^{N} a_j\underline{f}_j(\underline{r}) \tag{4.104}$$

where $\chi(\underline{r}) = 1 - \varepsilon_b/\varepsilon(\underline{r})$. The basis function is defined differently for different cell shapes. In general, a basis function $\underline{f}_j$ is defined on two adjacent cells, Ω_j^+ and Ω_j^- that share the same face S_j as shown in Fig. 4.11. The extended RWG function for tetrahedron mesh is described in [28]. Following the MoM procedure using similar testing functions as basis functions (Galerkin method), the integral eq. (4.103) can be converted into the matrix equation $[Z_{ij}][a_j] = [Vm_i]$ in which $Vm_i = \int_{V_i} \underline{f}_i \cdot \underline{E}^i dv'$. This matrix equation can be solved for the expansion coefficients a_j. The matrix elements are given by

$$Z_{ij} = \int_{V_i} \frac{\underline{f}_i \cdot \underline{f}_j}{j\omega\varepsilon(\underline{r}')} dv' + \int_{V_i} \underline{f}_i \cdot (j\omega \underline{A}_j + \nabla\Phi_j) dv' \tag{4.105}$$

where

$$\underline{A}_j(\underline{r}) = \mu_b \int_{V_j} G(\underline{r}, \underline{r}') \chi(\underline{r}') \underline{f}_j(\underline{r}') dv' \tag{4.106}$$

and

$$\Phi_j(\underline{r}) = -\frac{1}{j\omega\varepsilon_b} \int_{V_j} G(\underline{r}, \underline{r}') \nabla' \cdot [\chi(\underline{r}') \underline{f}_j(\underline{r}')] dv' \tag{4.107}$$

The unknown vector function $j\omega\varepsilon(\underline{r})\underline{E}(\underline{r})$ ensures the continuity of the normal component of the electric flux density $\underline{D}$, since the rooftop basis functions automatically satisfy this condition. The continuity conditions between two volume cells with different permittivities introduce surface charges proportional to $(\chi_j^+ - \chi_j^-)$. This does not lead to additional number of degrees of freedom. In fact, for piecewise continuous χ among the cells, the integral associated to the electric potential in (4.107) can be written as

$$\int_{V_j^\pm} G(\underline{r}, \underline{r}') \nabla' \cdot (\chi_j^\pm \underline{f}_j) dv'$$

$$= \int_{V_j^\pm} \chi_j^\pm G(\underline{r}, \underline{r}') \nabla' \cdot \underline{f}_j dv' + \int_{S_j} G(\underline{r}, \underline{r}') [\chi_j^+ - \chi_j^-] \underline{f}_j \cdot \underline{ds}' \tag{4.108}$$

In (4.108), the second integral on the right-hand side (RHS) is over the common face of the two adjacent cells. As is expected, if the two adjacent cells have the same permittivity, leading to the same χ for both cells, then there will be no net charge on the interface. This result is also obtained in [28] for the tetrahedron cells.

Another issue to be considered is the mesh truncation. In general, the vector $j\omega\varepsilon(\underline{r})\underline{E}(\underline{r})$ is nonzero in the entire background media. Hence, it is necessary to extend the mesh into the whole space. However, by (4.104), only the basis functions defined on cells belonging to the dielectric material region are used to represent the induced

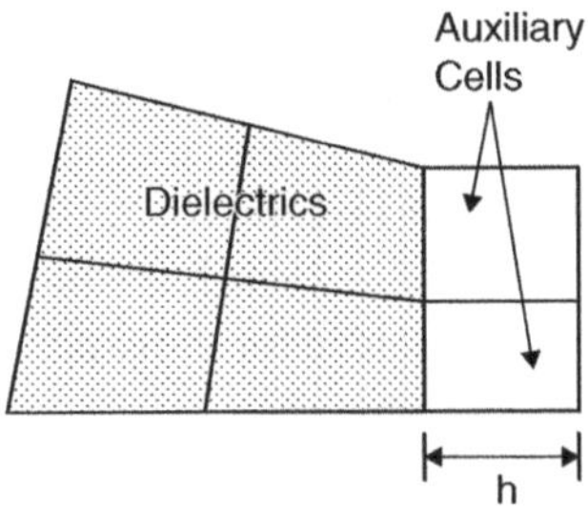

Figure 4.15 Cells in dielectrics (shaded) and auxiliary cells (unshaded). The auxiliary cells are introduced to terminate the mesh (the dimension *h* is arbitrary and is taken to be zero in the numerical implementation).

current $\underline{J}_V(\underline{r})$, because $\chi(\underline{r}) = 0$ if $\underline{r}$ is outside the dielectric region. Therefore, the mesh should be truncated such that only those volume cells that are either inside the dielectric region or have direct contact with the dielectric volume are retained. The cell that does not belong to the dielectric region is called an auxiliary cell, as shown in Fig. 4.15.

The existence of auxiliary cells introduces several difficulties: (a) it extends the source integration in (4.106) and (4.107), and the field integration in (4.105) to beyond the true dielectric object; (b) it increases the computation time because of the extra integral regions; (c) it needs extra mesh generating task which may not always be possible. As a result, in practical implementation of the algorithm, the auxiliary cells are eliminated by, for example, taking the limit of $h \to 0$. This is possible because there is no restriction on the size of the auxiliary cell as long as it shares a face with a cell in the dielectric region. As a result, a boundary basis function that is originally associated with a dielectric cell and an auxiliary cell now is defined on the dielectric cell only. The basis function is called a "half" basis function since it is associated with only one cell, as compared to a regular (or full) basis function that is defined on two cells.

4.3.1 Solution Using Fast Multipole Method

If the matrix equation is solved by an iterative method such as the conjugate gradient (CG) method described in Appendix D, the computational complexity for performing a matrix vector multiplication is $O(N^2)$ for an N-unknown problem. Apparently, because of this complexity, both memory requirement and CPU time can become prohibitively high when N is large. Hence, direct solution by an iterative solver for the set of linear equations is not possible for electrically large radomes. This is the most important reason that previous studies of dielectric scattering problems by MoM are limited to small-sized structures. However, the above-mentioned complexity for a matrix vector multiplication can be reduced to a lower order if the multilevel fast multipole algorithm (MLFMA) with a computational complexity of $O(N\log N)$ for performing a matrix vector multiplication is implemented. The MLFMA [30] and [31] is the multilevel implementation of the fast multipole method (FMM) described in [32], with a computational complexity of $O(N^{1.5})$ for performing a matrix vector multiplication. The principle of FMM is based on the addition theorem for spherical harmonics [17]. The following presents the process to implement FMM with focus on the application to solving VIE. For a given testing function, all the basis functions can be classified into two categories based on the distance between the testing function and the basis function: one as near-neighbor (*NN*) elements and one as well-separated, or far-field (*FF*) elements. As a result, the matrix-vector multiplication in (4.105) can be written as a sum of two terms:

$$\sum_{j} Z_{ij}a_j = \sum_{j\in NN} Z_{ij}a_j + \sum_{j\in FF} Z_{ij}a_j \tag{4.109}$$

The *NN* elements are calculated directly using (4.105) and are stored in a sparse matrix. For *FF* matrix elements, an alternative form of the integral operator can be used, as explained in the following. Moreover, eq. (4.105) can be rewritten in the form

$$Z_{ij} = \int_{V_i} \frac{\underline{f}_i \cdot \underline{f}_j}{j\omega\varepsilon(\underline{r})} dv + j\omega\mu_b \int_{V_i}\int_{V_j} \underline{f}_i(\underline{r}) \cdot \left(\bar{\bar{I}} + \frac{1}{k_b^2}\nabla\nabla\right) \cdot G(\underline{r},\underline{r}')\chi(\underline{r}')\underline{f}_j(\underline{r}')dv'\,dv \tag{4.110}$$

Assuming the dielectric region is subdivided into M small rectangular boxes, the basis functions belonging to one box form a group. Let basis function $\underline{f}_j(\underline{r})$ belong to group-n centered at $\underline{r}_n$, and testing function $\underline{f}_i(\underline{r})$ belong to group-m centered at $\underline{r}_m$. Then, the source to field vector can be rewritten in the form $\underline{r} - \underline{r}' = (\underline{r} - \underline{r}_m) + (\underline{r}_m - \underline{r}_n) + (\underline{r}_n - \underline{r}') = \underline{d} + \underline{r}_{mn}$, where $\underline{d} = (\underline{r} - \underline{r}_m) + (\underline{r}_n - \underline{r}')$ and $\underline{r}_{mn} = (\underline{r}_m - \underline{r}_n)$. For $r_{mn} > d$ in which $r_{mn} = |\underline{r}_{mn}|$ and $d = |\underline{d}|$, the addition theorem for spherical functions [7] can be used to expand the scalar Green's function $G(\underline{r},\underline{r}')$ into a multipole expression:

$$G(\underline{r},\underline{r}') = \frac{e^{-jk_b|\underline{r}-\underline{r}'|}}{4\pi|\underline{r}-\underline{r}'|} = \frac{e^{-jk_b|\underline{r}_{mn}+\underline{d}|}}{4\pi|\underline{r}_{mn}+\underline{d}|}$$

$$= \frac{jk_b}{4\pi}\sum_{l=0}^{\infty}(-1)^l(2l+1)j_l(k_b d)h_l^{(2)}(k_b r_{mn})P_l(\hat{r}_{mn}\cdot\hat{d}) \tag{4.111}$$

where j_l is the spherical Bessel function, $h_l^{(2)}$ is the spherical Hankel function of the second kind, P_l is the Legendre polynomial and $d<r_{mn}$. Expansion of $j_l P_l$ product by plane wave expansion [7] yields

$$4\pi j^l j_l(k_b d) P_l(\hat{r}_{mn} \cdot \hat{d}) = \int P_l(\hat{r}_{mn} \cdot \hat{k}) e^{j\underline{k}_b \cdot \underline{d}} dk_x dk_y \tag{4.112}$$

Substitution of (4.112) into (4.111) gives

$$\begin{aligned} G(\underline{r}, \underline{r}') &= \frac{e^{-jk_b|\underline{r}_{mn}+\underline{d}|}}{4\pi|\underline{r}_{mn}+\underline{d}|} = \frac{jk_b}{(4\pi)^2} \int T_{mn}(\hat{k}) e^{jk_b \hat{k} \cdot \underline{d}} dk_x dk_y \\ T_{mn}(\hat{k}) &= \sum_{l=0}^{\infty} j^l (2l+1) h_l^{(2)}(k_b r_{mn}) P_l(\hat{r}_{mn} \cdot \hat{k}) \end{aligned} \tag{4.113}$$

in which $\hat{k} = (\sin\theta\cos\phi,\ \sin\theta\sin\phi,\ \cos\theta)$. In practical implementations, the infinite series in (4.113)b is approximated by L terms. The selection of L depends on the accuracy desired and [32] gives some guidelines on how to choose this truncation parameter. It suffices, in order to obtain sufficient accuracy that $L > k_b D$, where D is the maximum value of d for which the expansion will be used. In practice, one chooses r_{mn} to make d relatively small, so that good accuracy can be obtained with a modest value of L. The K directions of $\hat{k}$, at which the angular functions are computed, must be sufficient to give a quadrature rule that is exact for all spherical harmonics of order $l < 2L$. A simple method for accomplishing this task is to pick polar angles θ such that they are zeros of $P_L(\cos\theta)$, and azimuthal angles ϕ to be $2L$ equally spaced points. Thus, for this choice $K = 2L^2$. Since $kD \propto \sqrt{N/M}$, this justifies the assertion that $K \propto N/M$. The acceleration of FMM comes from the fact that the summations in (4.113)b can be grouped together before the integral over k is performed. Substitution of (4.113) into (4.110) and the definition of $\underline{d} = (\underline{r} - \underline{r}_m) + (\underline{r}_n - \underline{r}')$ yields the impedance matrix element corresponding to well-separated testing and basis functions:

$$\begin{aligned} Z_{ij} = \int_{V_i} \frac{\underline{f}_i \cdot \underline{f}_j}{j\omega\varepsilon(\underline{r})} dv - \frac{k_b^2 \eta_b}{(4\pi)^2} \int dk_x dk_y \int_{V_i} (\bar{\bar{I}} - \overline{kk}) \underline{f}_i(\underline{r}) e^{jk_b \hat{k} \cdot (\underline{r} - \underline{r}_m)} dv \cdot T_{mn}(\hat{k}) \cdot \int_{V_j} \\ (\bar{\bar{I}} - \overline{kk}) \underline{f}_j(\underline{r}') \chi(\underline{r}') e^{jk_b \hat{k} \cdot (\underline{r}_n - \underline{r}')} dv' \end{aligned} \tag{4.114}$$

in which $\eta_b = \sqrt{\mu_b/\varepsilon_b}$. Eq. (4.114) can be rewritten in the short form,

$$Z_{ij} = \int_{V_i} \frac{\underline{f}_i \cdot \underline{f}_j}{j\omega\varepsilon(\underline{r})} dv - \frac{k_b^2 \eta_b}{(4\pi)^2} \int \mathbf{V}_{fmi}(\hat{k}) T_{mn}(\hat{k}) \mathbf{V}_{snj}(\hat{k}) dk_x dk_y \tag{4.115}$$

in which

$$\mathbf{V}_{fmi}(\hat{k}) = \int_{V_i} (\bar{\bar{I}} - \overline{kk}) \cdot \underline{f}_i(\underline{r}) e^{jk_b \hat{k}\cdot(\underline{r}-\underline{r}_m)} dv$$
$$\mathbf{V}_{snj}(\hat{k}) = \int_{V_j} (\bar{\bar{I}} - \overline{kk}) \cdot \underline{f}_j(\underline{r}') \chi(\underline{r}') e^{jk_b \hat{k}\cdot(\underline{r}_n-\underline{r}')} dv' \quad (4.116)$$

The proposed algorithm described in [32] proposes the following steps toward the solution. First, we divide the N basis functions into M localized groups, labeled by an index m, each supporting about N/M basis functions. The best choice is $M \sim \sqrt{N}$. Next, establish a correspondence between the basis-function index, n, and a pair of indices (m,a), where a denotes the particular basis function within the mth group. Then, denote the center of the smallest sphere enclosing each group $\underline{r}_m$. Next, compute the $\mathbf{V}_{fmi}(\hat{k})$, $\mathbf{V}_{snj}(\hat{k})$ and $T_{mn}(\hat{k})$ matrices at K directions ($L \propto \sqrt{K}$) using eq. (4.116) and (4.113)b and combine them into (4.115) to obtain the Z matrix.

The MLFMA is an extension of FMM, which reduces the complexity for a matrix vector multiplication to $O(N\log N)$ for performing a matrix vector multiplication. To implement the MLFMA, the entire object is first enclosed in a large cube, which is partitioned into eight smaller cubes. Each subcube is then recursively subdivided into smaller cubes until the edge length of the finest cube is about half of a wavelength. Cubes at all levels are indexed. At the finest level, we find the cube in which each basis resides by comparing the coordinates of the center of the basis function with the center of cube. We further find nonempty cubes by sorting. Only nonempty cubes are recorded using tree-structured data at all levels. Thus, the computational cost depends only on the nonempty cubes.

The basic algorithm for matrix-vector multiplication is broken down into two sweeps: The first sweep consists of constructing outer multipole expansions for each nonempty cube at all levels. The second sweep consists of constructing local multipole expansions contributed from the well-separated cubes at all levels. When the cube becomes larger as one progresses from the finest level to the coarsest level, the number of multipole expansions should increase. In the first sweep, the outer multipole expansions are computed at the finest level, and then the expansions for the larger cube are obtained using shifting and interpolation. Let $\underline{r}_{n_l}$ and $\underline{r}_{n_{l-1}}$ be the cube centers at level l and $l-1$, respectively; then the outer multipole expansions for the coarser level

$l-1$ should be

$$\mathbf{V}_{sn_{l-1}j}(\hat{k}) = e^{j\underline{k}\cdot\underline{r}_{n_l n_{l-1}}}\mathbf{V}_{sn_lj}(\hat{k}) \tag{4.117}$$

But $\mathbf{V}_{sn_lj}(\hat{k})$ has only K_l values, and we need $K_{l\text{-}1}$ values of $\mathbf{V}_{sn_{l-1}j}(\hat{k})$. Therefore, we will interpolate $\mathbf{V}_{sn_lj}(\hat{k})$ to $K_{l\text{-}1}$ values first. That is

$$\mathbf{V}_{sn_{l-1}j}(\hat{k}_{(l-1)p'}) = e^{j\underline{k}_{(l-1)p'}\cdot\underline{r}_{n_l n_{l-1}}}\sum_{p=1}^{K_l}\mathbf{W}_{p'p}\mathbf{V}_{sn_lj}(\hat{k}_{lp}) \tag{4.118}$$

where the interpolation matrix $\mathbf{W}$ is a sparse matrix. Similarly, we interpolate $\mathbf{V}_{fm_{l-1}i}(\hat{k}_{(l-1)p})$ using

$$\mathbf{V}_{fm_{l-1}i}(\hat{k}_{(l-1)p'}) = e^{j\underline{k}_{(l-1)p'}\cdot\underline{r}_{m_l m_{l-1}}}\sum_{p=1}^{K_l}\mathbf{W}_{p'p}\mathbf{V}_{fm_li}(\hat{k}_{lp}) \tag{4.119}$$

At the coarsest level, the local multipole expansions contributed from well-separated cubes are calculated using the second part of (4.115). At the second sweep, the local expansions for smaller cubes include the contributions from the parent cube using shifting and interpolation [33], and from the well-separated cubes at this level, but not the well-separated cubes at the parent level. If the local multipole expansions received by a cube center at level $l-1$ is $\mathbf{B}(\hat{k})$, then the contribution from all well-separated cubes can be written as

$$I = \int \mathbf{V}_{fm_{l-1}i}(\hat{k})\cdot\mathbf{B}(\hat{k})dk_x dk_y = \sum_{p'=1}^{K_{l-1}} w_{p'}\mathbf{V}_{fm_{l-1}i}(\hat{k}_{(l-1)p'})\cdot\mathbf{B}(\hat{k}_{(l-1)p'}) \tag{4.120}$$

where $w_{p'}$ is the weighting function. Substituting the interpolation expression (4.119) for $\mathbf{V}_{fm_{l-1}i}(\hat{k}_{(l-1)p'})$ into (4.120) and changing the order of two summations leads to

$$I = \sum_{p=1}^{K_l} w_p\mathbf{V}_{fm_li}(\hat{k}_{lp})\cdot\sum_{p'=1}^{K_{l-1}}\mathbf{W}_{p'p}\mathbf{B}(\hat{k}_{(l-1)p'})e^{j\underline{k}_{(l-1)p'}\cdot\underline{r}_{m_l m_{l-1}}} \tag{4.121}$$

At the finest level, the contributions from non-well-separated cubes are calculated directly. Because only nonempty cubes are considered, the complexity for MLFMA is reduced to $O(N\log N)$, and the memory requirements for MLFMA are of the same order.

The next example shows the calculated results for a circular parabolic reflector antenna $6\lambda_0$ in diameter, with f/D = 0.25 and gain of 22.5 dBi in the presence of three types of radomes:

1. A tangent ogive radome with $\nu = 2$, length $10.4\lambda_0$, and base diameter $9.3\lambda_0$ (modeled by 76914 tetrahedrons)
2. A conical radome with θ_a=20 deg (see Fig. 4.2), length $12.8\lambda_0$, and base diameter $9.3\lambda_0$ (modeled by 55308 tetrahedrons)

3. A hemisphere radome of base diameter $9.3\lambda_0$ and length $4.6\lambda_0$ (modeled by 45,000 tetrahedrons).

The thicknesses and the permitivities of all three radomes are the same: $t = 0.21\lambda_0$, $\varepsilon_r = 5.7$. The reflector antenna is oriented in the same direction as the radome axis. The gain and the calculated radiation patterns for linear polarization (with HFSS) for $\phi = 0$ deg and θ from 0 deg to 90 deg are shown in Fig. 4.16. As a reference, the radiation pattern of the reflector antenna in free-space is also plotted in the figure.

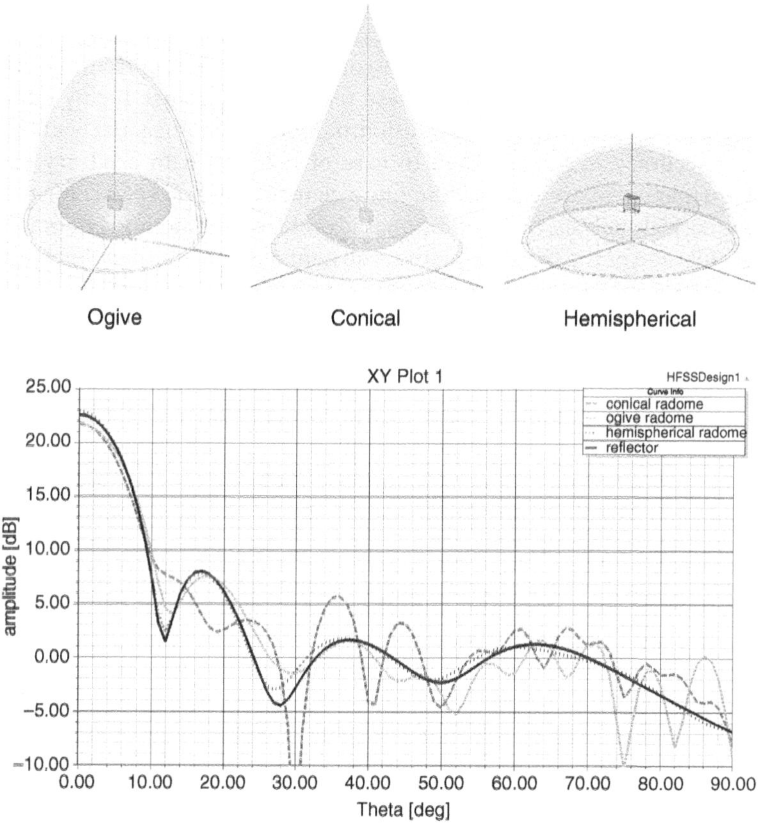

Figure 4.16 Comparison of the normalized radiation of a reflector antenna ($6\lambda_0$ in diameter) in the presence of three types of radome shapes: hemisphere shape (dot line), ogive shape (long dash line), conical shape (dash line), and radiation of the reflector in free space (solid line). All simulations performed with HFSS.

One can observe that the hemispherical radome gives the closest unperturbed pattern in comparison to the radiation pattern of the reflector without radome. The ogive and conical radome performance is almost similar with an insertion loss of 1.5 dB.

4.4 Differential Equation Formulation Methods

In the previous sections, we discussed the analysis of antennas enclosed in radomes using a variety of techniques such as ray optical technique combined with physical optics and numerical analysis using MoM for solving surface and volume integrals representing the multilayer radome structure. These techniques produce highly populated matrices with high condition number, which negatively affect the accuracy of the solution. In case of radomes with arbitrary inhomogeneities, we can use the finite element method (FEM) for the numerical analysis. Some additional advantages this method offers are: (1) fill time of the matrix equation substantially smaller than that with the MoM approach; (2) the resulting matrix is highly sparse and can be efficiently solved; (3) the radome shape can be conveniently described using an automatic mesh generation algorithm.

In this section, we present a general-purpose method for analyzing *axisymmetric* radomes based on the papers [34] and [35]. In the analysis, the reciprocity theorem is used in conjunction with an *axisymmetric* finite element approach that employs an absorbing boundary condition for mesh truncation. A two-step procedure is used, which initially employs the FEM to determine the near fields within the radome when it is illuminated by a distant source. In the second step, the reciprocity theorem is used to find the far-field pattern of an antenna with a given current distribution enclosed in the radome, from the near-field information determined earlier in the first step. The method presented assumes that the presence of the radome does not have a significant effect on the antenna currents.

We begin the first step by expressing $\underline{E}$ and $\underline{H}$ in terms of Fourier series expansions in ϕ as

$$\begin{aligned}\underline{E}(\rho,\phi,z) &= \sum_{m=-\infty}^{\infty}[E_{\rho,m}\hat{\rho}+E_{\phi,m}\hat{\phi}+E_{z,m}\hat{z}]e^{jm\phi}\\ \underline{H}(\rho,\phi,z) &= \sum_{m=-\infty}^{\infty}[H_{\rho,m}\hat{\rho}+H_{\phi,m}\hat{\phi}+H_{z,m}\hat{z}]e^{jm\phi}\end{aligned} \tag{4.122}$$

Let us introduce the scalar potentials u_m and v_m of the mth mode defined by

$$\begin{aligned} u_m &= E_{\phi,m} \\ v_m &= \eta_0 H_{\phi,m} \end{aligned} \tag{4.123}$$

Inserting (4.123) into (4.122), and substituting the resulting equations into Maxwell's two curl equations, six scalar equations are obtained for each value of m. The equations can be written in the form

$$\begin{aligned} E_{\rho,m} &= jf_m\left(m\frac{\partial(\rho u_m)}{\partial\rho} + k_0\rho^2\mu_r\frac{\partial v_m}{\partial z}\right) \\ E_{z,m} &= jf_m\left(m\rho\frac{\partial u_m}{\partial z} - k_0\rho\mu_r\frac{\partial(\rho v_m)}{\partial\rho}\right) \\ H_{\rho,m} &= jf_m\left(m\frac{\partial(\rho v_m)}{\partial\rho} - k_0\rho^2\varepsilon_r\frac{\partial u_m}{\partial z}\right) \\ H_{z,m} &= jf_m\left(m\rho\frac{\partial v_m}{\partial z} + k_0\rho\varepsilon_r\frac{\partial(\rho u_m)}{\partial\rho}\right) \end{aligned} \tag{4.124}$$

and

$$\begin{aligned} u_m\varepsilon_r + \nabla\cdot(f_m\rho\varepsilon_r\nabla(\rho u_m)) + (m/k_0)\nabla\cdot(f_m\hat{\phi}\times\nabla(\rho v_m)) &= 0 \\ v_m\mu_r + \nabla\cdot(f_m\rho\mu_r\nabla(\rho v_m)) - (m/k_0)\nabla\cdot(f_m\hat{\phi}\times\nabla(\rho u_m)) &= 0 \end{aligned} \tag{4.125}$$

in which

$$f_m = [k_0^2\mu_r(\rho,z)\varepsilon_r(\rho,z)\rho^2 - m^2]^{-1} \tag{4.126}$$

Eq. (4.124) demonstrates that $E_{\rho,m}$, $E_{z,m}$, $H_{\rho,m}$, $H_{z,m}$ can be expressed in terms of $E_{\phi,m}$, $H_{\phi,m}$ or alternatively in terms of u_{m}, v_{m}. Moreover, eq. (4.125) describes the basic two coupled differential equations for the FEM solution, which determine u_{m}, v_{m} using the problem boundary conditions. The total electric and magnetic fields for each mode are expressed as a summation of the mth mode scattered fields E_m^s, H_m^s, and the mth mode incident fields E_m^{inc}, H_m^{inc}. Furthermore, because of the azimuthal symmetry of the body, rather than a 3D mesh, the proposed representation requires only a 2D mesh in any ϕ = constant plane, such as the one shown in Fig. 4.17.

This representation results yield in a significant reduction in the number of unknowns, which, in turn, permits the analysis of significantly larger radomes. Multiplication of (4.125) by a testing function T, integration over the entire meshed region Γ, and using the divergence theorem in its 2D form $\iint_\Gamma \nabla\cdot(T\underline{A})ds = \iint_\Gamma$

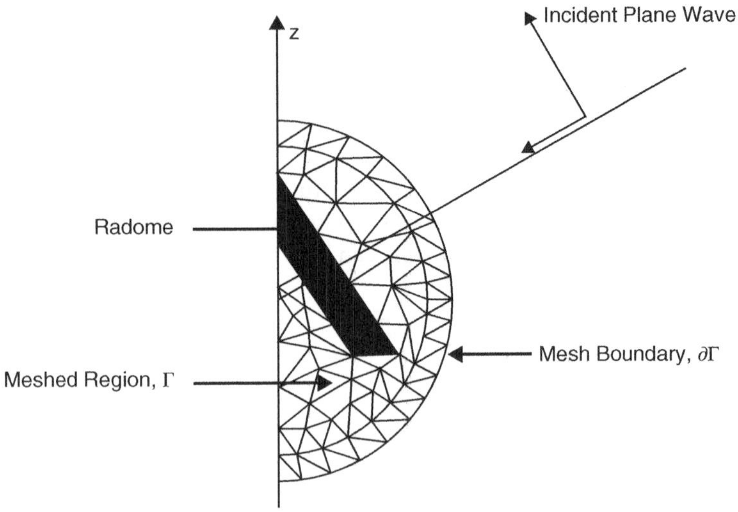

Figure 4.17 Finite element mesh for analysis of axisymmetric radome.

$[\nabla T \cdot \underline{A} + T\nabla \cdot \underline{A}]ds = \int_{\partial\Gamma} T\underline{A} \cdot \hat{n}dl$ yields

$$\iint_\Gamma Tu_m\varepsilon_r ds - \iint_\Gamma f_m\rho\varepsilon_r \nabla T \cdot \nabla(\rho u_m)ds - \frac{m}{k_0}\iint_\Gamma f_m \nabla T \cdot (\hat{\phi} \times \nabla(\rho v_m))ds$$
$$= -\int_{\partial\Gamma} Tf_m\rho\varepsilon_r \frac{\partial(\rho u_m)}{\partial n}dl - \frac{m}{k_0}\int_{\partial\Gamma} f_m T(\hat{\phi} \times \nabla(\rho v_m)) \cdot \hat{n}dl \tag{4.127}$$

and

$$\iint_\Gamma Tv_m\mu_r ds - \iint_\Gamma f_m\rho\mu_r \nabla T \cdot \nabla(\rho v_m)ds + \frac{m}{k_0}\iint_\Gamma f_m \nabla T \cdot (\hat{\phi} \times \nabla(\rho u_m))ds$$
$$= -\int_{\partial\Gamma} Tf_m\rho\mu_r \frac{\partial(\rho v_m)}{\partial n}dl + \frac{m}{k_0}\int_{\partial\Gamma} f_m T(\hat{\phi} \times \nabla(\rho u_m)) \cdot \hat{n}dl \tag{4.128}$$

where $\partial\Gamma$ is the boundary of the meshed region, Γ and $\hat{n}$ is the unit outward normal vector on $\partial\Gamma$ and dl is the arc length element along $\partial\Gamma$. These two coupled equations are implemented using FEM to determine u_m and v_m. The first term on the right-hand side of (4.127), (4.128) involves integrals of the normal derivative of an unknown along the outer boundary $\partial\Gamma$. These integrals are evaluated with the assistance of absorbing boundary conditions derived from Wilcox's expansion for the scattered fields [36], Bayliss-Gunzburger-Turkel boundary conditions [37], or perfect matching layers (PML) boundary conditions described in [38].

The two unknowns u_m and v_m can be represented as finite sums of basis functions weighted by unknown coefficients, as follows:

$$u_m = \sum_{j=1}^{N} a_j B_j(\rho, z) \tag{4.129}$$

and

$$v_m = \sum_{j=1}^{N} b_j B_j(\rho, z) \tag{4.130}$$

where a_j and b_j are unknown coefficients, and $B_j(\rho, z)$ is the basis function at node j in each of the triangular cells of the mesh shown in Fig. 4.12. The basis and testing functions at node j have the form

$$T_j = B_j = \rho^q (R_j \rho + Q_j z + K_j) \tag{4.131}$$

in which R_j, Q_j, and K_j depend on the position of node j and q is either 0 or 1 depending on which mode is considered. Note that if $q = 1$, then $T_j = B_j = 0$ if $\rho = 0$ (i.e., the z-axis). This is acceptable if the absolute value of the mode number is not equal to 1 because the boundary condition along the z-axis is $u_m = v_m = 0$. Thus, neither testing nor basis functions are required at nodes along the z-axis for these modes. However, when the absolute value of the mode number is equal to 1, the unknowns u_m and v_m are not required to be 0 along z-axis. Therefore, the value of q cannot be 1 for this case. Instead, when the absolute value of the mode number is 1, q is taken to be 0. In fact, $q = 0$ could be used for every mode except for the zero-order mode (i.e., $m = 0$). To see why q cannot be 0 in this case, it is necessary to consider that the third integral on the left-hand side of (4.118) will require evaluation of the integral

$$\iint_\Gamma \frac{\rho^{2q}}{k^2\rho^2 - m^2} ds \tag{4.132}$$

For $m = 0$ this becomes

$$\iint_\Gamma \frac{\rho^{2q}}{k^2\rho^2} ds \tag{4.133}$$

Since one edge of Γ coincides with the z-axis ($\rho = 0$), it is impossible to evaluate this integral unless q is larger than 0.5. Thus, q is taken to be equal to 1 when $m = 0$. Using the notation

$$I(n, m) = \int_{\rho_i}^{\rho_j} \frac{\rho^n}{k^2\rho^2 - m^2} d\rho \tag{4.134}$$

Morgan and Mei [39] observed that $I(0,m)$ and $I(1,m)$ are easily calculated and that, for $n > 1$, $I(n,m)$ can be found by using the recursive formula

$$I(n,m) = \frac{1}{k^2}\left(m^2 I(n-2,m) + \frac{\rho_j^{n-1} - \rho_i^{n-1}}{n-1}\right) \tag{4.135}$$

Note that in lossless regions, k will be real and the integrand of (4.134) will have a simple pole singularity at lines along which ρ is equal to $1/k$ multiplied by the absolute value of m. This pole will contribute to the integral a residue of $+\mathrm{j}\pi$ if $\rho_\mathrm{j} > \rho_\mathrm{i}$ and $-\mathrm{j}\pi$ if $\rho_\mathrm{j} < \rho_\mathrm{i}$. Thus, all terms on the left-hand side (LHS) of (4.127) and (4.128) can now be computed. The computation of the right-hand side (RHS) of (4.127) and (4.128) requires also special attention. While the second integral on the RHS of (4.127) and (4.128) involves only a tangential derivative of an unknown along the outer boundary and thus presents no difficulty, the first integral on the RHS requires knowledge of the normal derivative of an unknown along $\partial\Gamma$. Thus, we need to model the behavior of the normal derivatives of unknowns at the outer boundary of the mesh. This task can be achieved using an outer boundary condition derived by Wilcox's expansion for the scattered fields or PML boundary conditions.

The finite element solution for the near fields in the vicinity of the radome, when it is illuminated by a plane wave incident at some angle θ with respect to the z-axis, completes the first step of the two-step procedure.

Next, the reciprocity theorem is used to find the far field at the desired observation angle, for any antenna located anywhere within the radome. This is the second step of the two-step procedure, detailed underneath. Let $(\underline{J}^a, \underline{J}_m{}^a)$ and $(\underline{J}^b, \underline{J}_m{}^b)$ be two sets of sources existing in the same linear medium, and let $\underline{E}^a$ and $\underline{H}^a$ be the fields produced by $(\underline{J}^a, \underline{J}_m{}^a)$ alone, and $\underline{E}^b$ and $\underline{H}^b$ be the corresponding fields produced by $(\underline{J}^b, \underline{J}_m{}^b)$ alone. Then, by invoking the reciprocity theorem, we obtain

$$\iiint (\underline{E}^a \cdot \underline{J}^b - \underline{H}^a \cdot \underline{J}_m{}^b) dv = \iiint (\underline{E}^b \cdot \underline{J}^a - \underline{H}^b \cdot \underline{J}_m{}^a) dv \tag{4.136}$$

Thus, for instance, if we set $\underline{J}_m{}^a$ and $\underline{J}_m{}^b$ to be zero, $\underline{J}^a$ to be the electric current distribution on the antenna, and $\underline{J}^b$ to be an electric dipole located far from the radome, then (4.136) can be used to determine the electric far-field, $\underline{E}^a$, of the antenna radiating within the radome. In order to carry out this calculation, both $\underline{J}^a$ and $\underline{E}^b$ must be known. We assume that the presence of the radome has little effect on the antenna current distribution. Consequently, $\underline{J}^a$ can be determined through an analysis of the antenna radiating in free space. We also note that $\underline{E}^b$

is the electric near field in the vicinity of the radome, when it is illuminated by the electric dipole $\underline{J}^b$. This is precisely what has been determined in the first step—that is, the plane wave incident upon the radome by the electric dipole $\underline{J}^b$. Thus, the electric field determined in the first step is identical to $\underline{E}^b$. So, once the first step has been concluded for a particular location of $\underline{J}^b$, eq. (4.136) can be used to find, at this same location, the electric field radiated by the antenna within the radome. Similarly, the magnetic far field can be determined by setting $\underline{J}^b$ equal to zero and letting $\underline{J}_m{}^b$ be a magnetic dipole located far from the radome. We note at this point that it is unnecessary to recalculate the finite element matrix for each new location of $\underline{J}^b$ and $\underline{J}_m{}^b$; these sources affect only the incident field. Furthermore, we note that while the radome must have azimuthal symmetry, the antenna may have any shape and may be located anywhere within the radome.

The test case antenna chosen is that of a parabolic reflector antenna with $7.5\lambda_0$ diameter and $f/D = 0.28$ inclined $\Omega = 30$ deg with respect to radome axis and excited with a linear polarized waveguide feed. The reflector antenna is enclosed in a conical radome open at the bottom with length $L = 15.5\ \lambda_0$, base diameter $dc = 11.2\ \lambda_0$, apex angle $\theta_a = 20$ deg and thickness $0.21\lambda_0$ made of a material with dielectric constant $\varepsilon_r = 5.7$. Fig. 4.18 shows the gain radiation pattern in x-z plane with and without the radome.

The antenna radiation pattern is symmetric. One can observe that the radome shifts the radiation pattern boresight peak 1.5 deg to the left and breaks its symmetry. All that affects the direction-finding error of the system. In addition, the radome is causing a flash sidelobe due to the multiple reflections caused by the radome at approximately $\theta = 165$ deg.

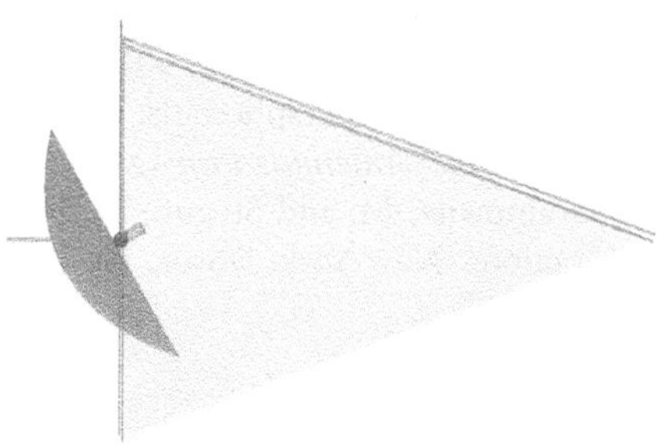

(a)

Figure 4.18 Radiation pattern in x-z plane of a reflector antenna enclosed in a conical radome: (a) the conical radome geometry with $\theta_a = 20$ deg, $\Omega = 30$ deg, (b) the radiation pattern of the reflector antenna with (dash line) and without the radome (solid line). All simulations performed with HFSS.

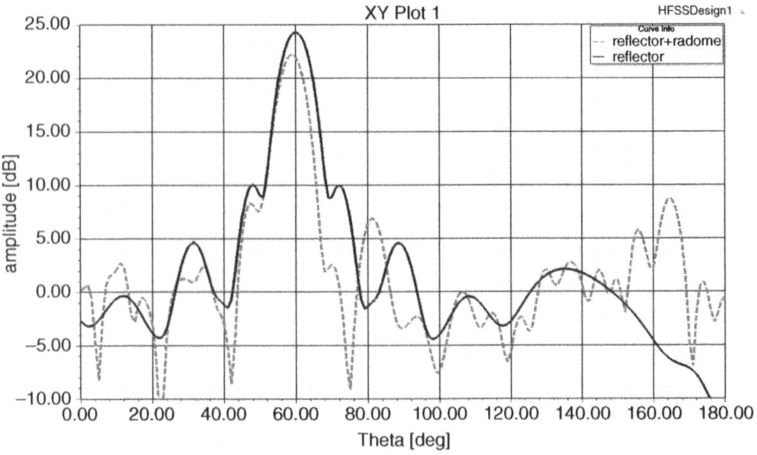

(b)

Figure 4.18 (*Continued*)

References

1 Paris, DT. Computer-aided radome analysis. *IEEE Trans. Antennas Propagat.*, 18(1), 7–15, 1970.
2 Cornbleet, S. Microwave optics. New York: Academic Press, 1976.
3 Wu, DCF, and Rudduck, RC. Plane wave spectrum-surface integration technique for radome analysis. *IEEE Trans. Antennas Propagat.*, 22(3), 497–500, 1974.
4 Boag A, and Letrou, C. Multilevel fast physical optics algorithm for radiation from non-planar apertures. *IEEE Trans. Antennas Propagat.*, 53(6), 2064–2075, 2005.
5 Arvas, E, Rahhalarabi, A, Pekel, U, and Gundogan, E. Electromagnetic transmission through a small radome of arbitrary shape. *IEE Proceedings, pt. H*, 137(6), 401–405, 1990.
6 Overfelt, PL. Superspheroids: A new family of radome shapes. *IEEE Trans. Antennas Propagat.*, 43(2), 215–220, 1995.
7 Abramowitz, M, and Stegun, IA. Handbook of mathematical functions. New York: Dover, 1964.

8 Gradshteyn, IS, and Ryzhik, IM. Table of integrals, series, and products. New York: Academic Press, 1980.

9 Silver, S. Microwave antenna theory and design. London: Peter Peregrinus, 1984.

10 Burks, DG, Graf, ER, and Fahey, MD. A high-frequency analysis of radome-induced radar pointing error. *IEEE Trans. Antennas Propagat.*, 30(5), 947–955, 1982.

11 Crone, GAE, Rudge, AW, and Taylor, GN. Design and performance of airborne radomes: a review." *IEE Proceedings, pt. F*, 128(7), 451–464, 1981.

12 Boag, A, and Letrou, C. Fast Radiation Pattern Evaluation for Lens and Reflector Antennas. *IEEE Trans. Antennas Propagat.*, 51(5), 1063–1069, 2003.

13 Balanis, CA. Antenna theory—analysis and design. Hoboken, NJ: John Wiley & Sons, 2005.

14 Bucci, OM, Gennarelli, C, and Savarese, C. Representation of electromagnetic fields over arbitrary surfaces by a finite and nonredundant number of samples. *IEEE Trans. Antennas Propag.*, 46(3), 351–359, 1998.

15 Bucci, OM, and Franceschetti, G. On the spatial bandwidth of scattered fields. *IEEE Trans. Antennas Propag.*, 35(12), 1445–1455, 1987.

16 Harrington, RF. Field computation by moment methods. New York: IEEE Press, 1968.

17 Harrington, RF. Time-harmonic electromagnetic fields. New York: McGraw-Hill, 1961.

18 Rao, SM, Wilton, DR, and Glisson, AW. Electromagnetic scattering by surfaces of arbitrary shape. *IEEE Trans. Antennas Propagat.*, 30(5), 409–418, 1982.

19 Umanshakar, K., Taflove, A, and Rao, S. M. Electromagnetic Scattering by Three-Dimensional Homogeneous Lossy Dielectric Objects. *IEEE Trans. Antennas Propagat.*, 34(6), 758–766, 1986.

20 Peterson, AF, Ray, SL, Chan, CH, and Mittra, R. Numerical implementations of the conjugate gradient method and the CG-FFT for electromagnetic scattering. In Applications of Conjugate Gra-

dient Method to Electromagnetics and Signal Analysis. New York, Elsevier, 1991, 125–145.
21 Bleszynski, E, Bleszynski, M, and Jaroszewicz, T. AIM: Adaptive integral method for solving large-scale electromagnetic scattering and radiation problems. *Radio Science*, 31(5), 1225–1251, 1996.
22 Chew, WC, Jin, JM, Lu, CC, Michielssen, E, and Song, JM. Fast solution methods in electromagnetics. *IEEE Trans. Antennas Propagat.*, 45(3), 533–543, 1997.
23 Moneum, MAA, Shen, Z, Volakis, JL, and Graham, O. Hybrid PO-MoM analysis of large axi-symmetric radomes. *IEEE Trans. Antennas Propagat.*, 49(12), 1657–1666, 2001.
24 Lu, CC. A fast algorithm based on volume integral equation for analysis of arbitrarily shaped dielectric radomes. *IEEE Trans. Antennas Propagat.*, 51(3), 606–612, 2003.
25 Livesay, DE, and Chen, KM. Electromagnetic fields induced inside arbitrary shaped biological bodies. *IEEE Trans. Microwave Theory Tech..*, 22(12), 1273–1280, 1974.
26 Catedra, MF, Gago, E, and Nulo, L. A numerical scheme to obtain the RCS of three dimensional bodies of resonant size using the conjugate gradient method and the fast fourier transform. *IEEE Trans. Antennas Propagat.*, 37(5), 528–537, 1989.
27 Gan, H, and Chew, W. A discrete BCG-FFT algorithm for solving 3D inhomogeneous scatterer problems. *J. Electromagnetic Waves Applicat.*, 9(10), 1339–1357, 1995.
28 Schaubert, DH, Wilton, DR, and Glisson, AW. A tetrahedral modeling method for electromagnetic scattering by arbitrary shaped inhomogeneous dielectric bodies. *IEEE Trans. Antennas Propagat.*, 32(1), 77-85, 1984.
29 Graglia, RD. The use of parametric elements in the moment method solution of static and dynamic volume integral equations. *IEEE Trans. Antennas Propagat.*, 36(5), 636–646, 1996.
30 Song, JM, and Chew, WC. Multilevel fast-multipole algorithm for solving combined field integral equations of electromagnetic scattering. *Microwave Opt. Tech. Lett.*, 10(1), 14–19, 1995.
31 Song, JM, Lu, CC, and Chew, WC. Multilevel fast-multipole algorithm for electromagnetic scattering by large complex objects. *IEEE Trans. Antennas Propagat.*, 45(10), 1488–1493, 1997.
32 Coifman, R, Rokhlin, V, and Wandzura, S. The fast multipole method for the wave equation: a pedestrian prescription. *IEEE Antennas Propagat. Magazine*, 35(3), 7–12, 1993.

33 Brandt, A. Multilevel computations of integral transforms and particle interactions with oscillatory kernels. *Computation Physical Commun.*, 65(2), 24–38, 1991.
34 Gordon, RK, and Mittra, R. Finite element analysis of axisymmetric radomes. *IEEE Trans. Antennas Propagat.*, 41(7), 975–981, 1993.
35 Gordon, RK, and Mittra, R. PDE techniques for solving the problem of radar scattering by a body of revolution. *Proc. of IEEE*, 79(10), 1449–1458, 1991.
36 Wilcox, CH. An expansion theorem for electromagnetic fields. *Communication Pure Appl. Math.*, IX, 115–134, 1956.
37 Bayliss, A, Gunzburger, M, and Turkel, E. Boundary conditions for the numerical solution of elliptic equations in exterior regions. *SIAM J. Appl. Math.*, 42(4), 430–451, 1982.
38 Sacks, ZS, Kingsland, DM, Lee, R, and Lee, JF. A perfectly matched anisotropic absorber for use as an absorbing boundary condition. *IEEE trans. Antennas Propagat.*, 43(12), 1460–1463, 1995.
39 Morgan, MA, and Mei, KK. Finite-element computation of scattrering by inhomogeneous penetrable bodies of revolution. *IEEE Trans. Antennas Propagat.*, 27(2), 202–215, 1979.
40 Siwiak, K, Dowling, TB, and Lewis, LR. Boresight errors induced by missile radomes. *IEEE Trans. Antennas Propagat.*, 27(11), 832–841, 1979.

Problems

P4.1 A circular aperture antenna with 10λ diameter and uniform distribution is enclosed in a conical radome with a length of 24.5λ, base diameter 12λ and apex angle $\theta_a = 20$ deg. The antenna is inclined $\Omega = 20$ deg with respect to the radome axis and transmits horizontal polarization (y) as shown in Fig. 4.2. The radome thickness is $t = \frac{\lambda}{2\sqrt{\epsilon_r}}$ with $\epsilon_\mathrm{r} = 5.7$. Compute:

(a) The field distribution (amplitude and phase) on the internal surface of the radome in the x-z plane, assuming that the incident field on the radome is locally plane wave and compare to the field distribution of the full spectrum of plane waves from the aperture.

(b) The field distribution on the external surface of the radome in the x-z plane, assuming a single incident plane wave.

(c) Compute the radiation pattern of the antenna in x-z plane using the multilevel algorithm using polynomial coefficients (−1/16, 9/16, 9/16, −1/16). Compare the radiation pattern with and without the radome. What is the expected insertion loss and boresight error?

P4.2 Replace the aperture antenna in P4.1 with a 2λ diameter antenna and compute:

(a) The field distribution (amplitude and phase) on the internal surface of the radome in the x-z plane assuming that the incident field on the radome is locally plane wave and compare to the field distribution of the full spectrum of plane waves from the aperture.

(b) Use ray-tracing technique to evaluate the field distribution on the exterior surface of the radome.

(c) Use the surface integral formulation and its MoM solution to determine the field distribution on the interior and exterior surfaces of the radome. Compare to the field distributions results using ray tracing in (a) and (b). Discuss the differences.

(d) Compute the far field pattern using ray tracing results in (a) and (b) and field distribution results using (c). Discuss differences.

(e) What are the values of the insertion loss and boresight error using the ray-tracing method and surface integral formulation?

P4.3 A circular aperture antenna with 10λ diameter and uniform distribution is enclosed in a conical radome with a length of 24.5λ, base diameter 12λ and apex angle $\theta_a = 20$ deg. The antenna is inclined $\Omega = 20$ deg with respect to the radome axis and transmits horizontal polarization (y) as shown in Fig. 4.2. The operational frequency is 14 GHz. The radome is made from a C-sandwich with exterior skins of 0.27 mm thickness and interior skin of 0.58 mm thickness and electrical parameters $\varepsilon_r = 4.6$, $\tan\delta = 0.028$. The C-sandwich core thickness is 5.8 mm and electrical parameters $\varepsilon_r = 1.3$, $\tan\delta = 0.003$.

(a) Formulate the radome analysis problem using the volume integral equation model and its numerical solution using the MoM.

(b) Compute the field distribution on the internal and external radome surfaces.
(c) Compute the radiation pattern of the enclosed antenna in the radome.
(d) What is the radome's insertion loss and its boresight error?

5

Scattering from Infinite Cylinders

Large radomes are assembled due to fabrication and mechanical considerations from many panels connected together with metallic or dielectric beams, in contrast to small- and medium-sized radomes, which are planar or conformal and made from one piece. The panels are made of thin membranes or sandwiches, which are optimized for minimum transmission loss over the operational frequency band of the radome, as explained in Chapters 2 and 3. The beams introduce scattering effects that degrade the overall electromagnetic performance of the antenna enclosed in the radome, in addition to the transmission losses through the panels. This chapter will describe various numerical methods to compute the scattering effect from a dielectric, conductive and a heterogeneous beam (dielectric beam with conductive strips) with arbitrary cross section, for orthogonal polarizations and for oblique incident angles. The scattering from a beam can be characterized by its scattering pattern and the induced field ratio (*IFR*) introduced by [1, 2]. The *IFR* is defined as the ratio of the forward scattered far-field to the hypothetical far-field radiated in the forward direction by a 2D aperture with a width equal to the optical shadow of the beam on the incident wavefront. The *IFR* depends on polarization, such that IFR_e is related to TM incident wave and IFR_h is related to TE incident wave. The total scattering effect from an array of beams in front of the enclosed antenna is described in Chapter 6.

The computation of the scattering from a heterogeneous beam made of dielectric and conductive materials is possible by considering the scattered field as a superposition of the radiation from equivalent electric polarization currents and conductive currents. Evaluation of these equivalent sources can be obtained by formulating two

Radome Electromagnetic Theory and Design, First Edition. Reuven Shavit.

volume-coupled integral equations: the electric field integral equation (EFIE) and the magnetic field integral equation (MFIE). These integral equations are formulated by recognizing that the total field is equal to the summation of the incident field and the scattered field by the beam equivalent electric and magnetic polarization currents. The two integral equations can be solved by the method of moments (MoM) [3] using for this purpose point matching or linear basis functions $B_n(x,y)$. The formulation and the numerical evaluation of the integral equations involved are described in section 5.1.

The volume discretization is necessary to properly model a heterogeneous scatterer and as such is computational intensive. In case of *homogeneous* or *layered geometries*, they can be treated with surface integral equation formulation and as such the number of unknowns in the numerical evaluation of the solution can be reduced. In a similar fashion to the volume integral equations, two coupled integral equations EFIE and MFIE are formulated using the equivalence theorem to define equivalent electric and magnetic sources on the beam surface. The formulation of the surface integral equations and their numerical evaluation is described in section 5.2.

In case of conductive beams, we can use the derivation of the surface integral equation with adaptation to the perfect electric conductor (PEC) boundary conditions $\hat{n} \times \underline{E} = 0$ on the beam surface. A further simplification of the surface integral equations is obtained if we recognize that the EFIE and MFIE are uncoupled for orthogonal incident field illumination. This observation simplifies and further reduces the computational complexity of the integral equations solution using the MoM. However, the numerical solution of the integral equations encounters singularity problems, due to internal resonances. These internal cavity resonances are the result of the homogeneous solution of the integral equations and they corrupt the solution at these frequencies. The root of the problem is that the surface integral equation involves only data on the mathematical surface of the scatterer and cannot distinguish between "inside" and "outside" in order to produce the desired external solution. Since the interior resonances are different for the EFIE and MFIE the effect of the interior resonances can be minimized if we combine these integral equations into the so called combined field integral equation (CFIE). The formulation of the surface integral equations and their numerical evaluation is described in section 5.3.

An important objective in the design of space frame radomes made of homogeneous dielectric beams is to reduce the forward scattering of the beams. A possible way to reduce the forward scattering from dielectric beams is to tune them with conductive strips inserted in the dielectric beam volume. The induced currents in the conductive strips offset the effect of the electric polarization currents induced in the homogeneous dielectric beam and reduce the forward scattering of the beam for a limited frequency bandwidth. The analysis of the scattering for these tuned beams is possible by formulating the EFIE of the problem and forcing the tangential electric field along the conductive strips of the beam to be equal to zero. The formulation and the numerical evaluation of scattering from a tuned dielectric beam with vertical and horizontal conductive strips to reduce their forward scattering is described in section 5.4.

Volume integral equation solved by MoM as outlined in section 5.1 is versatile, but involves fully populated matrices and requires relatively large computational effort to treat scatterers of moderate electrical size. On the other hand, surface integral equation matrices are populated as the volume integral equation matrices, but are smaller in size. However, they can accommodate only homogeneous, multilayer, and conductive problems. An alternative option to numerical computation of the scattering from a composite inhomogeneous beam is to solve its differential equation using the finite element method (FEM) [4] and [5]. Comparison of the FEM computational effort to the solution with volume integral equations reveals that it is less computationally intensive, since the matrices involved are sparse, but they are slightly larger in size. The differential equation methods often include an additional region of space outside the scatterer to ensure that the scattered fields represent outward propagating solutions. This additional region must be terminated with a radiation boundary condition, as will be discussed later. The formulation and the numerical evaluation of the scattering from a heterogenous beam using FEM are described in section 5.5.

5.1 Heterogeneous Beams—Volume Integral Equation Formulation

Oblique incidence volume integral equation formulation presented in this section is mainly based on [3]. Other relevant references for normal and oblique TM and TE incidence and volume formulation can be found in [6, 7, 8, 9, 10], and [11]. The cross section of the beam is shown in Fig. 5.1. Moreover, the incident field propagation vector

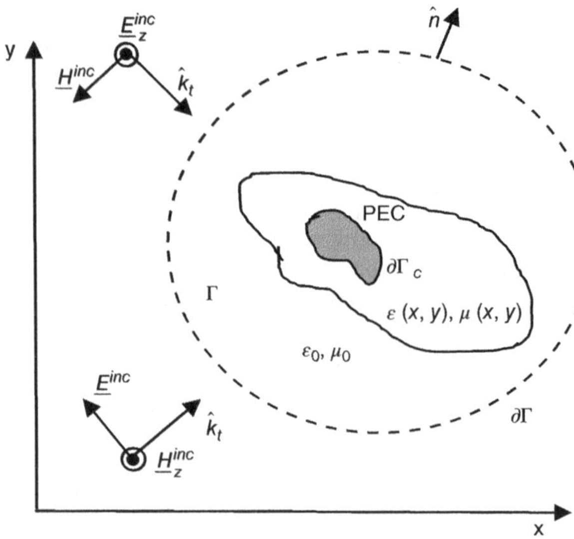

Figure 5.1 -Beam geometry.

components are $k_t = \sqrt{k_0^2 - k_z^2} = k_0 \sin\theta_0$ and $k_z = k_0 \cos\theta_0$, in which $k_0 = \omega\sqrt{\mu_0\varepsilon_0}$, θ_0 is the angle between the incident field propagation direction and the beam axis (for normal incidence $\theta_0 = \frac{\pi}{2}$) and ω is the angular frequency.

We start with the time harmonic $\{e^{j\omega t}\}$ Maxwell equations and introduce equivalent electric $\underline{J}$ and magnetic currents $\underline{J}_m$ to represent the polarization currents induced in a medium with relative permittivity $\varepsilon_r(x,y)$ and permeability $\mu_r(x,y)$.

$$\nabla \times \underline{E} = -j\omega\mu_0\mu_r\underline{H} = -j\omega\mu_0\underline{H} - \underbrace{j\omega\mu_0(\mu_r - 1)\underline{H}}_{\underline{J}_m} \tag{5.1}$$

$$\nabla \times \underline{H} = j\omega\varepsilon_0\varepsilon_r\underline{E} = j\omega\varepsilon_0\underline{E} + \underbrace{j\omega\varepsilon_0(\varepsilon_r - 1)\underline{E}}_{\underline{J}} \tag{5.2}$$

$$\nabla \cdot (\varepsilon_0\varepsilon_r\underline{E}) = 0 \tag{5.3}$$

$$\nabla \cdot (\mu_0\mu_r\underline{H}) = 0 \tag{5.4}$$

The z dependence of the oblique incident field is $e^{jk_z z}$ with k_z being the propagation constant in z direction and correspondingly, we assume that the z dependence of the total fields in the system is also

$e^{jk_z z}$. Thus, the total electric and magnetic fields can be described by

$$\underline{E}(x,y,z) = (\underline{E}_t(x,y) + E_z(x,y)\hat{z})e^{jk_z z} \quad (5.5)$$

$$\underline{H}(x,y,z) = (\underline{H}_t(x,y) + H_z(x,y)\hat{z})e^{jk_z z} \quad (5.6)$$

in which $(\underline{E}_t(x,y),\ \underline{H}_t(x,y))$ are the transverse electric and magnetic fields and $(\underline{E}_z(x,y),\ \underline{H}_z(x,y))$ are the z-components of the electric and magnetic fields. Suppose that the cylinder cross section is modeled by triangular cells of constant permittivity and permeability, but varying from cell to cell. The total scattering from the beam would be the superposition of the scattering from all these various homogeneous triangular cells. Let's define the operators $\nabla = \nabla_t + jk_z\hat{z}$ and $\nabla_t = \frac{\partial}{\partial x}\hat{x} + \frac{\partial}{\partial y}\hat{y}$ to obtain the differential equations for the magnetic field of each triangular cell:

$$\left.\begin{aligned} \nabla_t^2 \underline{H}_t + \left(k_0^2 \varepsilon_r \mu_r - k_z^2\right) \underline{H}_t &= 0 \\ \nabla_t^2 H_z + \left(k_0^2 \varepsilon_r \mu_r - k_z^2\right) H_z &= 0 \end{aligned}\right\} \quad (5.7)$$

Similarly, we obtain for the electric field

$$\left.\begin{aligned} \nabla_t^2 \underline{E}_t + \left(k_0^2 \varepsilon_r \mu_r - k_z^2\right) \underline{E}_t &= 0 \\ \nabla_t^2 E_z + \left(k_0^2 \varepsilon_r \mu_r - k_z^2\right) E_z &= 0 \end{aligned}\right\} \quad (5.8)$$

Substitution of (5.5) and (5.6) into (5.1)–(5.4) yields

$$\left.\begin{aligned} \nabla_t \times \underline{H}_t &= j\omega\varepsilon_0\varepsilon_r E_z\hat{z} \\ \nabla_t H_z \times \hat{z} + jk_z\hat{z} \times \underline{H}_t &= j\omega\varepsilon_0\varepsilon_r \underline{E}_t \end{aligned}\right\} \quad (5.9)$$

$$\left.\begin{aligned} \nabla_t \times \underline{E}_t &= -j\omega\mu_0\mu_r H_z\hat{z} \\ \nabla_t E_z \times \hat{z} + jk_z\hat{z} \times \underline{E}_t &= -j\omega\mu_0\mu_r \underline{H}_t \end{aligned}\right\} \quad (5.10)$$

$$\nabla_t \cdot \underline{H}_t = -jk_z H_z \quad (5.11)$$

$$\nabla_t \cdot \underline{E}_t = -jk_z E_z \quad (5.12)$$

Moreover, if we apply the operator $\nabla_t\times$ on (5.9) and substitute (5.7), we obtain after some algebraic manipulations,

$$\nabla_t \times \nabla_t \times \underline{H}_t = j\omega\varepsilon_0\varepsilon_r \nabla_t \times E_z\hat{z} \quad (5.13)$$

$$\nabla_t\left(\nabla_t \cdot \underline{H}_t\right) - \nabla_t^2 \underline{H}_t = j\omega\varepsilon_0\varepsilon_r \nabla_t E_z \times \hat{z} \quad (5.14)$$

$$-jk_z\nabla_t H_z + \left(k_0^2\varepsilon_r\mu_r - k_z^2\right)\underline{H}_t = j\omega\varepsilon_0\varepsilon_r \nabla_t E_z \times \hat{z} \quad (5.15)$$

$$\underline{H}_t = \frac{1}{k_0^2\varepsilon_r\mu_r - k_z^2}\left(j\omega\varepsilon_0\varepsilon_r \nabla_t E_z \times \hat{z} + jk_z\nabla_t H_z\right) \quad (5.16)$$

We can define the transverse and longitudinal equivalent polarization magnetic currents $\underline{J}_{mt}, J_{mz}$ and express them in terms of the z-directed electric and magnetic components of the fields.

$$\left.\begin{aligned} \underline{J}_{mt} &\triangleq j\omega\mu_0(\mu_r - 1)\underline{H}_t = \frac{\mu_r - 1}{k_0^2\varepsilon_r\mu_r - k_z^2}\left(k_0^2\varepsilon_r\hat{z}\times\nabla_t E_z - k_z\omega\mu_0\nabla_t H_z\right) \\ J_{mz} &\triangleq j\omega\mu_0(\mu_r - 1)H_z \end{aligned}\right\} \tag{5.17}$$

Similarly, one can show that

$$\underline{E}_t = \frac{1}{k_0^2\varepsilon_r\mu_r - k_z^2}\left(-j\omega\mu_0\mu_r\nabla_t H_z \times \hat{z} + jk_z\nabla_t E_z\right) \tag{5.18}$$

and the transverse and longitudinal electric polarization currents $\underline{J}_t$ and J_z are,

$$\left.\begin{aligned} \underline{J}_t &\triangleq j\omega\varepsilon_0(\varepsilon_r - 1)\underline{E}_t = \frac{\varepsilon_r - 1}{k_0^2\varepsilon_r\mu_r - k_z^2}\left(-k_0^2\mu_r\hat{z}\times\nabla_t H_z - k_z\omega\varepsilon_0\nabla_t E_z\right) \\ J_z &\triangleq j\omega\varepsilon_0(\varepsilon_r - 1)E_z \end{aligned}\right\} \tag{5.19}$$

The equivalent polarization currents radiate in free space, but are unknown quantities since they depend on the total field components E_z and H_z. If we denote by $(\underline{E}_1^s, \underline{H}_1^s)$ the electric and magnetic fields due to the electric currents and the fields $(\underline{E}_2^s, \underline{H}_2^s)$ due to the magnetic currents by superposition we can rewrite the total fields,

$$\underline{E}^s = \underline{E}_1^s + \underline{E}_2^s \tag{5.20}$$

$$\underline{H}^s = \underline{H}_1^s + \underline{H}_2^s \tag{5.21}$$

We define $\underline{H}_1^s \triangleq \nabla \times \underline{A}$ and $\underline{E}_2^s \triangleq -\nabla \times \underline{F}$ in which $\underline{A}$ is the vector magnetic potential and $\underline{F}$ is the vector electric potential. Thus, substitution of the vector potentials into Maxwell eqs. (5.1) and (5.2), and using some vector identities yields

$$\nabla \times \underline{E}_1^s = -j\omega\mu_0\underline{H}_1^s = -j\omega\mu_0\nabla \times \underline{A} \tag{5.22}$$

$$\underline{E}_1^s = -j\omega\mu_0\underline{A} - \nabla\Phi \tag{5.23}$$

in which Φ is the scalar electric potential

$$\nabla \times \underline{H}_1^s = j\omega\varepsilon_0\underline{E}_1^s + \underline{J} \tag{5.24}$$

$$\nabla \times \nabla \times \underline{A} = j\omega\varepsilon_0\underline{E}_1^s + \underline{J}$$

$$\nabla(\nabla \cdot \underline{A}) - \nabla^2\underline{A} = j\omega\varepsilon_0\left(-j\omega\mu_0\underline{A} - \nabla\Phi\right) + \underline{J} \tag{5.25}$$

If we use Lorentz gauge $\nabla \cdot \underline{A} \triangleq -j\omega\varepsilon_0\Phi$, we obtain

$$\nabla^2\underline{A} + k_0^2\underline{A} = -\underline{J} \tag{5.26}$$

and

$$\underline{E}_1^s = \frac{\eta_0}{jk_0}\left(\nabla(\nabla \cdot \underline{A}) + k_0^2\underline{A}\right) \tag{5.27}$$

Similarly, we obtain the Lorentz gauge for the magnetic currents $\nabla \cdot \underline{F} \triangleq -j\omega\varepsilon_0\Phi_m$ with Φ_m being the magnetic scalar potential and

$$\nabla^2\underline{F} + k_0^2\underline{F} = -\underline{J}_m \tag{5.28}$$

and

$$\underline{H}_2^s = \frac{1}{jk_0\eta_0}\left[\nabla(\nabla\cdot\underline{F}) + k_0^2\underline{F}\right] \tag{5.29}$$

Finally, we get the scattered fields in terms of vector magnetic and electric potentials:

$$\underline{E}^s = \frac{\eta_0}{jk_0}\left(\nabla(\nabla\cdot\underline{A}) + k_0^2\underline{A}\right) - \nabla\times\underline{F} \tag{5.30}$$

Similarly, for the magnetic field we obtain the expression

$$\underline{H}^s = \nabla\times\underline{A} + \frac{1}{jk_0\eta_0}\left(\nabla(\nabla\cdot\underline{F}) + k_0^2\underline{F}\right) \tag{5.31}$$

Substitution of (5.5) and (5.6) and the explicit expressions of the operators in terms of transverse and longitudinal components into (5.30), yields

$$\begin{aligned}\underline{E}^s &= -j\omega\mu_0\left(\underline{A}_t + A_z\hat{z}\right) + \frac{\left(\nabla_t + jk_z\hat{z}\right)\left[\left(\nabla_t + jk_z\hat{z}\right)\cdot\left(\underline{A}_t + A_z\hat{z}\right)\right]}{j\omega\varepsilon_0}\\ &\quad - \left(\nabla_t + jk_z\hat{z}\right)\times\left(\underline{F}_t + F_z\hat{z}\right)\end{aligned} \tag{5.32}$$

From (5.32), we obtain the explicit form of the z-component

$$\begin{aligned}E_z^s &= -j\omega\mu_0A_z + \frac{1}{j\omega\varepsilon_0}\left[jk_z\left(\nabla_t\cdot\underline{A}_t\right) - k_z^2A_z\right] - \left(\nabla_t\times\underline{F}_t\right)\cdot\hat{z}\\ &= \frac{k_0^2 - k_z^2}{j\omega\varepsilon_0}A_z + \frac{k_z}{\omega\varepsilon_0}\nabla_t\cdot\underline{A}_t - \hat{z}\cdot\left(\nabla_t\times\underline{F}_t\right)\end{aligned} \tag{5.33}$$

Similarly, for the magnetic scattered field we obtain:

$$\begin{aligned}\underline{H}^s &= -j\omega\varepsilon_0\left(\underline{F}_t + F_z\hat{z}\right) + \frac{\left(\nabla_t + jk_z\hat{z}\right)\left[\left(\nabla_t + jk_z\hat{z}\right)\cdot\left(\underline{F}_t + F_z\hat{z}\right)\right]}{j\omega\mu_0}\\ &\quad + \left(\nabla_t + jk_z\hat{z}\right)\times\left(\underline{A}_t + A_z\hat{z}\right)\end{aligned} \tag{5.34}$$

and its associated z-component:

$$\begin{aligned}H_z^s &= -j\omega\varepsilon_0F_z + \frac{1}{j\omega\mu_0}\left[jk_z\left(\nabla_t\cdot\underline{F}_t\right) - k_z^2F_z\right] + \left(\nabla_t\times\underline{A}_t\right)\cdot\hat{z}\\ &= \frac{k_0^2 - k_z^2}{j\omega\mu_0}F_z + \frac{k_z}{\omega\mu_0}\nabla_t\cdot\underline{F}_t + \hat{z}\cdot\left(\nabla_t\times\underline{A}_t\right)\end{aligned} \tag{5.35}$$

Given a z-directed line source with current $\underline{J} = \delta(x)\delta(y)e^{jk_z z}\hat{z}$, we can solve the differential equation for A_z in cylindrical coordinates (r, ϕ, z),

$$\frac{1}{r}\frac{\partial}{\partial r}\left(r\frac{\partial A_z}{\partial r}\right) + \frac{1}{r^2}\frac{\partial^2 A_z}{\partial \phi^2} + \frac{\partial^2 A_z}{\partial z^2} + k_0^2 A_z = -\delta(r)e^{jk_z z} \tag{5.36}$$

By symmetry considerations of the line source geometry, we can assume that $A_z(r, \phi, z) = a_z(r)e^{jk_z z}$. Thus,

$$\frac{1}{r}\frac{\partial}{\partial r}\left(r\frac{\partial a_z}{\partial r}\right) + \left(k_0^2 - k_z^2\right) a_z = -\delta(r) \tag{5.37}$$

and the solution [12] is,

$$a_z(r) = \frac{1}{4j}H_0^{(2)}\left(\sqrt{k_0^2 - k_z^2}\, r\right) \tag{5.38}$$

Solution for J_x and J_y yields similar results for $a_x(r)$ and $a_y(r)$. $H_0^{(2)}(x)$ is the second kind Hankel function of zero order. If we repeat the differential equation solution for a magnetic line source current, we obtain the same results but for the electric vector potential components $f_x(r)$, $f_y(r)$, $f_z(r)$. Moreover, displacement of the line source to r' location in (x,y) plane will change the solution to $G(r, r') = \frac{1}{4j}H_0^{(2)}\left(\sqrt{k_0^2 - k_z^2}\,|\vec{r} - \underline{r}'|\right)$ in which $|\vec{r} - \underline{r}'| = \sqrt{(x - x')^2 + (y - y')^2}$. If we extend the derivation to a 2D electric and magnetic current distribution and use superposition for the computation, we obtain the following result for the z-directed electric and magnetic fields,

$$E_z^s(x, y) = \frac{k_0^2 - k_z^2}{j\omega\varepsilon_0}\left(J_z * G\right) + \frac{k_z}{\omega\varepsilon_0}\nabla_t \cdot \left(\underline{J}_t * G\right) - \hat{z} \cdot \left(\nabla_t \times \left(\underline{J}_{mt} * G\right)\right) \tag{5.39}$$

and

$$H_z^s(x, y) = \frac{k_0^2 - k_z^2}{j\omega\mu_0}\left(J_{mz} * G\right) + \frac{k_z}{\omega\mu_0}\nabla_t \cdot \left(\underline{J}_{mt} * G\right) + \hat{z} \cdot \left(\nabla_t \times \left(\underline{J}_t * G\right)\right) \tag{5.40}$$

in which * denotes convolution. The electric field integral equation (EFIE) and magnetic field integral equation (MFIE) are formulated by recognizing that the total field $E_z(x, y)$ is equal to the summation of the incident field $E_z^{inc}(x, y)$ and the scattered field $E_z^s(x, y)$ excited by the electric and magnetic polarization currents $\underline{J}, \underline{J}_m$ such that

$$E_z^{inc}(x, y) = E_z(x, y) - \frac{k_0^2 - k_z^2}{j\omega\varepsilon_0}\left(J_z * G\right) - \frac{k_z}{\omega\varepsilon_0}\nabla_t \cdot \left(\underline{J}_t * G\right) + \hat{z} \cdot \left(\nabla_t \times \left(\underline{J}_{mt} * G\right)\right) \tag{5.41}$$

and

$$H_z^{inc}(x,y) = H_z(x,y) - \frac{k_0^2 - k_z^2}{j\omega\mu_0}\left(J_{mz} * G\right)$$
$$- \frac{k_z}{\omega\mu_0}\nabla_t \cdot \left(\underline{J}_{mt} * G\right) - \hat{z}\cdot\left(\nabla_t \times \left(\underline{J}_t * G\right)\right) \tag{5.42}$$

in which $E_z^{inc}(x,y) = E_0 \sin\theta_0 e^{-jk_0 \sin\theta_0 (x\cos\phi_0 + y\sin\phi_0)}$ and $H_z^{inc}(x,y) = H_0 \sin\theta_0 e^{-jk_0 \sin\theta_0 (x\cos\phi_0 + y\sin\phi_0)}$ with E_0 and H_0 being the amplitudes of the incident electric and magnetic fields, respectively. These two integral equations can be solved by method of moments (MoM) [3], using, for instance, point matching or linear basis functions $B_n(x,y)$. Thus, the z-component of the electric and magnetic field can be expressed by

$$E_z(x,y) \cong \sum_{n=1}^{N} e_n B_n(x,y) \tag{5.43}$$

and

$$H_z(x,y) \cong \sum_{n=1}^{N} h_n B_n(x,y) \tag{5.44}$$

in which N is the number of basis functions and $B_n(x,y)$ denotes a pyramid basis function centered at the nth corner or node within the triangular-cell model of the cylinder cross section as shown in Fig. 5.2.

The basis function connects the triangular cells grouped around a given vertex and vanishes at every other node in the model. Within the pth cell, there are three nodes $i = 1,2,3$ and the field can be described as

$$H_z(x,y) = a_1 + a_2 x + a_3 y \tag{5.45}$$

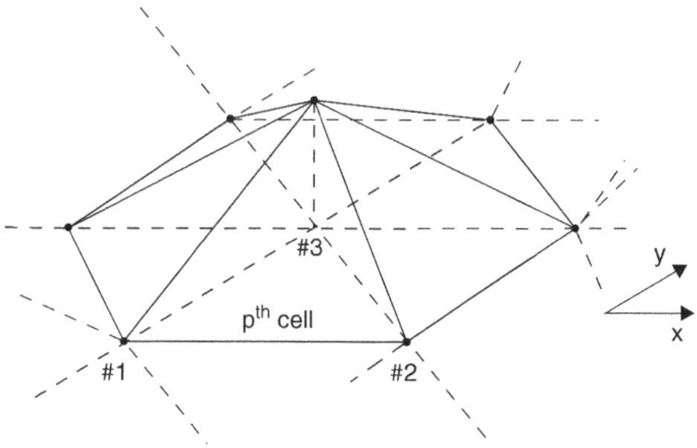

Figure 5.2 Linear pyramid basis function.

The coefficients a_1, a_2, and a_3 can be evaluated in terms of the field values at the cell vertices c_1, c_2, and c_3 with corresponding coordinates (x_1, y_1), (x_2, y_2), (x_3, y_3) such that

$$\begin{pmatrix} c_1 \\ c_2 \\ c_3 \end{pmatrix} = \begin{pmatrix} 1 & x_1 & y_1 \\ 1 & x_2 & y_2 \\ 1 & x_3 & y_3 \end{pmatrix} \begin{pmatrix} a_1 \\ a_2 \\ a_3 \end{pmatrix} \tag{5.46}$$

Solution of (5.46) for (a_1, a_2, a_3) yields

$$\begin{aligned} a_1 &= \frac{x_2y_3 - x_3y_2}{2S} \\ a_2 &= \frac{c_1(y_2 - y_3) + c_2(y_3 - y_1) + c_3(y_1 - y_2)}{2S} \\ a_3 &= \frac{c_1(x_3 - x_2) + c_2(x_1 - x_3) + c_3(x_2 - x_1)}{2S} \end{aligned} \tag{5.47}$$

with $S = \frac{1}{2}[y_1(x_3 - x_2) + y_2(x_1 - x_3) + y_3(x_2 - x_1)]$ representing the area of the triangle cell. Accordingly, the basis function as shown in Fig. 5.2 can be derived from (5.45) and (5.47) for $c_1 = 1$, $c_2 = 0$, $c_3 = 0$.

$$B_n(x, y) = \underbrace{\frac{x_2y_3 - x_3y_2}{2S}}_{a_1} + \underbrace{\frac{y_2 - y_3}{2S}}_{a_2}x + \underbrace{\frac{x_3 - x_2}{2S}}_{a_3}y \tag{5.48}$$

Substitution of (5.43), (5.44) into (5.41) and (5.42) and evaluation at discrete points (x_m, y_m) with $m = 1..N$ in the cylinder cross section yields the matrix equation

$$\begin{bmatrix} EE & EH \\ HE & HH \end{bmatrix} \begin{bmatrix} e_1 \\ \vdots \\ e_N \\ h_1 \\ \vdots \\ h_N \end{bmatrix} = \begin{bmatrix} E_z^{inc}(x_1, y_1) \\ \vdots \\ E_z^{inc}(x_N, y_N) \\ H_z^{inc}(x_1, y_1) \\ \vdots \\ H_z^{inc}(x_N, y_N) \end{bmatrix} \tag{5.49}$$

in which

$$\begin{aligned} EE_{mn} = {} & \delta_{mn} - (k_0^2 - k_z^2)\left[(\varepsilon_r - 1)\, B_n * G\right]\Big|_{x=x_m,\, y=y_m} \\ & + k_z^2 \left[\nabla_t \cdot \left(\left(\frac{\varepsilon_r - 1}{k_0^2\varepsilon_r\mu_r - k_z^2}\nabla_t' B_n\right) * G\right)\right]\Bigg|_{x=x_m,\, y=y_m} \\ & + k_0^2 \left[\hat{z} \cdot \nabla_t \times \left(\left(\frac{\varepsilon_r(\mu_r - 1)}{k_0^2\varepsilon_r\mu_r - k_z^2}\hat{z} \times \nabla_t' B_n\right) * G\right)\right]\Bigg|_{x=x_m,\, y=y_m} \end{aligned} \tag{5.50}$$

$$EH_{mn} = \frac{k_0^2 k_z}{\omega\varepsilon_0}\left[\nabla_t \cdot \left(\left(\frac{(\varepsilon_r - 1)\mu_r}{k_0^2\varepsilon_r\mu_r - k_z^2}\hat{z} \times \nabla_t' B_n\right) * G\right)\right]\Bigg|_{x=x_m,\, y=y_m}$$

$$- k_z \omega \mu_0 \left[\hat{z} \cdot \nabla_t \times \left(\left(\frac{(\mu_r - 1)}{k_0^2 \varepsilon_r \mu_r - k_z^2} \nabla'_t B_n \right) * G \right) \right] \Bigg|_{x=x_m,\, y=y_m} \quad (5.51)$$

$$HE_{mn} = -\frac{k_0^2 k_z}{\omega \mu_0} \left[\nabla_t \cdot \left(\left(\frac{(\mu_r - 1)\varepsilon_r}{k_0^2 \varepsilon_r \mu_r - k_z^2} \hat{z} \times \nabla'_t B_n \right) * G \right) \right] \Bigg|_{x=x_m,\, y=y_m}$$

$$+ k_z \omega \varepsilon_0 \left[\hat{z} \cdot \nabla_t \times \left(\left(\frac{(\varepsilon_r - 1)}{k_0^2 \varepsilon_r \mu_r - k_z^2} \nabla'_t B_n \right) * G \right) \right] \Bigg|_{x=x_m,\, y=y_m} \quad (5.52)$$

$$HH_{mn} = \delta_{mn} - (k_0^2 - k_z^2) \left[(\mu_r - 1) B_n * G \right] \Big|_{x=x_m,\, y=y_m}$$

$$+ k_z^2 \left[\nabla_t \cdot \left(\left(\frac{\mu_r - 1}{k_0^2 \varepsilon_r \mu_r - k_z^2} \nabla'_t B_n \right) * G \right) \right] \Bigg|_{x=x_m,\, y=y_m}$$

$$+ k_0^2 \left[\hat{z} \cdot \nabla_t \times \left(\left(\frac{\mu_r(\varepsilon_r - 1)}{k_0^2 \varepsilon_r \mu_r - k_z^2} \hat{z} \times \nabla'_t B_n \right) * G \right) \right] \Bigg|_{x=x_m,\, y=y_m} \quad (5.53)$$

Because the z-components of the fields are linear functions within this representation, transverse fields and transverse current densities are constant within a triangular cell. Let us denote

$$A_{mn}^{(p)} = \frac{1}{4j} \iint_{s_p} H_0^{(2)} \left(\sqrt{k_0^2 - k_z^2} \left| \underline{r}_m - \underline{r}' \right| \right) dx' dy'$$

$$B_{mn}^{(p)} = \frac{1}{4j} \iint_{s_p} x' H_0^{(2)} \left(\sqrt{k_0^2 - k_z^2} \left| \underline{r}_m - \underline{r}' \right| \right) dx' dy'$$

$$C_{mn}^{(p)} = \frac{1}{4j} \iint_{s_p} y' H_0^{(2)} \left(\sqrt{k_0^2 - k_z^2} \left| \underline{r}_m - \underline{r}' \right| \right) dx' dy'$$

$$D_{mn}^{(p)} = \frac{1}{4j} \iint_{s_p} \frac{\sqrt{k_0^2 - k_z^2}\,(x_m - x')}{\left| \underline{r}_m - \underline{r}' \right|} H_1^{(2)} \left(\sqrt{k_0^2 - k_z^2} \left| \underline{r}_m - \underline{r}' \right| \right) dx' dy'$$

$$E_{mn}^{(p)} = \frac{1}{4j} \iint_{s_p} \frac{\sqrt{k_0^2 - k_z^2}\,(y_m - y')}{\left| \underline{r}_m - \underline{r}' \right|} H_1^{(2)} \left(\sqrt{k_0^2 - k_z^2} \left| \underline{r}_m - \underline{r}' \right| \right) dx' dy' \quad (5.54)$$

in which s_p denotes the area of the pth triangle with a vertex at (x_n, y_n). $H_0^{(2)}$ and $H_1^{(2)}$ are the second kind Hankel functions of zero and first order, respectively. The convolutions in (5.54) are two-dimensional integrals. In general, all of the convolution integrals must be evaluated by numerical quadrature. Substitution of (5.48) into (5.50)–(5.53), evaluation of the vector operations, and substitution of (5.54) yields the following:

$$EE_{mn} = P_n \delta_{mn} - (k_0^2 - k_z^2) \sum_{p=1}^{P_n} \left(\varepsilon_r^{(p)} - 1 \right) \left(a_1^{(p)} A_{mn}^{(p)} + a_2^{(p)} B_{mn}^{(p)} + a_3^{(p)} C_{mn}^{(p)} \right)$$

$$
-\sum_{p=1}^{P_n} \frac{k_z^2\left(\varepsilon_r^{(p)}-1\right)+k_0^2\varepsilon_r^{(p)}\left(\mu_r^{(p)}-1\right)}{k_0^2\varepsilon_r^{(p)}\mu_r^{(p)}-k_z^2}\left(a_2^{(p)}D_{mn}^{(p)}+a_3^{(p)}E_{mn}^{(p)}\right) \tag{5.55}
$$

$$
EH_{mn}=\frac{k_0^2k_z}{\omega\varepsilon_0}\sum_{p=1}^{P_n}\frac{\left(\varepsilon_r^{(p)}\mu_r^{(p)}-1\right)}{\left(k_0^2\varepsilon_r^{(p)}\mu_r^{(p)}-k_z^2\right)}\left(a_3^{(p)}D_{mn}^{(p)}-a_2^{(p)}E_{mn}^{(p)}\right) \tag{5.56}
$$

$$
HE_{mn}=-\frac{k_0^2k_z}{\omega\mu_0}\sum_{p=1}^{P_n}\frac{\left(\varepsilon_r^{(p)}\mu_r^{(p)}-1\right)}{\left(k_0^2\varepsilon_r^{(p)}\mu_r^{(p)}-k_z^2\right)}\left(a_3^{(p)}D_{mn}^{(p)}-a_2^{(p)}E_{mn}^{(p)}\right) \tag{5.57}
$$

$$
\begin{aligned}
HH_{mn}&=P_n\delta_{mn}-(k_0^2-k_z^2)\sum_{p=1}^{P_n}\left(\mu_r^{(p)}-1\right)\left(a_1^{(p)}A_{mn}^{(p)}+a_2^{(p)}B_{mn}^{(p)}+a_3^{(p)}C_{mn}^{(p)}\right)\\
&\quad-\sum_{p=1}^{P_n}\frac{k_z^2\left(\mu_r^{(p)}-1\right)+k_0^2\mu_r^{(p)}\left(\varepsilon_r^{(p)}-1\right)}{k_0^2\varepsilon_r^{(p)}\mu_r^{(p)}-k_z^2}\left(a_2^{(p)}D_{mn}^{(p)}+a_3^{(p)}E_{mn}^{(p)}\right)
\end{aligned} \tag{5.58}
$$

in which P_n is the total number of triangles with vertex at (x_n, y_n). Each entry in (5.55)–(5.58) represents a contribution from a source that is distributed over several adjacent triangular cells. The electric and magnetic fields vectors in (5.49) can be evaluated by matrix inversion.

An important parameter, which characterizes the scattering from a beam, is the *IFR*. The *IFR* is equal to the ratio of the maximum forward far-field scattering field to the hypothetical forward maximum radiated far field from a 2D aperture with a width equal to the optical projected width of the beam on the incident plane wave front. The aperture field distribution is equal to the incident field [12]. The *IFR* depends on polarization, such that IFR_e is related to TM incident wave and IFR_h is related to TE incident wave. Usually, the maximum forward scattering occurs in the propagation direction of the incident wave, as shown in Fig. 5.3, with the maximum forward-scattering direction $(\theta = \pi - \theta_0, \phi = \phi_0)$.

Eq. (5.39) can be rewritten in the form

$$
\begin{aligned}
E_z^s &= \frac{k_t^{\,2}}{j\omega\varepsilon_0}\frac{1}{4j}\int_S J_z H_0^{(2)}\left(k_t\left|\underline{r}-\underline{r}'\right|\right)dx'dy'\\
&+\frac{k_z}{\omega\varepsilon_0}\frac{1}{4j}\int_S \underline{J}_t\cdot\nabla_t H_0^{(2)}\left(k_t\left|\underline{r}-\underline{r}'\right|\right)dx'dy'\\
&+\hat{z}\cdot\frac{1}{4j}\int_s \underline{J}_{mt}\times\nabla_t H_0^{(2)}\left(k_t\left|\underline{r}-\underline{r}'\right|\right)dx'dy'
\end{aligned} \tag{5.59}
$$

in which $k_t = \sqrt{k_0^2 - k_z^2} = k_0 \sin\theta_0$ and $k_z = k_0\cos\theta_0$. S denotes the cross-section area of the beam and the electric and magnetic current distributions have been computed numerically using the MoM, as

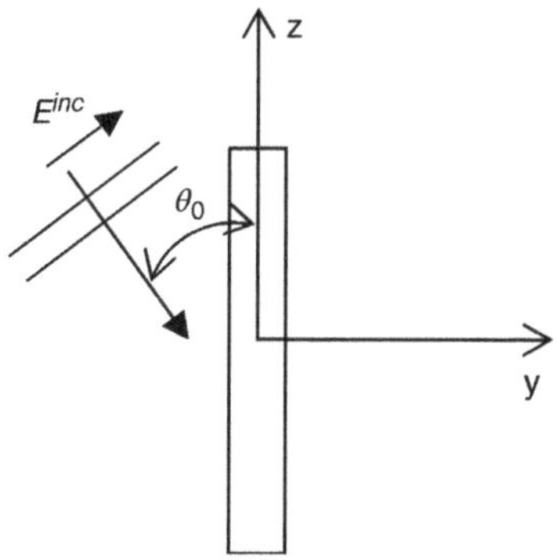

Figure 5.3 The geometry of the oblique incident beam.

explained above. Next, we need to evaluate the scattered field for far-field approximation and use the expressions of Hankel functions for large arguments [13]:

$$
\begin{aligned}
H_0^{(2)}\left(k_t\left|\underline{r}-\underline{r}'\right|\right)\big|_{|\underline{r}\to\infty|} &\approx \sqrt{\frac{2j}{\pi k_t r}}\, e^{-jk_{t'} r} e^{j\underline{k}_{t'}\cdot\underline{r}'} \\
\nabla_t H_0^{(2)}\left(k_t\left|\underline{r}-\underline{r}'\right|\right)\big|_{|\underline{r}\to\infty|} &\approx -jk_t\sqrt{\frac{2j}{\pi k_t r}}\, e^{-jk_{t'} r} e^{j\underline{k}_{t'}\cdot\underline{r}'}\,\hat{r}
\end{aligned}
\tag{5.60}
$$

Evaluation of the phase term in (5.60) for $\theta = \pi - \theta_0$, $\phi = \phi_0$ yields

$$
\begin{aligned}
\underline{k}_{t'}\cdot\underline{r}'\Big|_{\theta=\pi-\theta_0,\,\phi=\phi_0} &= k_0 r'\sin\theta_0\left(\cos\phi_0\cos\phi' + \sin\phi_0\sin\phi'\right) \\
&= k_0 r'\sin\theta_0\cos(\phi_0-\phi')
\end{aligned}
\tag{5.61}
$$

Evaluation of (5.59) for far field ($r \to \infty$) approximation using (5.60),(5.61) and the representation of $\hat{r}$ in Cartesian coordinates $\hat{r} = \sin\theta\cos\phi\hat{x} + \sin\theta\sin\phi\hat{y} + \cos\theta\hat{z}$ yields

$$
E_z^s(r,\pi-\theta_0,\phi_0) = \frac{1}{4j}\sqrt{\frac{2j}{\pi k_t r}}\,e^{-jk_{t'} r}
$$
$$
\times\left[\begin{aligned}
&\frac{k_t^{\,2}}{j\omega\varepsilon_0}\int_S J_z e^{jk_0 r'\sin\theta_0\cos(\phi_0-\phi')}dx'dy' \\
&+\frac{k_z k_t}{j\omega\varepsilon_0}\sin\theta_0\int_S\left(J_{tx}\cos\phi_0 + J_{ty}\sin\phi_0\right)e^{jk_0 r'\sin\theta_0\cos(\phi_0-\phi')}dx'dy' \\
&+jk_t\sin\theta_0\int_S\left(J_{mtx}\sin\phi_0 - J_{mty}\cos\phi_0\right)e^{jk_0 r'\sin\theta_0\cos(\phi_0-\phi')}dx'dy'
\end{aligned}\right]
\tag{5.62}
$$

Similar derivation enables to compute the maximum radiated electric field in the direction $\theta = \pi - \theta_0$, $\phi = \phi_0$ by an infinite aperture in z direction with width w positioned at the origin embedded in an infinite conductive plane and illuminated by the same TM polarization oblique incident field as the beam:

$$E_z^a\left(r,\pi-\theta_0,\phi_0\right)=\frac{1}{4j}\sqrt{\frac{2j}{\pi k_t r}}e^{-jk_{t'}r}\frac{k_t^{\,2}}{j\omega\varepsilon_0}\frac{2E_0 w}{\eta_0} \tag{5.63}$$

in which $\eta_0=\sqrt{\frac{\mu_0}{\varepsilon_0}}$. The ratio of (5.62) and (5.63) results in the expression of IFR_e, such that

$$IFR_e=\frac{\eta_0}{2E_0 w}\int_s\left(J_z+\cos\theta_0\left(J_{tx}\cos\phi_0+J_{ty}\sin\phi_0\right)-\frac{1}{\eta_0}\left(J_{mtx}\sin\phi_0-J_{mty}\cos\phi_0\right)\right)$$
$$\times e^{jk_0r'\sin\theta_0\cos(\phi_0-\phi')}dx'dy' \tag{5.64}$$

By a similar derivation eq. (5.40) can be rewritten for far field ($r\to\infty$) approximation for the maximum forward scattered magnetic field by,

$$H_z^s(r,\pi-\theta_0,\phi_0)=\frac{1}{4j}\sqrt{\frac{2j}{\pi k_t r}}e^{-jk_{t'}r}$$
$$\times\begin{bmatrix}\frac{k_t^{\,2}}{j\omega\mu_0}\int_S J_{mz}e^{jk_0r'\sin\theta_0\cos(\phi_0-\phi')}dx'dy' \\ +\frac{k_zk_t}{j\omega\mu_0}\sin\theta_0\int_S\left(J_{mtx}\cos\phi_0+J_{mty}\sin\phi_0\right)e^{jk_0r'\sin\theta_0\cos(\phi_0-\phi')}dx'dy' \\ -jk_t\sin\theta_0\int_S\left(J_{tx}\sin\phi_0-J_{ty}\cos\phi_0\right)e^{jk_0r'\sin\theta_0\cos(\phi_0-\phi')}dx'dy'\end{bmatrix} \tag{5.65}$$

In a similar fashion, we consider the maximum radiated TE polarized magnetic field in the direction $\theta=\pi-\theta_0$, $\phi=\phi_0$ by an infinite aperture with width w,

$$H_z^a(r,\pi-\theta_0,\phi_0)=\frac{1}{4j}\sqrt{\frac{2j}{\pi k_t r}}e^{-jk_{t'}r}\frac{k_t^{\,2}}{j\omega\mu_0}2H_0\eta_0 w \tag{5.66}$$

The ratio of (5.65) and (5.66) results in the expression of IFR_h, such that

$$IFR_h=\frac{1}{2H_0 w}\int_s\left(\frac{1}{\eta_0}J_{mz}+\frac{\cos\theta_0}{\eta_0}\left(J_{mtx}\cos\phi_0+J_{mty}\sin\phi_0\right)+\left(J_{tx}\sin\phi_0-J_{ty}\cos\phi_0\right)\right)$$
$$\times e^{jk_0r'\sin\theta_0\cos(\phi_0-\phi')}dx'dy' \tag{5.67}$$

An alternative figure of merit to describe the scattering from the scatterer is the radar cross section (RCS). For TM polarization the RCS is defined by

$$\sigma_e(\phi,\theta,\phi_0,\theta_0)=\lim_{r\to\infty}2\pi r\left|\frac{E_z^s(r,\phi,\theta)}{E_z^{inc}(\phi_0,\theta_0)}\right|^2 \tag{5.68}$$

and for TE polarization,

$$\sigma_h(\phi,\theta,\phi_0,\theta_0)=\lim_{r\to\infty}2\pi r\left|\frac{H_z^s(r,\phi,\theta)}{H_z^{inc}(\phi_0,\theta_0)}\right|^2 \tag{5.69}$$

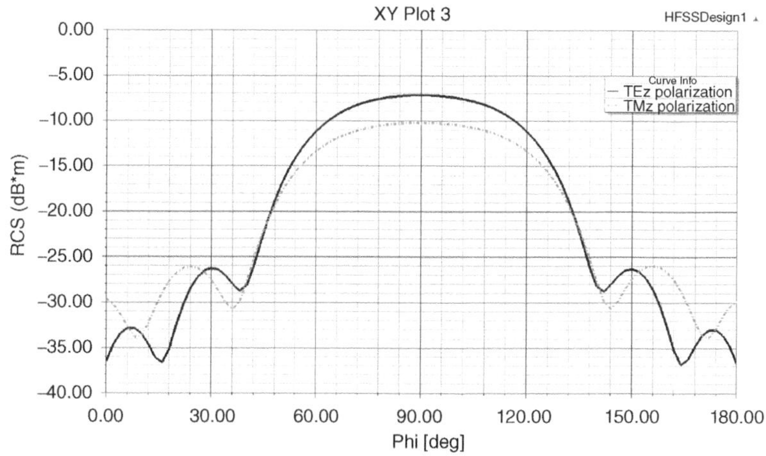

Figure 5.4 Bistatic scattering of a rectangular beam 1.38 × 6.2 in.^{2} with $\varepsilon_r = 4.6$, $\tan\delta = 0.014$ normal incidence illumination by a plane wave on its narrow side at 5 GHz for TE_z polarization (solid line) and TM_z polarization (dash line) simulated with HFSS.

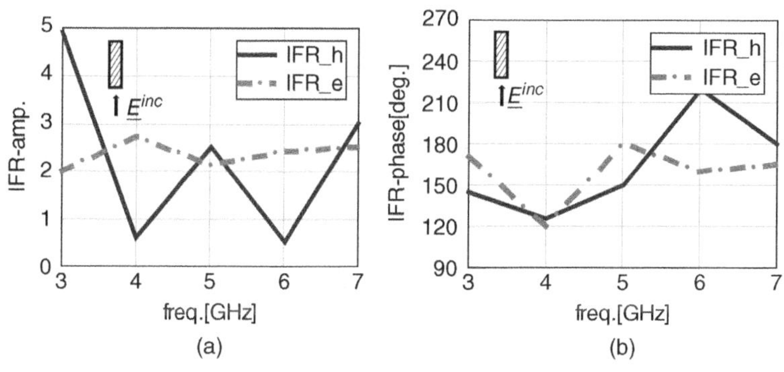

Figure 5.5 *IFR* as function of frequency for a dielectric beam with dimensions 1.38 × 6.2 in.^{2} with $\varepsilon_r = 4.6$, $\tan\delta = 0.014$ and illuminated by a normal plane wave on its narrow side: (a) *IFR*-amplitude; (b) *IFR*-phase.

Fig. 5.4 shows the bistatic scattering of a rectangular beam 1.38 × 6.2 in.^{2} with $\varepsilon_r = 4.6$, $\tan\delta = 0.014$ illuminated normally by a plane wave on its narrow side at 5 GHz for TE_z and TM_z polarizations. The simulation was performed with HFSS commercial software from ANSYS. One can observe a higher scattering of the TE_z polarization.

Fig. 5.5 shows the *IFR* (amplitude and phase) as a function of frequency for two orthogonal polarizations and an incident plane wave on the beam narrow side. The beam is dielectric with mechanical and electrical properties as shown in Fig. 5.4. It is interesting to note that the IFR_h is higher and more oscillating compared to IFR_e.

5.2 Homogeneous Beams—Surface Integral Equation Formulation

A volume discretization is necessary to model properly a heterogeneous scatterer and as such is computational intensive. In case of *homogeneous, layered geometries,* and *conductive cylinders,* they can be treated with surface integral equation formulation and consequently reduce the required number of unknowns in the MoM numerical algorithm [14]. Therefore, this method is considered as a more efficient alternative to the volume integral equation discussed in the previous section. In the derivation, the equivalence principle [15] is used. Fig. 5.6 shows the beam geometry and the equivalent sources for the external and internal methods.

For the EFIE, we consider the tangential electric field at S^+ and S^- such that

$$\begin{aligned}\hat{n}\times\underline{E}^{inc} &= -\underline{J}_m - \hat{n}\times\left\{\frac{\eta_0}{jk_0}\left[\nabla\left(\nabla\cdot\underline{A}^{(0)}\right)+k_0^2\underline{A}^{(0)}\right]-\nabla\times\underline{F}^{(0)}\right\}_{S^+}\\ 0 &= \underline{J}_m - \hat{n}\times\left\{\frac{\eta_d}{jk_d}\left[\nabla\left(\nabla\cdot\underline{A}^{(d)}\right)+k_d^2\underline{A}^{(d)}\right]-\nabla\times\underline{F}^{(d)}\right\}_{S^-}\end{aligned} \tag{5.70}$$

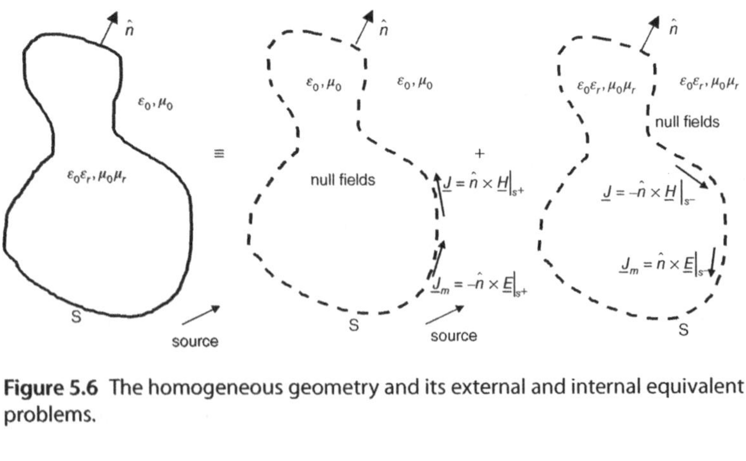

Figure 5.6 The homogeneous geometry and its external and internal equivalent problems.

in which $k_d = k_0\sqrt{\mu_r\varepsilon_r}$, $\eta_d = \eta_0\sqrt{\frac{\mu_r}{\varepsilon_r}}$ and $\left(\underline{A}^{(0)}, \underline{F}^{(0)}\right)$, $\left(\underline{A}^{(d)}, \underline{F}^{(d)}\right)$ are the magnetic and electric vector potentials in free space and in the homogeneous beam, respectively. Similarly, for the MFIE we consider the tangential magnetic field at S^+ and S^- such that

$$\hat{n}\times\underline{H}^{inc} = \underline{J} - \hat{n}\times\left\{\frac{1}{jk_0\eta_0}\left[\nabla\left(\nabla\cdot\underline{F}^{(0)}\right) + k_0^2\underline{F}^{(0)}\right] + \nabla\times\underline{A}^{(0)}\right\}_{S^+}$$

$$0 = -\underline{J} - \hat{n}\times\left\{\frac{1}{jk_d\eta_d}\left[\nabla\left(\nabla\cdot\underline{F}^{(d)}\right) + k_d^2\underline{F}^{(d)}\right] + \nabla\times\underline{A}^{(d)}\right\}_{S^-} \tag{5.71}$$

The z dependence of the oblique incident field is $e^{jk_z z}$ and correspondingly, we assume that the z dependence of the total fields in the system is also $e^{jk_z z}$, in which $k_z = k_0\cos\theta_0$. The electric and magnetic fields can be described by (5.5), (5.6) and the incident fields for the TM and TE cases are $E_z^{inc}(x, y) = E_0\sin\theta_0 e^{-jk_0 r\sin\theta_0\cos(\phi-\phi_0)}$ and $H_z^{inc}(x, y) = H_0\sin\theta_0 e^{-jk_0 r\sin\theta_0\cos(\phi-\phi_0)}$ with E_0 and H_0 being the incident electric and magnetic field intensity, respectively. Let's define the operators $\nabla = \nabla_t + jk_z\hat{z}$ and $\nabla_t = \frac{\partial}{\partial x}\hat{x} + \frac{\partial}{\partial y}\hat{y}$. Substitution of these operators in (5.70) and (5.71) results in

$$E_z^{inc} = J_{mt} + \left\{-\frac{\eta_0 k_z}{k_0}\nabla_t\cdot A_t^{(0)} + j\frac{\eta_0\left(k_0^2 - k_z^2\right)A_z^{(0)}}{k_0} + \hat{z}\cdot\nabla_t\times\underline{F}_t^{(0)}\right\}_{S^+}$$

$$0 = -J_{mt} + \left\{-\frac{\eta_d k_z}{k_d}\nabla_t\cdot A_t^{(d)} + j\frac{\eta_d\left(k_d^2 - k_z^2\right)A_z^{(d)}}{k_d} + \hat{z}\cdot\nabla_t\times\underline{F}_t^{(d)}\right\}_{S^-} \tag{5.72}$$

and

$$H_z^{inc} = -J_t + \left\{-\frac{k_z}{k_0\eta_0}\nabla_t\cdot\underline{F}_t^{(0)} + j\frac{\left(k_0^2 - k_z^2\right)F_z^{(0)}}{k_0\eta_0} - \hat{z}\cdot\nabla_t\times\underline{A}_t^{(0)}\right\}_{S^+}$$

$$0 = J_t + \left\{-\frac{k_z}{k_d\eta_d}\nabla_t\cdot\underline{F}_t^{(d)} + j\frac{\left(k_d^2 - k_z^2\right)F_z^{(d)}}{k_d\eta_d} - \hat{z}\cdot\nabla_t\times\underline{A}_t^{(d)}\right\}_{S^-} \tag{5.73}$$

in which

$$A_z^{(i)} = \frac{1}{4j}\int_S J_z(t')H_0^{(2)}\left(\sqrt{k_i^2 - k_z^2}\,|\underline{r} - \underline{r}'|\right)dt'$$

$$\underline{A}_t^{(i)} = \frac{1}{4j}\int_S \hat{t}(t')J_t(t')H_0^{(2)}\left(\sqrt{k_i^2 - k_z^2}\,|\underline{r} - \underline{r}'|\right)dt'$$

$$F_z^{(i)} = \frac{1}{4j}\int_S J_{mz}(t')H_0^{(2)}\left(\sqrt{k_i^2 - k_z^2}\,|\underline{r} - \underline{r}'|\right)dt'$$

$$\underline{F}_t^{(i)} = \frac{1}{4j}\int_S \hat{t}(t')J_{mt}(t')H_0^{(2)}\left(\sqrt{k_i^2 - k_z^2}\,|\underline{r} - \underline{r}'|\right)dt' \tag{5.74}$$

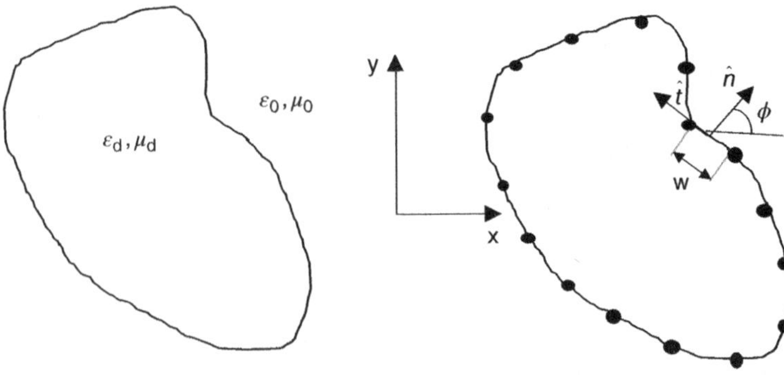

Figure 5.7 Flat-strip model of the surface of a homogeneous dielectric cylinder.

where $i = 0,d$ and $|\underline{r} - \underline{r}'| = \sqrt{(x(t) - x(t'))^2 + (y(t) - y(t'))^2}$. $\hat{t}$ is the unit vector tangent to the cylinder contour, as shown in Fig. 5.7.

$$\begin{aligned} \hat{n}(t) &= \cos\phi(t)\hat{x} + \sin\phi(t)\hat{y} \\ \hat{t}(t) &= -\sin\phi(t)\hat{x} + \cos\phi(t)\hat{y} \end{aligned} \tag{5.75}$$

Eq. (5.72)a and (5.73)a should be evaluated with the observer an infinitesimal distance outside S, while (5.72)b and (5.73)b should be evaluated with the observer an infinitesimal distance inside S. Equations (5.72) and (5.73) can be solved with MoM with N pulse basis functions on the cylinder contour used to represent the unknowns J_z, J_t, J_{mz}, J_{mt} and delta functions as testing functions enforced at the center of each of the cells of the model. The result is a matrix 4 × 4 block structure:

$$\begin{bmatrix} A & B & 0 & C \\ D & E & 0 & F \\ 0 & G & H & I \\ 0 & J & K & L \end{bmatrix} \begin{bmatrix} j_{z1} \\ \vdots \\ j_{zN} \\ j_{t1} \\ \vdots \\ j_{tN} \\ j_{mz1} \\ \vdots \\ j_{mzN} \\ j_{mt1} \\ \vdots \\ j_{mtN} \end{bmatrix} = \begin{bmatrix} E_z^{inc}(x_1, y_1) \\ \vdots \\ E_z^{inc}(x_N, y_N) \\ 0 \\ \vdots \\ 0 \\ H_z^{inc}(x_1, y_1) \\ \vdots \\ H_z^{inc}(x_N, y_N) \\ 0 \\ \vdots \\ 0 \end{bmatrix} \tag{5.76}$$

where

$$
\begin{aligned}
A_{mn} &= \frac{k_t^2}{4\omega\varepsilon_0}\int_{cell\ n} H_0^{(2)}\left(k_t R_{mn}\right) dt' \\
B_{mn} &= \frac{k_z k_t}{4j\omega\varepsilon_0}\int_{cell\ n}\left(-\sin\phi_m \frac{x_m - x_n - x(t')}{R_{mn}} + \cos\phi_m \frac{y_m - y_n - y(t')}{R_{mn}}\right) \\
&\quad \times H_1^{(2)}\left(k_t R_{mn}\right) dt' \\
C_{mn} &= 1 + \frac{k_t}{4j}\int_{cell\ n}\left(\cos\phi_m \frac{x_m - x_n - x(t')}{R_{mn}} + \sin\phi_m \frac{y_m - y_n - y(t')}{R_{mn}}\right) \\
&\quad \times H_1^{(2)}\left(k_t R_{mn}\right) dt' \\
D_{mn} &= \frac{k_{dt}^2}{4\omega\varepsilon_d}\int_{cell\ n} H_0^{(2)}\left(k_{dt} R_{mn}\right) dt' \\
E_{mn} &= -\frac{k_z k_{dt}}{4j\omega\varepsilon_d}\int_{cell\ n}\left(-\sin\phi_m \frac{x_m - x_n - x(t')}{R_{mn}} + \cos\phi_m \frac{y_m - y_n - y(t')}{R_{mn}}\right) \\
&\quad \times H_1^{(2)}\left(k_{dt} R_{mn}\right) dt' \\
F_{mn} &= -1 + \frac{k_{dt}}{4j}\int_{cell\ n}\left(\cos\phi_m \frac{x_m - x_n - x(t')}{R_{mn}} + \sin\phi_m \frac{y_m - y_n - y(t')}{R_{mn}}\right) \\
&\quad \times H_1^{(2)}\left(k_{dt} R_{mn}\right) dt' \\
G_{mn} &= -1 - \frac{k_t}{4j}\int_{cell\ n}\left(\cos\phi_m \frac{x_m - x_n - x(t')}{R_{mn}} + \sin\phi_m \frac{y_m - y_n - y(t')}{R_{mn}}\right) \\
&\quad \times H_1^{(2)}\left(k_t R_{mn}\right) dt' \\
H_{mn} &= \frac{k_t^2}{4\omega\mu_0}\int_{cell\ n} H_0^{(2)}\left(k_t R_{mn}\right) dt' \\
I_{mn} &= \frac{k_z k_t}{4j\omega\mu_0}\int_{cell\ n}\left(-\sin\phi_m \frac{x_m - x_n - x(t')}{R_{mn}} + \cos\phi_m \frac{y_m - y_n - y(t')}{R_{mn}}\right) \\
&\quad \times H_1^{(2)}\left(k_t R_{mn}\right) dt' \\
J_{mn} &= 1 - \frac{k_{dt}}{4j}\int_{cell\ n}\left(\cos\phi_m \frac{x_m - x_n - x(t')}{R_{mn}} + \sin\phi_m \frac{y_m - y_n - y(t')}{R_{mn}}\right) \\
&\quad \times H_1^{(2)}\left(k_{dt} R_{mn}\right) dt' \\
K_{mn} &= \frac{k_{dt}^2}{4\omega\mu_d}\int_{cell\ n} H_0^{(2)}\left(k_{dt} R_{mn}\right) dt' \\
L_{mn} &= \frac{k_z k_{dt}}{4j\omega\mu_d}\int_{cell\ n}\left(-\sin\phi_m \frac{x_m - x_n - x(t')}{R_{mn}} + \cos\phi_m \frac{y_m - y_n - y(t')}{R_{mn}}\right) \\
&\quad \times H_1^{(2)}\left(k_{dt} R_{mn}\right) dt'
\end{aligned}
\tag{5.77}
$$

in which $k_t = \sqrt{k_0^2 - k_z^2}$; $k_{dt} = \sqrt{k_0^2\varepsilon_d\mu_d - k_z^2}$; $R_{mn} = \left|\underline{r}_m - \underline{r}_n - \underline{r}'\right| = \sqrt{(x_m - x_n - x(t'))^2 + (y_m - y_n - y(t'))^2}$, and ϕ_m is the polar angle defining the outward normal vector to the nth strip in the model shown in Fig. 5.7.

In the next step, the *IFR* of the cylinder can be evaluated. For this task, we need to evaluate the forward-scattered far-field approximation using the expressions of Hankel functions for large arguments [13]. The maximum forward-scattering occurs in the propagation direction of the incident wave, $(\theta = \pi - \theta_0, \phi = \phi_0)$. In the TM case based on (5.72), the z-directed scattered electric field is

$$\begin{aligned} E_z^s(x,y) = {} & \frac{k_z}{4j\omega\varepsilon_0}\int_s \underline{J}_t(t')\cdot\nabla_t H_0^{(2)}\left(k_t|\underline{r}-\underline{r}'|\right)dt' \\ & - \frac{k_t^2}{4\omega\varepsilon_0}\int_s J_z(t')H_0^{(2)}\left(k_t|\underline{r}-\underline{r}'|\right)dt' + \dots \\ & + \frac{1}{4j}\hat{z}\cdot\int_s \underline{J}_{mt}(t')\times\nabla_t H_0^{(2)}\left(k_t|\underline{r}-\underline{r}'|\right)dt' \end{aligned} \tag{5.78}$$

Evaluation of the scattered field for far-field approximation, using the expressions of Hankel functions for large arguments in eq. (5.60) and the phase argument for $(\theta = \pi - \theta_0\ \phi = \phi_0)$ described in eq. (5.78) yields

$$\begin{aligned} E_z^s\Big|_{\substack{r\to\infty \\ \theta=\pi-\theta_0 \\ \phi=\phi_0}} \cong {} & \frac{1}{4j}\sqrt{\frac{2j}{\pi k_t r}}e^{-jk_{t'}r} \\ & \times\left[\begin{array}{l} \dfrac{k_t^2}{j\omega\varepsilon_0}\displaystyle\int_S J_z(t')e^{jk_0 r'\sin\theta_0\cos(\phi_0-\phi')}dt' + \dots \\ \dfrac{k_z k_t}{j\omega\varepsilon_0}\sin\theta_0\displaystyle\int_S J_t(t')\cos(\phi_0-\phi')e^{jk_0 r'\sin\theta_0\cos(\phi_0-\phi')_0}dt' + \dots \\ +jk_t\sin\theta_0\displaystyle\int_S J_{mt}(t')\sin(\phi_0-\phi')e^{jk_0 r'\sin\theta_0\cos(\phi_0-\phi')}dt' \end{array}\right] \end{aligned} \tag{5.79}$$

Derivation of the maximum radiated electric field in the direction $\theta = \pi - \theta_0$, $\phi = \pi/2$ by an infinite aperture with width w, as shown in eq. (5.63) and taking the ratio of (5.79) and (5.63) yields the IFR_e for the homogeneous dielectric cylinder,

$$\begin{aligned} IFR_e \cong {} & \frac{\eta_0}{2E_0 w}\left[\int_S\left(J_z(t') + J_t\cos\theta_0\cos(\phi_0-\phi') - \frac{1}{\eta_0}J_{mt}(t')\sin(\phi_0-\phi')\right)\right. \\ & \left.\times\, e^{jk_0 r'\sin\theta_0\cos(\phi_0-\phi')}dt'\right] \end{aligned} \tag{5.80}$$

By similar derivation for the TE case we obtain the forward scattered magnetic field:

$$H_z^s\Big|_{\substack{r\to\infty \\ \theta=\pi-\theta_0 \\ \phi=\pi/2}} \cong \frac{1}{4j}\sqrt{\frac{2j}{\pi k_t r}}e^{-jk_{t'}r}$$

$$\times\begin{bmatrix} \dfrac{k_t^2}{j\omega\mu_0}\displaystyle\int_S J_{mz}(t')e^{jk_0r'\sin\theta_0\cos(\phi_0-\phi')}dt' \\ +\dfrac{k_zk_t}{j\omega\mu_0}\sin\theta_0\displaystyle\int_S J_{mt}(t')\cos(\phi_0-\phi')e^{jk_0r'\sin\theta_0\cos(\phi_0-\phi')}dt' - \dots \\ -jk_t\sin\theta_0\displaystyle\int_S J_t(t')\sin(\phi_0-\phi')e^{jk_0r'\sin\theta_0\cos(\phi_0-\phi')}dt' \end{bmatrix} \tag{5.81}$$

and the IFR_h is

$$IFR_h \cong \frac{1}{2H_0w}\left[\int_S\left(\frac{1}{\eta_0}J_{mz}(t')+\frac{1}{\eta_0}J_{mt}\cos\theta_0\cos(\phi_0-\phi')+J_t(t')\sin(\phi_0-\phi')\right)\right.$$
$$\left.\times\, e^{jk_0r'\sin\theta_0\cos(\phi_0-\phi')}dt'\right] \tag{5.82}$$

Fig. 5.8 shows the back-scattering RCS from a circular dielectric cylinder with radius $a = 24$ mm, for normal incidence TM polarized as a function of frequency. The cylinder is a lossy dielectric with $\varepsilon_r = 4$ and $\sigma = 0$ (solid line), 0.017 S/m (dot line), 0.05 (dash-dot line) S/m.

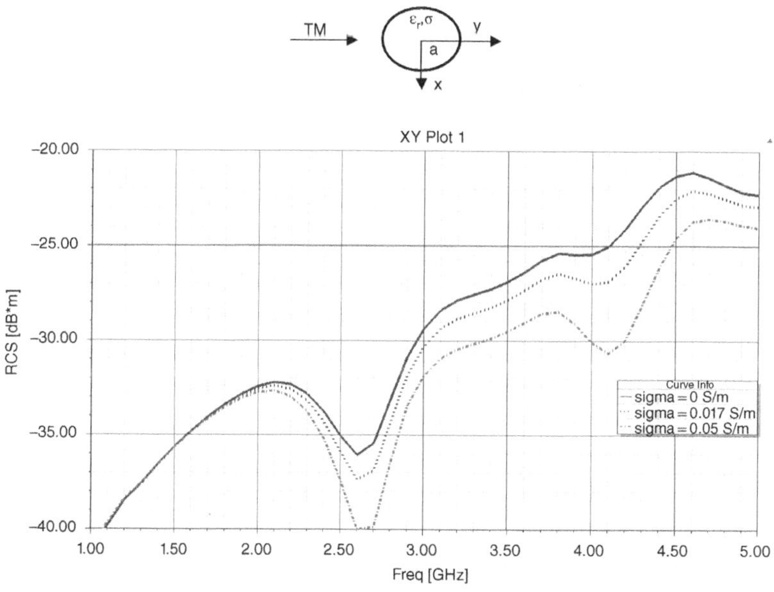

Figure 5.8 The back scattering RCS from a circular dielectric cylinder with radius $a = 24$ mm, for normal incidence TM polarized as a function of frequency. The cylinder is a lossy dielectric with $\varepsilon_r = 4$ and conductivity $\sigma = 0$ (solid line), 0.017 S/m (dot line), 0.05 S/m (dash-dot line). All simulations are performed with HFSS.

One can observe that increase in the cylinder conductivity reduces the back scattering.

Fig. 5.9 shows the RCS of a lossless elliptic cylinder ($\varepsilon_r = 2$) with major axis radius $a = 38$ mm and minor radius axis $b = 19$ mm illuminated by normal incidence TM polarized wave at angle $\phi_{\text{inc}} = 60$ deg with respect to x-axis.

One can observe the forward and backward scattering as a function of frequency in the range 1–5 GHz. In the scattering pattern at 5 GHz, one can observe the forward scattered lobe with peak at $\phi = 60$ deg and a minor lobe in the direction of the specular reflection at ϕ=120 deg. Fig. 5.10 shows the comparison of the numerical SIE solution (MoM with $N = 20$ contour divisions) and the exact solution of the TE bistatic RCS of three types of circular cylinders with parameters ($\varepsilon_r = 1$, $\mu_r = 10$, $k_0 a = 0.7$), ($\varepsilon_r = 9$, $\mu_r = 5$, $k_0 a = 0.7$), and ($\varepsilon_r = 9$, $\mu_r = 1$, $k_0 a = 2$).

The agreement of the numerical solution with the exact solution is good but slightly inferior to that of the TM polarization, as shown in Fig. 5.8.

5.3 Conductive Beams—Surface Integral Equation Formulation

In case of conductive beams, we can use the derivation of the surface integral equation with adaptation to the perfect electric conductor (PEC) boundary conditions $\hat{n} \times \underline{E} = 0$ on the beam surface [16]. Thus, for oblique incidence as shown in Fig. 5.3 and TM case, we can consider only the z-directed scattered electric field due to z-directed induced currents and the corresponding EFIE based on (5.72). In this case, no transverse electric and magnetic currents are induced; therefore,

$$E_z^{inc} = \frac{(k_0^2 - k_z^2)}{j\omega\varepsilon_0} A_z = \frac{(k_0^2 - k_z^2)}{j\omega\varepsilon_0} \frac{1}{4j} \int_S J_z(t') H_0^{(2)} \left(\sqrt{k_0^2 - k_z^2} |\underline{r} - \underline{r}'| \right) dt' \tag{5.83}$$

Similarly, based on (5.71)a, one can derive the MFIE. We start with the boundary condition $\hat{n} \times (\underline{H}^{inc} + \underline{H}^s) = \underline{J} \rightarrow H_t^{inc} + H_t^s = J_z$, which simplifies to

$$\begin{aligned} H_t^{inc} &= J_z - \hat{t} \cdot \nabla \times \int_s \hat{z} J_z(t') \frac{1}{4j} H_0^{(2)} \left(\sqrt{k_0^2 - k_z^2} |\underline{r} - \underline{r}'| \right) dt' \\ &= J_z + \hat{n} \cdot \nabla \int_s J_z(t') \frac{1}{4j} H_0^{(2)} \left(\sqrt{k_0^2 - k_z^2} |\underline{r} - \underline{r}'| \right) dt' \end{aligned} \tag{5.84}$$

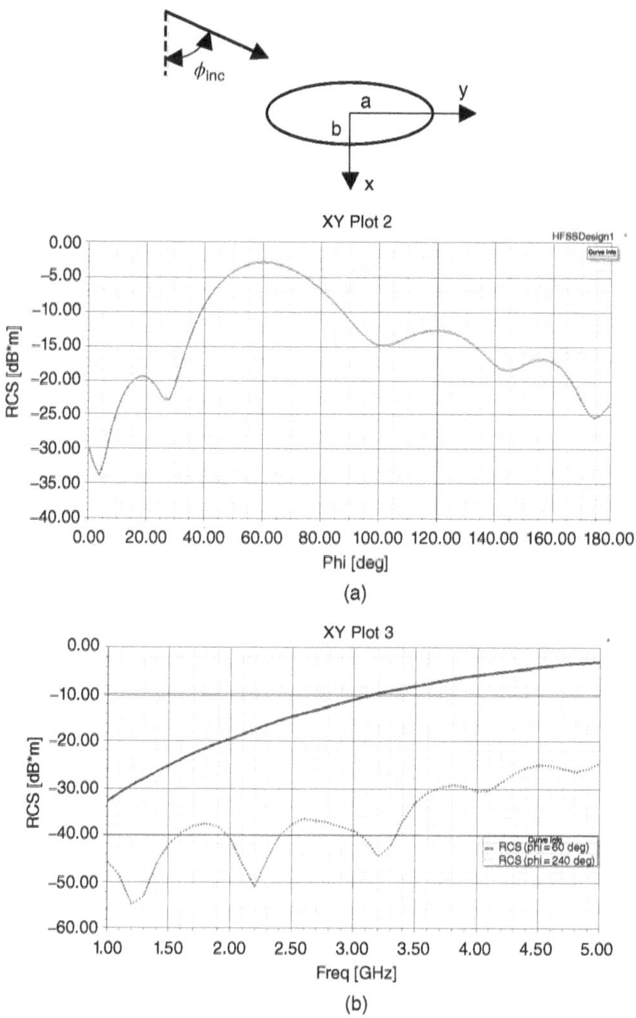

Figure 5.9 RCS of lossless elliptic cylinder ($\varepsilon_r = 2$) with a = 38 mm and b = 19 mm illuminated by normal incidence TM polarized wave at angle $\phi_{inc} = 60$ deg with respect to *x*-axis: (a) scattering pattern at 5GHz; (b) forward ($\phi = 60$ deg, solid line) and backward ($\phi = 240$ deg, dot line) RCS as a function of frequency. All simulations are performed with HFSS.

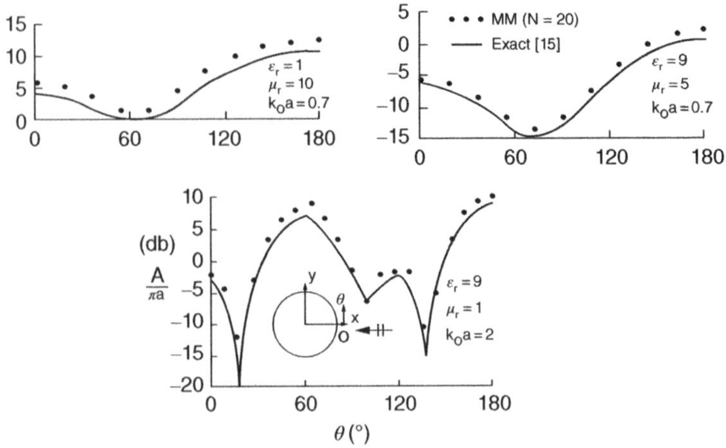

Figure 5.10 Comparison of bistatic RCS between SIE and exact solutions for three penetrable circular cylinders with normal incidence TE excitations [14].

Numerical solution of the integral equations (5.83) and (5.84) encounters singularity problems, due to internal resonances. These internal cavity resonances are the result of the homogeneous solution of the integral equations and they corrupt the solution at these frequencies. The root of the problem is that the surface integral equation involves only data on the mathematical surface of the scatterer and cannot distinguish between "inside" and "outside" in order to produce the desired external solution. Since the interior resonances are different for the EFIE and MFIE, the effect of the interior resonances can be minimized if we combine these integral equations into the so-called combined field integral equation (CFIE) [5]:

$$\begin{aligned}\alpha E_z^{inc} + (1-\alpha)\eta_0 H_t^{inc} &= (1-\alpha)\eta_0 J_z + j\alpha\eta_0 \frac{(k_0^2 - k_z^2)}{k_0} \\ &\times \int_S J_z(t')\frac{1}{4j}H_0^{(2)}(\sqrt{k_0^2 - k_z^2}|\underline{r} - \underline{r}'|)dt' \\ &+ (1-\alpha)\eta_0 \hat{n}\cdot\nabla \int_s J_z(t')\frac{1}{4j}H_0^{(2)}(\sqrt{k_0^2 - k_z^2}|\underline{r} - \underline{r}'|)dt'\end{aligned} \quad (5.85)$$

In eq. (5.85), we linearly combine eqs. (5.83) and (5.84) using the scale factors α and $(1-\alpha)\eta_0$ to employ the same units, where α is a real value between 0 and 1 and η_0 is the characteristic impedance of the host medium. However, this CFIE remains valid if $k_t = \sqrt{k_0^2 - k_z^2} = k_0 \sin\theta_0$ is purely imaginary, which would occur if $k_z > k$. There are no *interior resonances* frequencies with $k_z>k$, and consequently, the conventional EFIE and MFIE alone would be more efficient choice throughout the invisible region of the spectrum. Eq. (5.85) can be solved numerically using the point matching MoM. The incident field for the TM case is $E_z^{inc} = E_0 \sin\theta_0 e^{-jk_t(x\cos\phi_0+y\sin\phi_0)}e^{jk_z z}$ and $H_t^{inc} = -\frac{E_0}{\eta_0}\cos(\phi_0-\phi)e^{-jk_t(x\cos\phi_0+y\sin\phi_0)}e^{jk_z z}$. Using N pulse basis functions and N Dirac testing functions produces the matrix equation

$$\begin{bmatrix} C_{11} & C_{12} & \cdots & C_{1N} \\ C_{21} & C_{22} & \cdots & C_{2N} \\ \vdots & \vdots & \ddots & \vdots \\ C_{N1} & \cdots & \cdots & C_{NN} \end{bmatrix}\begin{bmatrix} j_1 \\ j_2 \\ \vdots \\ j_N \end{bmatrix} = \begin{bmatrix} e_1 \\ e_2 \\ \vdots \\ e_N \end{bmatrix} \tag{5.86}$$

in which $j_1 \ldots j_N$ are the z-directed currents and $e_1 \ldots e_N$ are the z-directed incident fields on the beam contour. Using the convention outlined in Fig. 5.7, the definition of $\hat{n}$ and $\hat{t}$ in eq. (5.75) and the approximations made for $R_{\mathrm{mm}}\rightarrow 0$ in (5.77), the diagonal and off-diagonal entries in (5.86) are determined by

$$\begin{aligned} C_{mm} &= (1-\alpha)\eta_0 + \alpha\eta_0\frac{k_t^2}{4k_0}\int_{cell\ m} H_0^{(2)}(k_t R_{mm})\,dt' - \frac{(1-\alpha)\eta_0}{2} \\ &= \frac{(1-\alpha)\eta_0}{2} + \alpha\eta_0\frac{k_t^2}{4k_0}\int_{cell\ m} H_0^{(2)}(k_t R_{mm})\,dt' \end{aligned} \tag{5.87}$$

and

$$\begin{aligned} C_{mn} &= \alpha\eta_0\frac{k_t^2}{4k_0}\int_{cell\ n} H_0^{(2)}(k_t R_{mn})\,dt' + \frac{j(1-\alpha)\eta_0 k_t}{4} \\ &\times \int_{cell\ n}\left(\cos\phi_m \frac{x_m - x_n - x(t')}{R_{mn}} + \sin\phi_m \frac{y_m - y_n - y(t')}{R_{mn}}\right) \\ &\times H_1^{(2)}(k_t R_{mn})\,dt' \end{aligned} \tag{5.88}$$

in which $R_{mn} = \left|\underline{r}_m - \underline{r}_n - \underline{r}'\right| = \sqrt{(x_m - x_n - x(t'))^2 + (y_m - y_n - y(t'))^2}$ and the entries of the excitation column vector are

$$e_m = E_0\left[\alpha\sin\theta_0 - (1-\alpha)\cos(\phi_0-\phi_m)\right]e^{-jk_t(x_m\cos\phi_0+y_m\sin\phi_0)} \tag{5.89}$$

Matrix inversion of (5.87) results in the z-directed currents on the beam contour. The knowledge of the currents enable to compute the IFR_e based on (5.80), given that in this case only z-directed electric currents are excited,

$$IFR_e \cong \frac{\eta_0}{2E_0 w}\left[\int_S J_z(t')e^{jk_0 r'\sin\theta_0\cos(\phi_0-\phi')}dt'\right] \tag{5.90}$$

where w is the optical projected width of the conductive beam on the incident wave plane and J_z is defined in Fig. 5.3 with the maximum forward-scattering direction ($\theta = \pi - \theta_0$, $\phi = \phi_0$).

Similar derivation can be applied for the TE case. The incident field for the TE case is $H_z^{inc} = H_0 \sin\theta_0 e^{-jk_t(x\cos\phi_0 + y\sin\phi_0)} e^{jk_z z}$ and $E_t^{inc} = H_0\eta_0 \cos(\phi_0 - \phi) e^{-jk_t(x\cos\phi_0 + y\sin\phi_0)} e^{jk_z z}$. In this case, we encounter only $\hat{t}$ components of the excited electric current $\underline{J}$. The boundary conditions on the beam surface is $\hat{n}\times\underline{E} = 0 \rightarrow \hat{n}\times(\underline{E}^{inc} + \underline{E}^s) = 0 \rightarrow \hat{t}\cdot(\underline{E}^{inc} + \underline{E}^s) = 0 \rightarrow E_t^{inc} = -\hat{t}\cdot\underline{E}^s$. Substitution of (5.27) for $\underline{E}^s$ and (5.35) for $\underline{A}$ yields the EFIE; in this case, i.e.,

$$\begin{aligned} E_t^{inc} &= \hat{t}\cdot\left(j\omega\mu_0\underline{A} - \frac{\nabla(\nabla\cdot\underline{A})}{j\omega\varepsilon_0}\right) \\ &= -\frac{\eta_0}{jk_0}\hat{t}\cdot\left(\nabla_t\nabla_t\cdot + k_t^2\right)\int_s \hat{t}(t')J_t(t')\frac{1}{4j}H_0^{(2)}\left(k_t|\underline{r}-\underline{r}'|\right)dt' \end{aligned} \tag{5.91}$$

The MFIE is based on the boundary condition $\hat{n}\times(\underline{H}^{inc} + \underline{H}^s) = \underline{J} \rightarrow H_z^{inc} + H_z^s = -J_t$, which can be simplified using (5.35) to

$$\begin{aligned} H_z^{inc} &= -J_t - \hat{z}\cdot\nabla\times\int_s \hat{t}(t')J_t(t')\frac{1}{4j}H_0^{(2)}(k_t|\underline{r}-\underline{r}'|)dt' \\ &= -J_t - \frac{1}{4j}\int_s J_t(t')\hat{n}(t')\cdot\nabla H_0^{(2)}(k_t|\underline{r}-\underline{r}'|)dt' \end{aligned} \tag{5.92}$$

To avoid the interior resonances excited in the EFIE and MFIE formulations in a similar fashion to the TM case, we linearly combine the two integral equations into the CFIE for the TE case:

$$\begin{aligned} &\alpha E_t^{inc} - (1-\alpha)\eta_0 H_z^{inc} = (1-\alpha)\eta_0 J_t + (1-\alpha)\eta_0 \\ &\times\int_s J_t(t')\hat{n}(t')\cdot\nabla\frac{1}{4j}H_0^{(2)}\left(k_t|\underline{r}-\underline{r}'|\right)dt' + j\alpha\frac{\eta_0}{k_0}\hat{t}\cdot\left(\nabla_t\nabla_t\cdot + k_t^2\right) \\ &\times\int_s \hat{t}(t')J_t(t')\frac{1}{4j}H_0^{(2)}\left(k_t|\underline{r}-\underline{r}'|\right)dt' \end{aligned} \tag{5.93}$$

In (5.93) the scale factors α and $(1-\alpha)\eta_0$ are used to employ the same units, where α is a real value between 0 and 1 and η_0 is the characteristic impedance of the host medium. Eq. (5.93) can be solved numerically using the MoM with N pulse basis functions and N Dirac delta-testing functions to obtain the matrix eq. (5.94):

$$\begin{bmatrix} C_{11} & C_{12} & \cdots & C_{1N} \\ C_{21} & C_{22} & \cdots & C_{2N} \\ \vdots & \vdots & \ddots & \vdots \\ C_{N1} & \cdots & \cdots & C_{NN} \end{bmatrix} \begin{bmatrix} j_{t1} \\ j_{t2} \\ \vdots \\ j_{tN} \end{bmatrix} = \begin{bmatrix} e_1 \\ e_2 \\ \vdots \\ e_N \end{bmatrix} \tag{5.94}$$

in which $j_{t1}\ldots j_{tN}$ are the tangential currents on the beam contour and $e_1\ldots e_N$ are the tangential incident fields on the beam contour. Using the convention outlined in Fig. 5.7, the definition of $\hat{n}$ and $\hat{t}$ in eq. (5.75) and the approximations made for $R_{mm}\to 0$ in (5.77), the diagonal and off-diagonal entries in (5.94) are determined:

$$\begin{aligned}
C_{mm} &= \frac{(1-\alpha)}{2} + j\alpha\frac{1}{k_0}\hat{t}\cdot\left(\nabla_t\nabla_t\cdot + k_t^2\right)\int_{cell\ m}\hat{t}(t')\frac{1}{4j}H_0^{(2)}\left(k_tR_{mm}\right)dt' \\
&= \frac{(1-\alpha)}{2} + \alpha\frac{1}{4k_0}\int_{cell\ m}\left[\frac{\partial^2 H_0^{(2)}\left(k_tR_{mm}\right)}{\partial t^2} + k_t^2 H_0^{(2)}\left(k_tR_{mm}\right)\right]dt' \\
&= \frac{(1-\alpha)}{2} + j\alpha\frac{1}{4k_0}\int_{cell\ m} \\
&\quad\times\left\{\begin{aligned}&\frac{[x(t')]^2}{R_{mm}^3}k_tH_1^{(2)}\left(k_tR_{mm}\right) + \ldots \\ &\frac{[y(t')]^2}{R_{mn}^2}\left[k_t^2H_0^{(2)}\left(k_tR_{mm}\right) - k_tH_1^{(2)}\left(k_tR_{mm}\right)\right] + k_t^2H_0^{(2)}\left(k_tR_{mm}\right)\end{aligned}\right\}dt'
\end{aligned} \tag{5.95}$$

and

$$\begin{aligned}
C_{mn} &= \frac{j(1-\alpha)k_t}{4}\int_{cell\ n}\left(\cos\phi_m\frac{x_m - x_n - x(t')}{R_{mn}} + \sin\phi_m\frac{y_m - y_n - y(t')}{R_{mn}}\right) \\
&\quad\times H_1^{(2)}\left(k_tR_{mn}\right)dt' \\
&\quad + j\alpha\frac{1}{k_0}\hat{t}\cdot\left(\nabla_t\nabla_t\cdot + k_t^2\right)\int_{cell\ n}\hat{t}(t')\frac{1}{4j}H_0^{(2)}\left(k_tR_{mn}\right)dt' \\
&= \frac{j(1-\alpha)k_t}{4}\int_{cell\ n}\left(\cos\phi_m\frac{x_m - x_n - x(t')}{R_{mn}} + \sin\phi_m\frac{y_m - y_n - y(t')}{R_{mn}}\right) \\
&\quad\times H_1^{(2)}\left(k_tR_{mn}\right)dt' + \ldots + j\alpha\frac{1}{4k_0} \\
&\quad\times\int_{cell\ n}\left\{\begin{aligned}&\frac{\left[(x_m - x_n - x(t'))\cos\phi_m + (y_m - y_n - y(t'))\sin\phi_m\right]^2}{R_{mn}^3} \\ &k_tH_1^{(2)}\left(k_tR_{mn}\right) + \ldots \\ &\frac{\left[(x_m - x_n - x(t'))\sin\phi_m - (y_m - y_n - y(t'))\cos\phi_m\right]^2}{R_{mn}^2} \\ &\left[k_t^2H_0^{(2)}\left(k_tR_{mn}\right) - k_tH_1^{(2)}\left(k_tR_{mn}\right)\right] + \ldots k_t^2H_0^{(2)}\left(k_tR_{mn}\right)\end{aligned}\right\}dt'
\end{aligned} \tag{5.96}$$

in which $R_{mn} = |\underline{r}_m - \underline{r}_n - \underline{r}'| = \sqrt{(x_m - x_n - x(t'))^2 + (y_m - y_n - y(t'))^2}$ and the entries of the excitation column vector are

$$e_m = H_0\left[\alpha\cos(\phi_0 - \phi_m) - (1-\alpha)\sin\theta_0\right]e^{-jk_t(x_m\cos\phi_0 + y_m\sin\phi_0)} \tag{5.97}$$

Matrix inversion of (5.94) results in the tangential currents on the beam contour on the beam contour. The knowledge of the currents enables to compute the IFR_h based on (5.82) with the maximum forward-scattering direction coinciding with axis ($\theta = \pi - \theta_0, \phi = \phi_0$). Given that in this case only transverse electric currents are excited,

$$IFR_h \cong \frac{1}{2H_0 w} \int_S J_t(t') \sin(\phi_0 - \phi') e^{jkr' \sin\theta_0 \cos(\phi_0 - \phi')} dt' \tag{5.98}$$

where w is the optical projected width of the conductive beam on the incident wave plane and J_t is defined in Fig. 5.3.

Fig. 5.11 shows the comparison of the numerical solutions of the EFIE, CFIE, and the exact solutions for the TM current density on a circular PEC cylinder of radius $0.82\lambda_0$ [5].

One can observe the distortion in the current distribution computed with EFIE due to the internal singularities compared to the exact solution. All that is in contrast to the numerical solution using CFIE, in which exists a nice agreement with the exact solution. Fig. 5.12 shows the comparison of the EFIE, CFIE, and exact solutions for the TM bistatic radar cross section scattering patterns from the circular PEC cylinder of radius $a = 0.82\lambda_0$ [5]. In this case, good agreement is obtained for both numerical solutions (EFIE and CFIE) and the exact solution. This observation implies that the internal singularities affect the current distribution on the scatterer and its near field but almost have no effect on the scattered far field. This can be attributed to the fact that the internal singularities affect the reactive scattered fields of the beam and have almost no effect on the real scattered fields, which determine the scattered far field.

The exact solution of the normalized scattered far field ($r \to \infty$) from a circular PEC cylinder is derived using cylindrical wave functions and is described in [15] by

$$\left|\frac{E_z^s}{E_z^i}\right| = \sqrt{\frac{2}{\pi k_0 r}} \left| \sum_{n=-\infty}^{\infty} \frac{J_n(k_0 a)}{H_n^{(2)}(k_0 a)} e^{jn\phi} \right| \tag{5.99}$$

with its corresponding IFR_e given by

$$IFR_e = -\frac{1}{k_0 a \cos\theta_0} \sum_{n=-\infty}^{\infty} J_n(k_0 a \cos\theta_0) / H_n^{(2)}(k_0 a \cos\theta_0) \tag{5.100}$$

where J_n and $H_n^{(2)}$ are the Bessel and outgoing Hankel functions and θ_0 is the tilt angle between the incident wavefront and the cylinder axis. Similar derivation using the cylindrical wave functions for the TE case

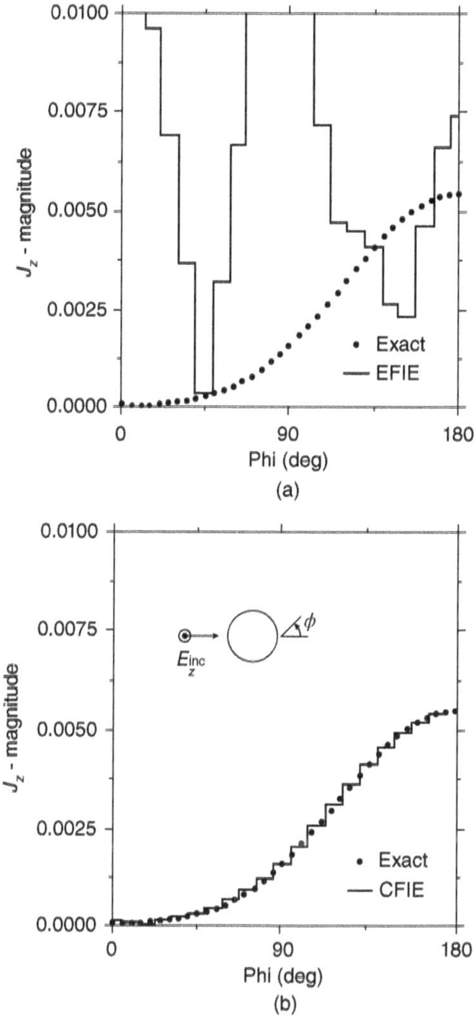

Figure 5.11 Comparison of the EFIE, CFIE, and exact solutions for the TM current density on a circular PEC cylinder of radius $0.82\lambda_0$. The numerical result was obtained using 40 equal sized cells: (a) EFIE and exact current distribution results; (b) CFIE and exact current distribution results [5].

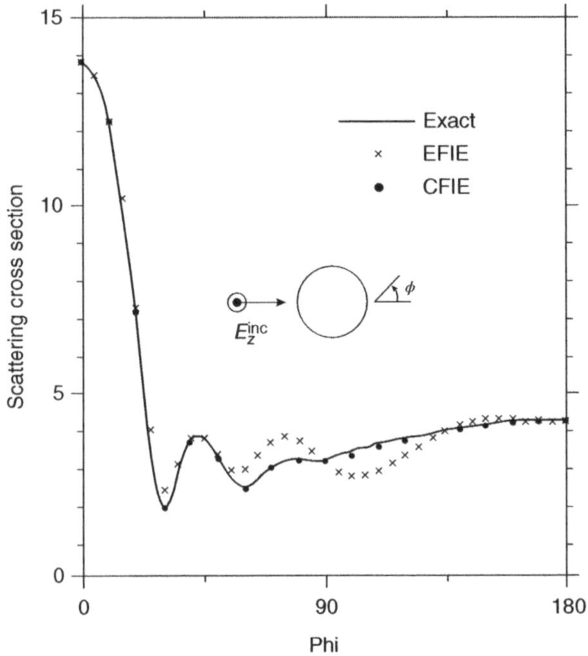

Figure 5.12 Comparison of the EFIE, CFIE, and exact solutions for the TM bistatic radar cross section scattering patterns from a circular PEC cylinder of radius a = $0.82\lambda_0$ [5].

yields

$$\left|\frac{H_z^s}{H_z^i}\right| = \sqrt{\frac{2}{\pi k_0 r}}\left|\sum_{n=-\infty}^{\infty}\frac{J_n'(k_0a)}{H_n^{(2)\prime}(k_0a)}e^{jn\phi}\right| \tag{5.101}$$

with the IFR_h given by

$$IFR_h = -\frac{1}{k_0a\cos\theta_0}\sum_{n=-\infty}^{\infty} J_n'(k_0a\cos\theta_0)/H_n^{(2)\prime}(k_0a\cos\theta_0) \tag{5.102}$$

In general, the IFR_e is larger in magnitude than IFR_h and has a positive phase angle compared to a negative phase angle for the *HP*. Both *IFR*'s approach the value of −1.0 + j0.0 as the radius increases. IFR_h approaches this limit from below and IFR_e approaches the limit from above.

Another example that demonstrates the effect of the internal resonances on the induced current distribution on a pie-shaped PEC cylinder for TM polarization is shown in Fig. 5.13. In a similar fashion

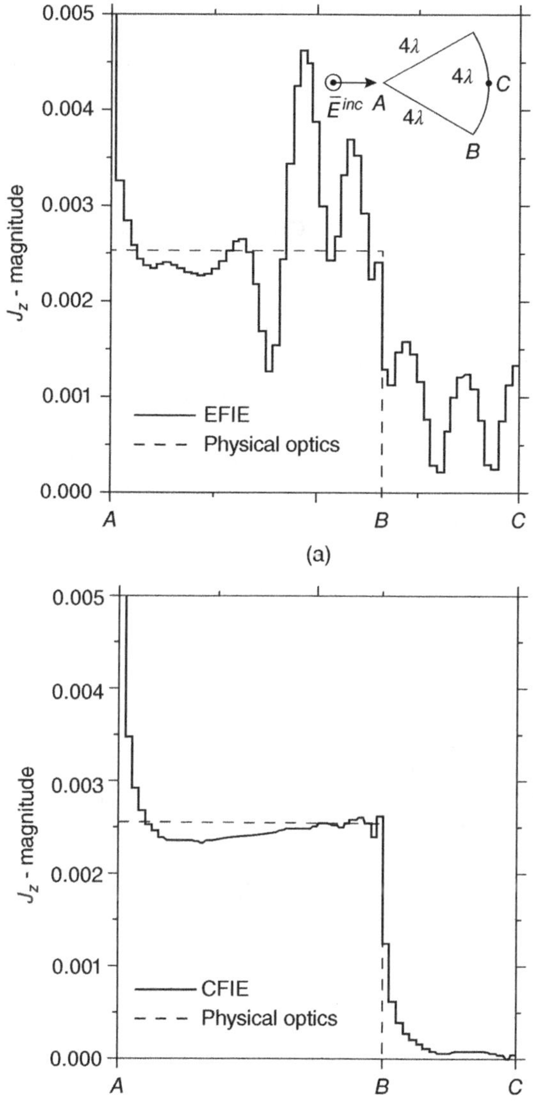

Figure 5.13 Comparison of the EFIE, CFIE, and physical optics solutions for the TM current density on a pie-shaped PEC cylinder: (a) EFIE and physical optics distribution results; (b) CFIE and physical optics distribution results [5].

to the circular PEC cylinder computation of the current distribution using the CFIE formulation solves the problem almost entirely. In Fig. 5.13 the numerical solution is compared to the induced current computed using physical optics approximation, which is nonzero only on the illuminated surfaces.

5.4 Tuned Beams—Surface Integral Equation Formulation

The main objective in the design of space frame radomes made of homogeneous dielectric beams is to reduce their forward scattering. This objective can be accomplished by inserting periodic vertical and horizontal conductive strips in the dielectric beam with periodicity Δ_z in z direction, as shown in Fig. 5.14. Consequently, the strips can be arranged into unit cells and it is sufficient to study the behavior of the electromagnetic field within a single reference unit cell. The induced currents in the conductive strips offset the effect of the electric polarization currents induced in the homogeneous dielectric beam

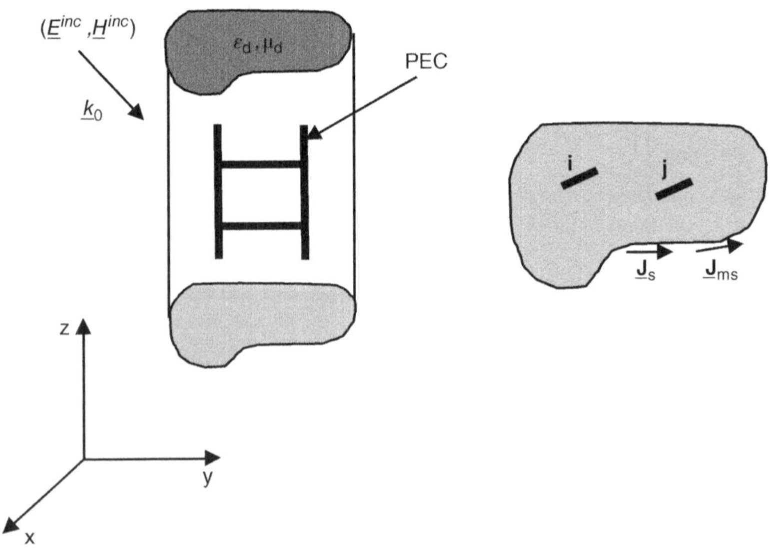

Figure 5.14 Tuned-beam geometry with a periodic vertical and horizontal wire loading.

and reduce the forward scattering of the beam for a limited frequency bandwidth. The analysis of the scattering for these tuned beams is possible by formulating the EFIE of the problem by forcing the tangential electric field along the conductive strips of the beam to be equal to zero. Evaluation of the total electric field at the strip locations is done in two steps. In the first step using the coupled surface integral equation of the unloaded homogeneous beam as outlined in Chapter 5.2, the equivalent surface electric currents $\underline{J}_s$ and surface magnetic currents $\underline{J}_{ms}$ on the beam contour S, as shown in Fig. 5.6, are evaluated using the MoM using eq. (5.72) for the TM case and eq. (5.73) for the TE case. In the second step, an EFIE is constructed in terms of the unknown induced currents on the conductive strips and enforcing the total electric field on the strips to vanish. The total electric field at the strip locations is equal to the field radiated by the surface equivalent sources on the beam contour S for the unloaded homogeneous beam and the scattered electric fields by the vertical and horizontal conductive strips. The scattered fields by the conductive strips are formulated in terms of a z-directed periodic array of vertical and horizontal current elements $\underline{J}$ as shown in Fig. 5.14.

Without loss of generality, we assume that the horizontal wires are y-directed with center coordinates (x_r,y_r,z_r) and the vertical wires are z-directed with center coordinates (x_q,y_q,z_q). The length of the vertical strips in the unit cell is L_v and the length of the horizontal strips is L_h with the strips width being w. The current singularity near the strip edges [17] is taken in consideration for the vertical strips by assuming that the transverse current distribution is $\frac{1}{\sqrt{1-\left(\frac{2(y-y_q)}{w}\right)^2}}$ and $\frac{1}{\sqrt{1-\left(\frac{2(z-z_r)}{w}\right)^2}}$ for the horizontal strips. Accordingly, the z-periodic currents on the conductive strips of the beam can be represented by a Fourier series:

$$\underline{J} = \sum_{n=-\infty}^{+\infty} \underline{J}_n(x,y,z) \tag{5.103}$$

$$= \sum_{n=-\infty}^{+\infty} \left[\hat{y} \sum_{r=1}^{R} f_n(x_r, y-y_r, z-z_r) e^{jk_{zn}z} + \hat{z} \sum_{q=1}^{Q} g_n(x_q, y-y_q, z-z_q) e^{jk_{zn}z} \right]$$

Here, $k_{zn} = k_d \cos\theta + \frac{2\pi}{\Delta_z} n$ with $n = 0,1,\ldots,$ denoting the Floquet harmonics [17] index, Δ_z being the unit cell length in z direction and $k_d = k_0\sqrt{\mu_r \varepsilon_r}$. The center locations of the horizontal strips are (x_r,y_r,z_r) with $r = 1,\ldots,R$ and R being the total number of horizontal strips in the unit cell. The center locations of the vertical strips are

(x_q, y_q, z_q) with $q = 1,\ldots,Q$ and Q being the total number of vertical strips in the unit cell. The electric field scattered by this current distribution can be described by

$$\underline{E}(\underline{J}) = \frac{\eta_d}{jk_d}\left[\nabla(\nabla\cdot\underline{A}^{(d)}) + k_d^2\underline{A}^{(d)}\right] \tag{5.104}$$

in which $\eta_d = \eta_0\sqrt{\frac{\mu_r}{\varepsilon_r}}$ and $\underline{A}^{(d)} = \frac{1}{4j\Delta_z}\sum_{n=-\infty}^{\infty}\int_v \underline{J}_{-n}(x',y',z')H_0^{(2)}(k_{tn}|\underline{r} - \underline{r}'|)e^{jk_{zn}(z-z')}dv'$ where $|\underline{r} - \underline{r}'| = \sqrt{(x-x')^2 + (y-y')^2}$, $k_{tn} = \sqrt{k_d^2 - k_{zn}^2}$, and v being the unit cell volume. The EFIE of the problem is set by forcing the tangential components of the total electric field to vanish along the conductive strips. The location along the strips is denoted by r_c.

$$\hat{n}\times\left\{\underline{E}(\underline{J}) + \frac{\eta_d}{jk_d}\left[\nabla(\nabla\cdot\underline{A}_s^{(d)}) + k_d^2\underline{A}_s^{(d)}\right] - \nabla\times\underline{F}_s^{(d)}\right\}\Bigg|_{r_c} = 0 \tag{5.105}$$

in which

$$A_{sz}^{(d)} = \frac{1}{4j\Delta_z}\sum_{n=-\infty}^{\infty}\int_S J_{sz,n}(t_s)H_0^{(2)}(k_{tn}|\underline{r} - \underline{r}_s|)e^{jk_{zn}(z-z')}dt_s dz'$$

$$\underline{A}_{st}^{(d)} = \frac{1}{4j\Delta_z}\sum_{n=-\infty}^{\infty}\int_S \hat{t}(t_s)J_{st,n}(t')H_0^{(2)}(k_{tn}|\underline{r} - \underline{r}_s|)e^{jk_{zn}(z-z')}dt_s dz'$$

$$F_{sz}^{(d)} = \frac{1}{4j\Delta_z}\sum_{n=-\infty}^{\infty}\int_S J_{msz,n}(t_s)H_0^{(2)}(k_{tn}|\underline{r} - \underline{r}_s|)e^{jk_{zn}(z-z')}dt_s dz'$$

$$\underline{F}_{st}^{(d)} = \frac{1}{4j\Delta_z}\sum_{n=-\infty}^{\infty}\int_S \hat{t}(t_s)J_{mst,n}(t')H_0^{(2)}(k_{tn}|\underline{r} - \underline{r}_s|)e^{jk_{zn}(z-z')}dt_s dz' \tag{5.106}$$

where $\underline{r}_s$ denotes points on the beam contour. The electric currents, $\underline{J}_{s,n}$ and magnetic currents, $\underline{J}_{ms,n}$ are the Fourier series of the currents $\underline{J}_s$ and $\underline{J}_{ms}$ evaluated in advance, as explained in section 5.2. The currents $\underline{J}$ on the strips can be determined by solving (5.105) using the MoM with pulses as basis functions and delta functions as test functions. The matrix inversion used in the MoM is performed for each harmonics n. Only the zeroth order harmonics ($n = 0$) is propagating, but few harmonics are needed to evaluate the total field in the beam. The next essential step is evaluation of the scattered far field of the tuned beam. As an interim step, we need to compute the total electric and magnetic field in the beam medium given the beam contour surface currents $\underline{J}_s$, $\underline{J}_{ms}$ and the evaluated conductive strip currents $\underline{J}$ using

$$\underline{E}_{tot} = \underline{E}(\underline{J}) + \frac{\eta_d}{jk_d}\left[\nabla(\nabla\cdot\underline{A}_s^{(d)}) + k_d^2\underline{A}_s^{(d)}\right] - \nabla\times\underline{F}_s^{(d)} \tag{5.107}$$

Using (5.107), the induced equivalent volumetric electric and magnetic currents in the beam volume can be computed using (5.17) and (5.19) to represent the induced polarization currents in the beam. Next, the forward far field can be computed using the induced electric and magnetic polarization currents in the beam and the electric currents on the strips. For this step, only the $n = 0$ propagating harmonics is needed. Without loss of generality, in the special case of incident field propagating in $\phi = \pi/2$ direction and maximum forward-scattered electric field also in $\phi = \pi/2$ direction, we can use (5.107) to compute the far-field approximation and the *IFR* values using (5.64) and (5.67).

Fig. 5.15 shows the geometry of a rectangular dielectric cylinder ($\varepsilon_r = 3$) loaded periodically with two horizontally and two vertically oriented PEC strips in the unit cell [17] with parameters $s^v = 0.6$, $s^h = 0.8$, and $p = 0.75$. The widths of the strips are $w^v = w^h = 0.01L$.

Fig. 5.16 shows the simulated RCS of the loaded and unloaded cylinder as a function of frequency for TE and TM polarizations. One can observe that for normal incidence and TE/TM polarizations, the RCS of the loaded dielectric cylinder is reduced by approximately 30 dB

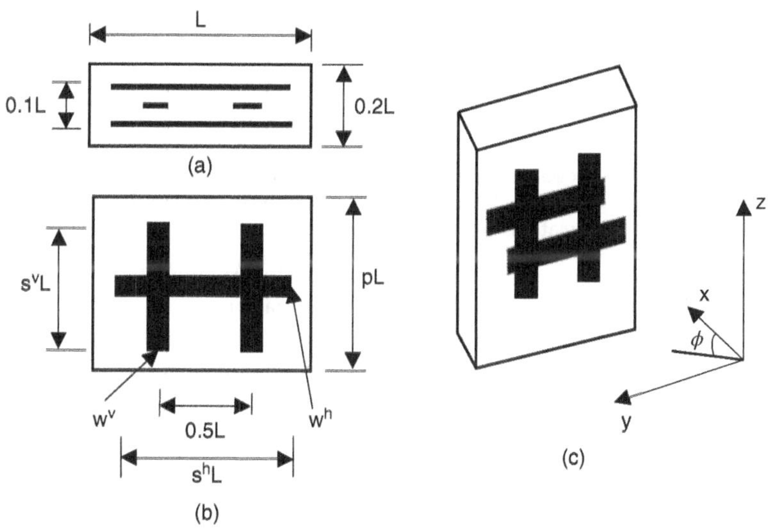

Figure 5.15 Rectangular dielectric cylinder, loaded periodically with two horizontally and two vertically oriented PEC strips: (a) cross section: (b) front view; (c) perspective view.

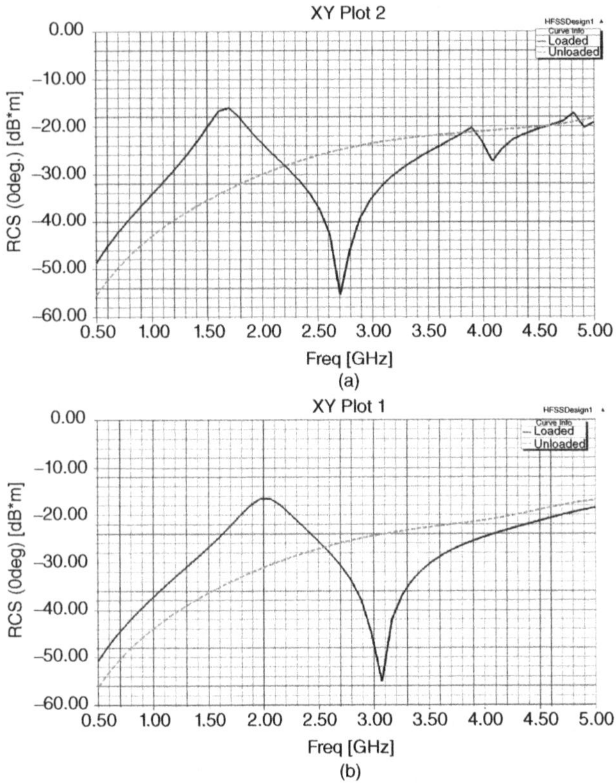

Figure 5.16 RCS (0 deg) of the unloaded and loaded cylinder as a function of frequency, for $s^v = 0.6$, $s^h = 0.8$, and $p = 0.75$: (a) TE polarization; (b) TM polarization. All simulations are performed with HFSS.

in a narrow frequency bandwidth, with center frequency 2.7 GHz for TE polarization and 3.1 GHz for TM polarization.

In a different example, a rectangular dielectric beam with electrical parameters $\varepsilon_r = 4.6$, $\tan\delta = 0.014$, and dimensions 2×0.4 in.^{2} is considered. The beam is tuned with a grid of eight vertical conductive strips located on its center line, spaced 0.277 in. apart and with a strip width of 0.062 in. Fig. 5.17 shows the IFR dependence of the tuned and untuned cylinder on frequency. The data presented is for the TM polarization and normal incidence. The computed and measured data are compared.

At the tuning frequency of 5.6 GHz, the magnitude of the IFR drops significantly in comparison to that of the untuned beam. In addition,

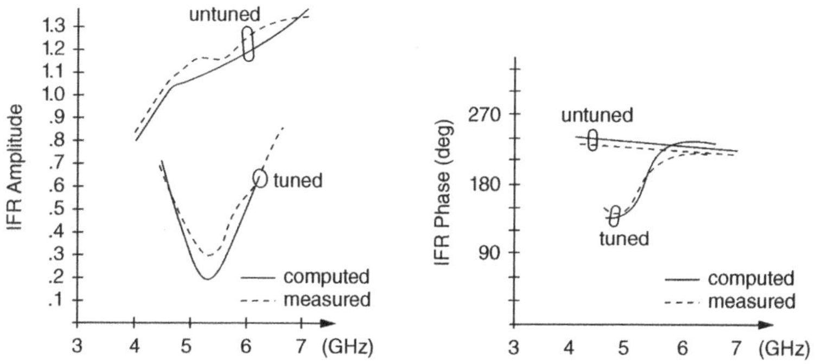

Figure 5.17 IFR as a function of frequency for a tuned/untuned dielectric beam 2×0.4 in.^{2} for vertical polarization and normal incidence.

the IFR phase goes through the 180 deg value close to the tuning frequency, a feature common to resonant devices. Fig. 5.18 shows the field distribution (amplitude and phase) of the untuned and tuned beams inside the cross section of the dielectric beam at 5.6 GHz.

As expected the total electric field goes to zero at the locations of the strips. Moreover, the phase of the electric field is relatively more uniform as compared to that of the untuned beam. The uniformity of the phase distribution is a good indicator of the transparency of the beam because it resembles the uniform phase front of the incident field. Fig. 5.19 shows a comparison between the computed scattering pattern of the untuned and tuned dielectric beams at the tuning frequency of 5.6 GHz.

One can observe a significant reduction of the scattering levels in the forward direction, as compared to that of the untuned beam.

A different approach to the dual polarization tuning problem (forward scattering reduction) is described in [19]. The concept is based on replacement of soft surfaces by hard surfaces in the dielectric beam for both TE_z and TM_z polarizations. The presence of hard surfaces on a dielectric or conductive beam is desirable to reduce the beam blockage. Generally, a hard surface supports waves with maximum value of the E-field, like normal electric field on a conductive surface, whereas the soft surface makes the amplitude of the tangential E-field zero at a conductive surface. This effect is not allowing propagation and increases scattering. In case of propagation along a conductive surface, if the E-field excited is normal to the surface it can propagate along the surface with a strong intensity as the boundary

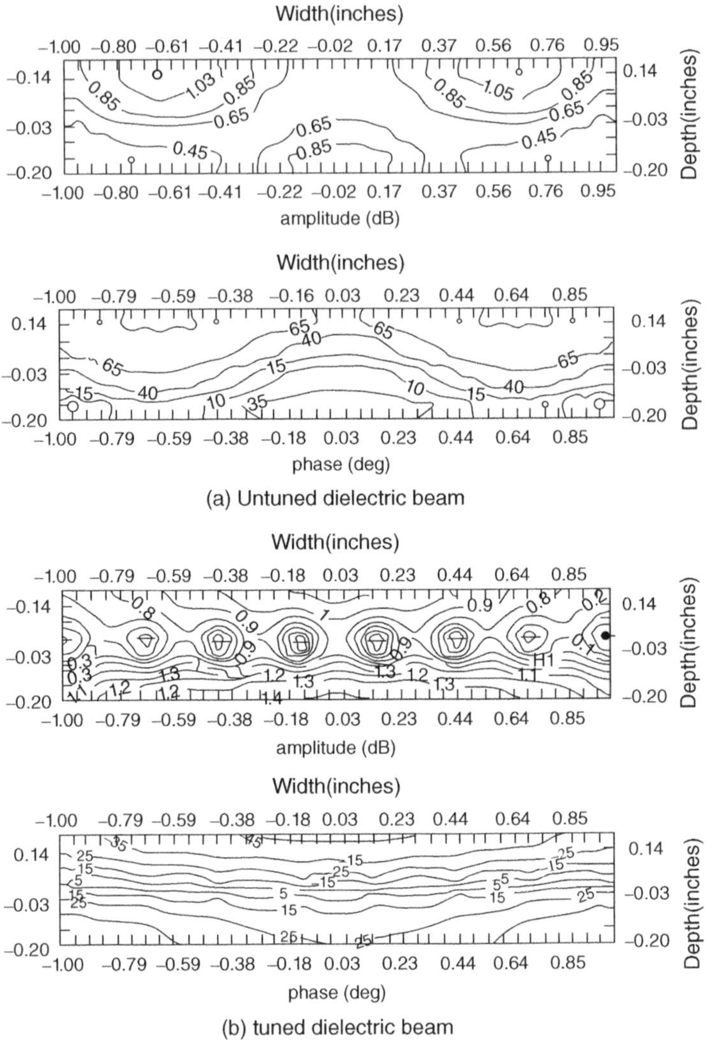

Figure 5.18 Field distribution in an untuned and tuned dielectric beam at the tuning frequency (5.6 GHz) at normal incidence: (a) untuned; (b) tuned.

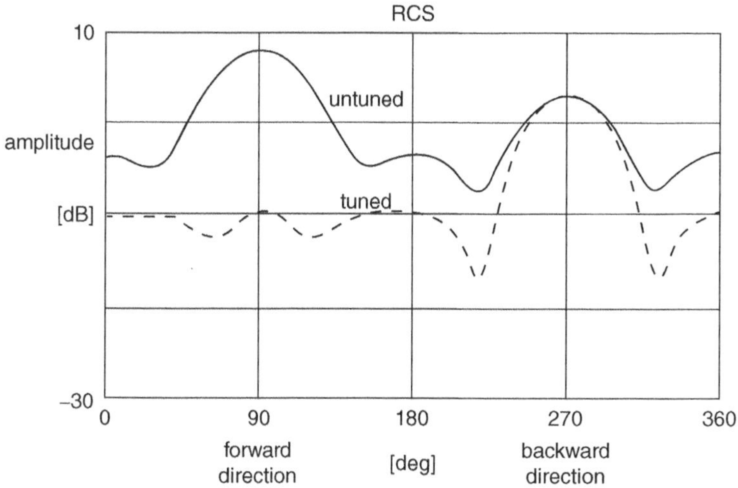

Figure 5.19 Scattering pattern for tuned and untuned beam at 5.6 GHz.

condition is hard with $\partial E_n/\partial n = 0$. On the other hand, if the E-field is transverse to the direction of propagation and tangential to the surface, the wave is effectively stopped from propagating along the conductive surface and this increases scattering. Thus, the electric conductor has a strongly polarization dependent boundary condition for waves propagating along it. Accordingly, we are interested to transform the beam surfaces to hard surfaces to reduce scattering. This desire for a dielectric beam for TE_z polarization can be obtained by changing the profile of the beam to an oblong shape and strip loading its outer surface with inter-element strip spacing of less than $\lambda/2$ to transform the strip periodic structure to a full conductive surface. However, in case we don't have the flexibility to change the cross section of the beam due to mechanical constraints and for instance, still need to reduce the forward scattering of a rectangular cross section dielectric beam, a different approach should be adopted. In this case, two scattering reduction mechanisms are considered, as shown in Fig. 5.20a.

In the first mechanism, two metal plates are inserted in the dielectric beam to form an internal parallel plate waveguide and the phase propagation through the waveguide is equalized to the phase through air, such that

$$\sqrt{\varepsilon_r} k_0 l = k_0 l + 2\pi n; \quad n = 1,2,\ldots \tag{5.108}$$

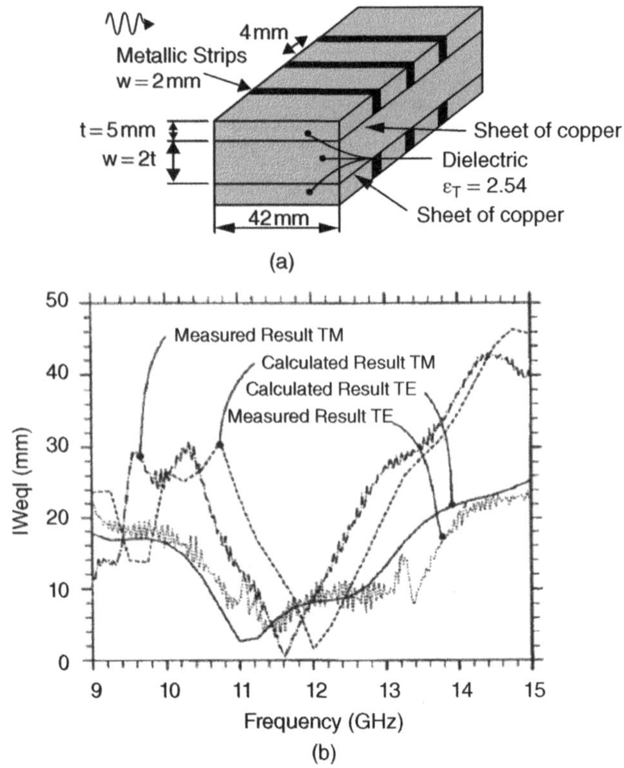

Figure 5.20 Measured and calculated blockage widths w_{eq} of cylinder realized as parallel-plate waveguide and outer walls of the cylinder coated with strip-loaded dielectric material to obtain the hard boundary condition for both TE and TM cases: (a) geometry; (b) absolute value of w_{eq} for TE and TM cases [19].

in which ε_r is the dielectric constant of the beam, k_0 is the propagation constant in free space, l is the beam length, and n is an integer. The second tuning mechanism is strip loading of the outer surface with inter-element strip spacing of less than $\lambda/2$, which transforms the outer surface to a conductive surface or a hard surface boundary for TE_z polarization. The hard surface boundary condition for TM_z polarization is obtained by coating the metal plates of the beam by a thin dielectric layer of thickness t, as shown in Fig. 5.20a. Inside the dielectric coating, the propagation constant will be $k_d = \sqrt{\varepsilon_r}k_0$, which is larger than the k_0 outside the coating. The component of k_x

along the dielectric surface must be equal to k_0. Therefore, we will get a normal component of the propagation vector, k_y inside the dielectric coating equal to $k_y = \sqrt{k_d^2 - {k_0}^2} = k_0\sqrt{\varepsilon_r - 1}$. This component of k_d will transform the boundary condition at the metal surface to the boundary condition of an artificial magnetic conductor at the surface of the dielectric provided that,

$$k_y t = \frac{\pi}{2} \rightarrow t = \frac{\lambda}{4\sqrt{\varepsilon_r - 1}} \tag{5.109}$$

Fig. 5.20b shows the absolute value of $w_{eq} = w \cdot IFR$ as a function of frequency. One can observe that for the tuned beam with dimensions and parameters outlined in Fig. 5.20a, a minimum in w_{eq}, which is proportional to the forward scattering, is obtained at 11.5 GHz.

Comparison of the two tuning techniques described above reveals that they both can achieve dual polarization forward-scattering reduction in a limited frequency band, but the tuning with vertical and horizontal conductive strips is more flexible from the manufacturing point of view, without the need to use special dielectric materials given a predetermined beam geometry, as may be required based on eqs. (5.108) and (5.109).

5.5 Scattering from Infinite Cylinders—Differential Equation Formulation

Volume integral equation solved by MoM as outlined in previous sections are versatile, but involve fully populated matrices and require relatively large computational effort to treat scatterers of moderate electrical size. On the other hand, surface integral equation matrices are populated as the volume integral equation matrices, but are smaller in size. However, they can accommodate only homogeneous, multilayer, and conductive problems. An alternative option to compute numerically the scattering from a composite inhomogeneous beam is to solve its differential equation using the finite element method (FEM) [4] and [5]. Comparison of the FEM computational effort to the solution with volume integral equations reveals that it is less computationally intensive, since the matrices involved are sparse, but they are slightly larger in size. The differential equation methods often include an additional region of space outside the scatterer

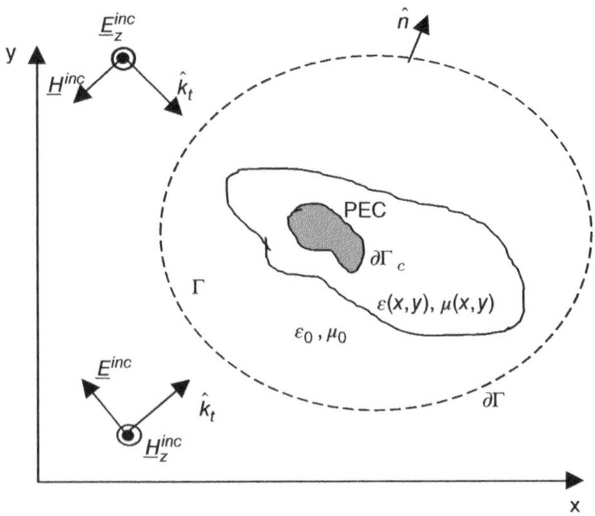

Figure 5.21 Cross section of the cylinder geometry showing the imbedded PEC region whose surface is denoted $\partial\Gamma_c$.

to ensure that the scattered fields represent outward propagating solutions. This additional region must be terminated with a radiation boundary condition as will be discussed later. Fig. 5.21 shows the cylinder geometry confined to the volume Γ made of a composite material with permittivity $\varepsilon_r(x,y)$, permeability $\mu_r(x,y)$ and PEC conductive surfaces denoted by $\partial\Gamma_c$. The surface that confines the computational domain is $\partial\Gamma$.

In the following, we will consider the general case of oblique scattering as discussed for the integral equation solution and described in [5]. The z dependence of the oblique incident field is $e^{jk_z z}$ and, correspondingly, we assume that the z dependence of the total fields in the system is also $e^{jk_z z}$. In a similar fashion to the volume integral equation, the electric and magnetic fields in the system can be described by eqs. (5.5) and (5.6). Moreover, we assume that the cylinder cross section is modeled by triangular cells of constant permittivity and permeability, but varying from cell to cell. Then, the total scattering from the beam would be the superposition of the scattering from all these various homogeneous triangular cells. Furthermore, like in the case of the volume integral equation derivation, we employ the total fields E_z and H_z as the primary unknowns to be determined and obtain the transverse magnetic and electric field components within a homogeneous triangular cell via equations (5.16) and (5.18), respectively.

The electromagnetic fields generated in the vicinity of the scatterer can be determined by the Helmholtz equations for the TM case:

$$\nabla \cdot \left(\frac{1}{\mu_r} \nabla E_z \right) + k_0^2 \varepsilon_r E_z = 0 \tag{5.110}$$

and similarly, for the TE case

$$\nabla \cdot \left(\frac{1}{\varepsilon_r} \nabla H_z \right) + k_0^2 \mu_r H_z = 0 \tag{5.111}$$

Eqs. (5.110) and (5.111) illustrate the "strong" form of Helmholtz equations with the unknown being in a second-order differential operator, which are very sensitive to numerical solution. To make these equations numerically more stable, they can be converted into the so-called "weak" form. The fields in the vicinity of the scatterer must satisfy Maxwell's equations:

$$\begin{aligned} &\nabla \times \underline{H} = j\omega\varepsilon_0\varepsilon_r \underline{E} \\ &\nabla \times \underline{E} = -j\omega\mu_0\mu_r \underline{H} \end{aligned} \tag{5.112}$$

Multiplying both sides of (5.112) with the testing function $\underline{T}(x, y, z) = \hat{z}T(x, y)$ and integrating over the domain Γ, we obtain

$$\begin{aligned} &\iint_\Gamma (\underline{T} \cdot \nabla \times \underline{H} - j\omega\varepsilon_0\varepsilon_r \underline{T} \cdot \underline{E})\, ds = 0 \\ &\iint_\Gamma (\underline{T} \cdot \nabla \times \underline{E} + j\omega\mu_0\mu_r \underline{T} \cdot \underline{H})\, ds = 0 \end{aligned} \tag{5.113}$$

Now, recall the two-dimensional divergence theorem

$$\iint_\Gamma \nabla \cdot (\underline{T} \times \underline{H})\, ds = \int_{\partial\Gamma + \partial\Gamma_c} (\underline{T} \times \underline{H}) \cdot \hat{n} dl \tag{5.114}$$

with $\hat{n}$ being the normal to the surfaces $\partial\Gamma$ and $\partial\Gamma_c$ and the vector identities

$$\begin{aligned} &\nabla \cdot (\underline{T} \times \underline{H}) = \nabla \times \underline{T} \cdot \underline{H} - \underline{T} \cdot \nabla \times \underline{H} \\ &(\underline{T} \times \underline{H}) \cdot \hat{n} = -\underline{T} \cdot (\hat{n} \times \underline{H}) \end{aligned} \tag{5.115}$$

Substitution of (5.114) and (5.115) into (5.113) enables us to rewrite it in the form,

$$\begin{aligned} &\iint_\Gamma (\nabla \times \underline{T} \cdot \underline{H}_t - j\omega\varepsilon_0\varepsilon_r T E_z)\, ds = -\int_{\partial\Gamma + \partial\Gamma_c} \underline{T} \cdot (\hat{n} \times \underline{H})\, dl \\ &\iint_\Gamma (\nabla \times \underline{T} \cdot \underline{E}_t + j\omega\mu_0\mu_r T H_z)\, ds = -\int_{\partial\Gamma + \partial\Gamma_c} \underline{T} \cdot (\hat{n} \times \underline{E})\, dl \end{aligned} \tag{5.116}$$

The nabla operator can be rewritten in the form $\nabla = \nabla_t + jk_z\hat{z}$ with $\nabla_t = \frac{\partial}{\partial x}\hat{x} + \frac{\partial}{\partial y}\hat{y}$. Substitution of the transverse fields from equations (5.16) and (5.18) into (5.116) and utilizing the vector identity $\nabla \times \underline{T} = -\hat{z} \times \nabla_t T$ yields, after some algebraic manipulations,

$$\iint_\Gamma \left(\frac{\varepsilon_r}{\varepsilon_r\mu_r - (k_z^2/k_0^2)} \nabla_t T \cdot \nabla_t E_z - k_0^2 \varepsilon_r T E_z - \frac{k_z}{\omega\varepsilon_0} \frac{1}{\varepsilon_r\mu_r - (k_z^2/k_0^2)} \hat{z} \cdot \nabla_t T \times \nabla_t H_z \right) ds$$
$$= -\int_{\partial\Gamma} (\underline{T} \cdot \hat{n} \times (\nabla \times \underline{E}))\, dl \tag{5.117}$$

and

$$\iint_\Gamma \left(\frac{\mu_r}{\varepsilon_r\mu_r - (k_z^2/k_0^2)} \nabla_t T \cdot \nabla_t H_z - k_0^2 \mu_r T H_z + \frac{k_z}{\omega\mu_0} \frac{1}{\varepsilon_r\mu_r - (k_z^2/k_0^2)} \hat{z} \cdot \nabla_t T \times \nabla_t E_z \right) ds$$
$$= -\int_{\partial\Gamma} (\underline{T} \cdot \hat{n} \times (\nabla \times \underline{H}))\, dl \tag{5.118}$$

Eqs. (5.117) and (5.118) constitute coupled "weak" differential equations describing the field throughout region Γ and is at the basis of the FEM. For normal incidence $k_z = 0$, we obtain two uncoupled equations one for the TM case (5.117) and the other for the TE case (5.118). Since the tangential electric field must vanish on the conductive surface, E_z is a known function on $\partial\Gamma_c$ and satisfies the Dirichlet boundary condition. Consequently, all testing functions used to discretize (5.118) will vanish on $\partial\Gamma_c$, and the boundary integral will contribute nothing to the matrix equation. Similarly, in the boundary integral $\partial\Gamma_c$ of eq. (5.116), Newman boundary condition is enforced by ignoring the integral over $\partial\Gamma_c$. In order to represent the complete scattering problem, it is necessary to augment these equations with additional information about the fields on $\partial\Gamma$, such as radiation boundary conditions (RBC), which will be explained in the following.

For simplicity, we assume that the boundary $\partial\Gamma$ is circular with radius a, $\hat{n} = \hat{r}$ and is located in free space. These assumptions simplify the right-hand side (RHS) of (5.117) and (5.118) such that

$$-\int_{\partial\Gamma} (\underline{T} \cdot \hat{n} \times (\nabla \times \underline{E}))\, dl = \int_0^{2\pi} T(a,\phi) \left(\frac{k_0^2}{k_t^2} \frac{\partial E_z}{\partial r} - \frac{k_z\omega\mu_0}{k_t^2} \frac{\partial H_z}{\partial \phi} \right) a d\phi \tag{5.119}$$

and similarly,

$$-\int_{\partial\Gamma} (\underline{T} \cdot \hat{n} \times (\nabla \times \underline{H}))\, dl = \int_0^{2\pi} T(a,\phi) \left(\frac{k_0^2}{k_t^2} \frac{\partial H_z}{\partial r} + \frac{k_z\omega\varepsilon_0}{k_t^2} \frac{\partial E_z}{\partial \phi} \right) a d\phi \tag{5.120}$$

in which $k_t = \sqrt{k_0^2 - k_z^2}$. To comply with the radiation boundary conditions on $\partial\Gamma$, the total electric and magnetic fields are expressed as a summation of the scattered fields E_z^{s}, H_z^{s} and the incident fields E_z^{inc}, H_z^{inc}. The scattered electric field on a circular boundary of radius $r = a$ can be expressed in terms of cylindrical harmonics [15] by

$$E_z^s(r,\phi) = \sum_{n=-\infty}^{\infty} e_n H_n^{(2)}(k_t r) e^{jn\phi} \quad ; \quad r \geq a \tag{5.121}$$

where

$$e_n = \frac{1}{2\pi H_n^{(2)}(k_t a)} \int_0^{2\pi} E_z^s(a,\phi') e^{-jn\phi'} d\phi' \tag{5.122}$$

and $H_n^{(2)}(x)$ is the nth order Hankel function of second type. Accordingly,

$$\left.\frac{\partial E_z^s}{\partial r}\right|_{r=a} = \int_0^{2\pi} E_z^s(a,\phi') \left[\frac{k_t}{2\pi} \sum_{n=-\infty}^{\infty} \frac{H_n^{(2)\prime}(k_t a)}{H_n^{(2)}(k_t a)} e^{jn(\phi-\phi')} \right] d\phi' \tag{5.123}$$

Similarly, the plane wave incident field can be expressed in terms of cylindrical harmonics [15]

$$E_z^{inc}(r,\phi) = \sum_{n=-\infty}^{\infty} e_n^{inc} J_n(k_t r) e^{jn\phi} \tag{5.124}$$

where $J_n(x)$ is the nth order Bessel function and

$$e_n^{inc} = \frac{1}{2\pi J_n(k_t a)} \int_0^{2\pi} E_z^{inc}(a,\phi') e^{-jn\phi'} d\phi' \tag{5.125}$$

By a similar procedure to that of the scattered field as described in (5.123), we obtain for the incident field

$$\left.\frac{\partial E_z^{inc}}{\partial r}\right|_{r=a} = \int_0^{2\pi} E_z^{inc}(a,\phi') \left[\frac{k_t}{2\pi} \sum_{n=-\infty}^{\infty} \frac{J_n'(k_t a)}{J_n(k_t a)} e^{jn(\phi-\phi')} \right] d\phi' \tag{5.126}$$

Using the Wronskian relationship [15]:

$$J_n'(k_t a) H_n^{(2)}(k_t a) - J_n(k_t a) H_n^{(2)\prime}(k_t a) = \frac{j2}{\pi k_t a} \tag{5.127}$$

and addition of (5.123) with (5.126), yields

$$\begin{aligned} \left.\frac{\partial E_z}{\partial r}\right|_{r=a} &= \frac{1}{2\pi} \int_0^{2\pi} E_z(a,\phi') \left[k_t \sum_{n=-\infty}^{\infty} \frac{H_n^{(2)\prime}(k_t a)}{H_n^{(2)}(k_t a)} e^{jn(\phi-\phi')} \right] d\phi' \\ &+ \frac{1}{2\pi} \int_0^{2\pi} E_z^{inc}(a,\phi') \left[\frac{j2}{\pi a} \sum_{n=-\infty}^{\infty} \frac{1}{J_n(k_t a) H_n^{(2)}(k_t a)} e^{jn(\phi-\phi')} \right] d\phi' \end{aligned} \tag{5.128}$$

Similar derivation for H_z, results in

$$\left.\frac{\partial H_z}{\partial r}\right|_{r=a} = \frac{1}{2\pi}\int_0^{2\pi} H_z(a,\phi')\left[k_t \sum_{n=-\infty}^{\infty} \frac{H_n^{(2)\prime}(k_t a)}{H_n^{(2)}(k_t a)} e^{jn(\phi-\phi')}\right] d\phi' + \frac{1}{2\pi}\int_0^{2\pi} H_z^{inc}(a,\phi')\left[\frac{j2}{\pi a}\sum_{n=-\infty}^{\infty}\frac{1}{J_n(k_t a)H_n^{(2)}(k_t a)} e^{jn(\phi-\phi')}\right] d\phi' \quad (5.129)$$

Next we derive $\left.\frac{\partial E_z}{\partial \phi}\right|_{r=a}$ from (5.121) and (5.122)

$$\left.\frac{\partial E_z}{\partial \phi}\right|_{r=a} = \int_0^{2\pi} E_z(a,\phi')\left[\frac{j}{2\pi}\sum_{n=-\infty}^{\infty} n e^{jn(\phi-\phi')}\right] d\phi' \quad (5.130)$$

and similarly, we obtain

$$\left.\frac{\partial H_z}{\partial \phi}\right|_{r=a} = \int_0^{2\pi} H_z(a,\phi')\left[\frac{j}{2\pi}\sum_{n=-\infty}^{\infty} n e^{jn(\phi-\phi')}\right] d\phi' \quad (5.131)$$

Substituting (5.128)–(5.131) into (5.119) yields the RBC for circular boundary with radius a:

$$\int_{\partial\Gamma} \underline{T}\cdot\hat{n}\times(\nabla\times\underline{E})dl = \frac{a}{2\pi}\int_0^{2\pi}\int_0^{2\pi} T(a,\phi)E_z(a,\phi')G_1(\phi-\phi')d\phi' d\phi + \frac{a\eta_0}{2\pi}\int_0^{2\pi}\int_0^{2\pi} T(a,\phi)H_z(a,\phi')G_2(\phi-\phi')d\phi' d\phi + \frac{a}{2\pi}\int_0^{2\pi}\int_0^{2\pi} T(a,\phi)E_z^{inc}(a,\phi')G_3(\phi-\phi')d\phi' d\phi \quad (5.132)$$

in which $\eta_0 = \sqrt{\frac{\mu_0}{\varepsilon_0}}$ and

$$G_1(\phi) = -\frac{k_0^2}{k_t}\sum_{n=-\infty}^{\infty}\frac{H_n^{(2)\prime}(k_t a)}{H_n^{(2)}(k_t a)}e^{jn\phi}$$

$$G_2(\phi) = \frac{jk_0 k_z}{k_t^2 a}\sum_{n=-\infty}^{\infty} n e^{jn\phi}$$

$$G_3(\phi) = -\frac{j2k_0^2}{\pi k_t^2 a}\sum_{n=-\infty}^{\infty}\frac{1}{J_n(k_t a)H_n^{(2)}(k_t a)}e^{jn\phi} \quad (5.133)$$

Similarly, (5.120) can be rewritten in the form

$$\int_{\partial\Gamma} \underline{T}\cdot\hat{n}\times(\nabla\times\underline{H})\,dl = \frac{a}{2\pi}\int_0^{2\pi}\int_0^{2\pi} T(a,\phi)H_z(a,\phi')G_1(\phi-\phi')d\phi' d\phi$$
$$-\frac{a}{2\pi\eta_0}\int_0^{2\pi}\int_0^{2\pi} T(a,\phi)E_z(a,\phi')G_2(\phi-\phi')d\phi' d\phi$$
$$+\frac{a}{2\pi}\int_0^{2\pi}\int_0^{2\pi} T(a,\phi)H_z^{inc}(a,\phi')G_3(\phi-\phi')d\phi' d\phi \quad (5.134)$$

Although the summations in (5.133)a and (5.133)b are divergent, the required calculations in (5.132) and (5.134) are well behaved on $\partial\Gamma$, since the integrals involved and the cylindrical harmonic content of the functions $E_z(\phi)$ and $H_z(\phi)$ ensures that the integrals are convergent and relatively easy to compute. To ensure the convergence, the integration should be performed term by term before the summation. Eqs. (5.132) and (5.134) should be substituted into (5.117) and (5.118) to complete the formulation. The resulting equations can be discretized following the FEM procedure using a piecewise-linear representation $B_n(x,y)$ for $E_z(\phi)$ and $H_z(\phi)$ and piecewise-linear testing functions $B_m(x,y)$, as shown in Fig. 5.2 and described in eq. (5.48). Each basis function has unity amplitude at one node and vanishes at all other nodes in the mesh. Three basis functions overlap each triangular cell to provide a continuous piecewise-linear representation. Along the circular boundary $\partial\Gamma$ the basis and testing functions, $B_m(\phi)$ may be expressed as

$$B_m(\phi) = \begin{cases} \dfrac{\phi-\phi_{m-1}}{\phi_m-\phi_{m-1}} & \phi_{m-1}<\phi<\phi_m \\ \dfrac{\phi_{m+1}-\phi}{\phi_{m+1}-\phi_m} & \phi_m<\phi<\phi_{m+1} \\ 0 & otherwise \end{cases} \quad (5.135)$$

where for convenience, the ϕ coordinates of the three nodes associated with the mth basis function are denoted as ϕ_{m-1}, ϕ_m, and ϕ_{m+1}. The incident uniform plane wave field with amplitude e_0 and at an angle (ϕ_0,θ_0) on a circular boundary of radius $r = a$ and in terms of cylindrical harmonics can be expressed in terms of

$$E_z^{inc}\Big|_{r=a} = e_0 e^{-jk_t(a\cos\phi\cos\phi_0 + a\sin\phi\sin\phi_0)} e^{jk_z z} = e_0 e^{-jk_t a\cos(\phi-\phi_0)} e^{jk_z z}$$
$$= e_0 \sum_{q=-\infty}^{\infty} j^{-q} J_q(k_t a) e^{jq(\phi-\phi_0)} e^{jk_z z} \quad (5.136)$$

Similarly,

$$H_z^{inc}|_{r=a} = h_0 \sum_{q=-\infty}^{\infty} j^{-q} J_q(k_t a) e^{jq(\phi-\phi_0)} e^{jk_z z} \tag{5.137}$$

Substitution of (5.43), (5.44), (5.132), (5.134), (5.136), and (5.137) into (5.117) and (5.118) results in the $N \times N$ matrix equation with N being the number of nodes in the solution domain:

$$\begin{bmatrix} A & B \\ C & D \end{bmatrix} \begin{bmatrix} e \\ h \end{bmatrix} = \begin{bmatrix} e^{inc} \\ h^{inc} \end{bmatrix} \tag{5.138}$$

in which

$$A_{mn} = \iint_\Gamma \left(\frac{\varepsilon_r}{\varepsilon_r \mu_r - (k_z^2/k_0^2)} \nabla_t B_m \cdot \nabla_t B_n - k_0^2 \varepsilon_r B_m B_n \right) ds + \frac{k_0^2 a}{2\pi k_t} \sum_{q=-\infty}^{\infty} \frac{H_q^{(2)\prime}(k_t a)}{H_q^{(2)}(k_t a)} I_m(q) I_n(-q) \tag{5.139}$$

$$B_{mn} = \iint_\Gamma \left(\frac{-k_z}{\omega \varepsilon_0} \frac{1}{\varepsilon_r \mu_r - (k_z^2/k_0^2)} \hat{z} \cdot \nabla_t B_m \times \nabla_t B_n \right) ds + \frac{j k_0 k_z \eta_0}{2\pi k_t^2} \sum_{q=-\infty}^{\infty} q I_m(q) I_n(-q) \tag{5.140}$$

$$C_{mn} = \iint_\Gamma \left(\frac{k_z}{\omega \mu_0} \frac{1}{\varepsilon_r \mu_r - (k_z^2/k_0^2)} \hat{z} \cdot \nabla_t B_m \times \nabla_t B_n \right) ds - \frac{j k_0 k_z}{2\pi k_t^2 \eta_0} \sum_{q=-\infty}^{\infty} q I_m(q) I_n(-q) \tag{5.141}$$

$$D_{mn} = \iint_\Gamma \left(\frac{\mu_r}{\varepsilon_r \mu_r - (k_z^2/k_0^2)} \nabla_t B_m \cdot \nabla_t B_n - k_0^2 \mu_r B_m B_n \right) ds - \frac{k_0^2 a}{2\pi k_t} \sum_{q=-\infty}^{\infty} \frac{H_q^{(2)\prime}(k_t a)}{H_q^{(2)}(k_t a)} I_m(q) I_n(-q) \tag{5.142}$$

where

$$I_m(q) = \int_{\partial\Gamma} B_m(\phi) e^{jq\phi} d\phi \tag{5.143}$$

which only contributes to the preceding expressions if node m is located on the circular boundary $\partial\Gamma$. In addition, the excitation vector is given by

$$e_m^{inc} = \frac{j2k_0^2 e_0}{\pi k_t^2} \sum_{q=-\infty}^{\infty} \frac{j^{-q} e^{-jq\phi_0}}{H_q^{(2)}(k_t a)} I_m(q) \tag{5.144}$$

$$h_m^{inc} = \frac{j2k_0^2 h_0}{\pi k_t^2} \sum_{q=-\infty}^{\infty} \frac{j^{-q} e^{-jq\phi_0}}{H_q^{(2)}(k_t a)} I_m(q) \tag{5.145}$$

Matrix inversion of (5.138) enables to evaluate the electric and magnetic field distributions ($e_1, e_2, \ldots .e_N$) and ($h_1, h_2, \ldots, h_N$) in the beam. The equivalent induced electric and magnetic polarization currents can be computed using (5.17) and (5.19). Next, the scattered far field and IFR_e and IFR_h can be computed using (5.64) and (5.67).

Fig. 5.22 shows a comparison of the E_z field produced on the surface of a circular dielectric cylinder with $k_0a = 0.25$ and $\varepsilon_r = 50 - j20$ by a

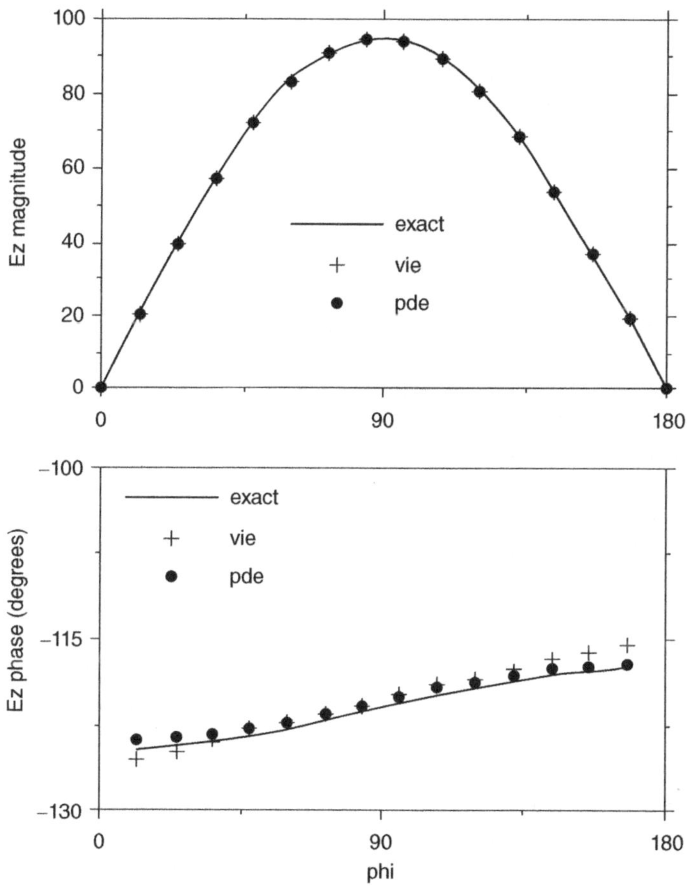

Figure 5.22 Comparison of the E_z field produced on the surface of a circular dielectric cylinder with $k_0a = 0.25$ and $\varepsilon_r = 50 - j20$ by a TE wave incident at an oblique angle of 30 deg computed analytically, numerically with volume integral equation (VIE) and with partial differential equation (PDE) or FEM [20].

TE wave incident at an oblique angle of 30 deg computed analytically, numerically with volume integral equation (VIE) and with partial differential equation (PDE) or FEM [20]. The agreement between the analytical and numerical results is very good and also between the VIE and FEM solutions.

References

1 Kay, AF. Electrical design of metal space frame radomes. *IEEE Trans. Antennas Propagat.*,13(2), 188–202, 1965.

2 Kennedy, PD. An analysis of the electrical characteristics of structurally supported radomes. The Ohio State University, Columbus, 1958.

3 Michielssen, E, Peterson, AF, and Mittra, R. Oblique scattering from inhomogeneous cylinders using a coupled integral formulation with triangular cells. *IEEE Trans. Antennas Propagat.*, 39(4), 485–490, 1991.

4 Wu, RB, and Chen, CH. Variational Reaction Formulation of Scattering Problem for Anisotropic Dielectric Cylinders. *IEEE Trans. Antennas Propagat.*, 34(5), 640–645, 1986.

5 Peterson, AF, Ray, SL, and Mittra, R. Computational methods for electromagnetics. New York: IEEE Press, 1998.

6 Richmond, JH. TE-wave scattering by a dielectric cylinder of arbitrary cross-section shape. *IEEE Trans. Antennas Propagat.*, 14(4), 460–464, 1966.

7 Richmond, JH. Scattering by a dielectric cylinder of arbitrary cross section shape. *IEEE Trans. Antennas Propagat.*, 13(3), 334–341, 1965.

8 Peterson, AF, and Klock, PW. An improved MFIE formulation for TE-wave scattering from Lossy, inhomogeneous dielectric cylinders. *IEEE Trans. Antennas Propagat.*, 36(1), 45–49, 1988.

9 Ricoy, MA, Kilberg, SM, and Volakis, JL. Simple integral equations for two-dimensional scattering with further reduction in unknowns. *IEE Proceedings*, 136(4), 298–304, 1989.

10 Rojas, RG. Scattering by an inhomogeneous dielectric/ferrite cylinder of arbitrary cross-section shape-oblique incidence case," *IEEE Trans. Antennas Propagat.*, 36(2), 238–246, 1988.

11 Gupta, IJ, Lai, AKY, and Burnside, WD. Scattering by dielectric straps with potential application as target support structure. *IEEE Trans. Antennas Propagat.*, 37(9), 1164–1171, 1989.
12 Rusch, WVT, Appel-Hansen, J, Klein, CA, and Mittra, R. Forward scattering from square cylinders in the resonance region with application to aperture blockage. *IEEE Trans. Antennas Propagat.*, 24(2), 182–189, 1976.
13 Abramowitz, M, and Stegun, IA. Handbook of mathematical functions. New York: Dover, 1964.
14 Wu, TK, and Tsai, LL. Scattering by arbitrarily cross-sectioned layered lossy dielectric cylinders. *IEEE Trans. Antennas Propagat.*, 25(4), 518–524, 1977.
15 Harrington, RF. Time-harmonic electromagnetic fields. New York: McGraw-Hill, 1961.
16 Harrington, RF. Field computation by moment methods. New York: IEEE Press, 1968.
17 Michielssen, E, and Mittra, R. Electromagnetic scattering from arbitrary strip loaded cylinders. In IEEE Ant. Propagat. Symposium, Chicago, Illinois, 1992.
18 Shavit, R, Smolski, AP, Michielssen, E, and Mittra, R. Scattering analysis of high performance large sandwich radomes. *IEEE Trans. Antennas Propagat.*, 40(2), 126–133, 1992.
19 Kildal, PS, Kishk, AA, and Tengs, A. Reduction of forward scattering from cylindrical objects using hard surfaces. *IEEE Trans. Antennas Propagat.*, 44(11), 1509–1520, 1996.
20 Peterson, AF. Application of volume discretization methods to oblique scattering from high-contrast penetrable cylinders. *IEEE Trans. Mic. Theory Tech.*, 42(4), 686–689, 1994.

Problems

P5.1 A rectangular PEC beam with cross section $1 \times 4\,\text{in.}^2$ is illuminated by a TM_z plane wave with amplitude 1V/m at 60 deg azimuth angle, as shown in the sketch. The operational frequency is 10 GHz.

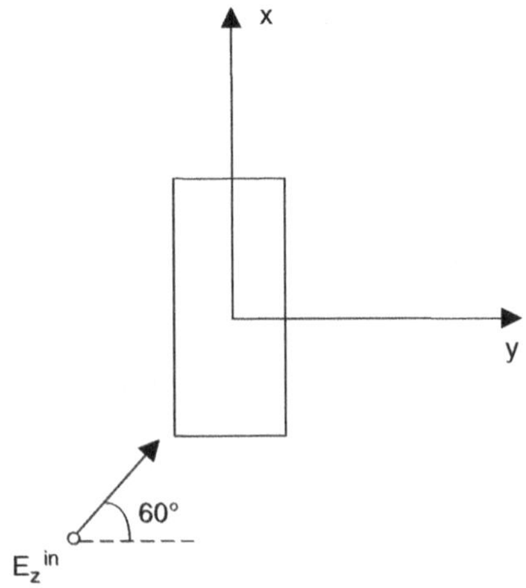

(a) Write the problem characteristic EFIE with the unknown current distribution on the beam.
(b) Formulate the numerical solution using MoM (point matching). Compute and plot the current distribution on the beam for subdivisions $\Delta = \lambda/10$, $\lambda/30$.
(c) Compute and plot the beam RCS $\sigma_{TM}(\phi)$ for subdivisions $\Delta = \lambda/10$, $\lambda/30$.
(d) Repeat items (a) – (c) for TE_z incident plane wave and formulation through a MFIE.
(e) Repeat items (a) – (c) for TM_z incident plane wave using Galerkin MoM with linear basis functions as shown in the sketch:

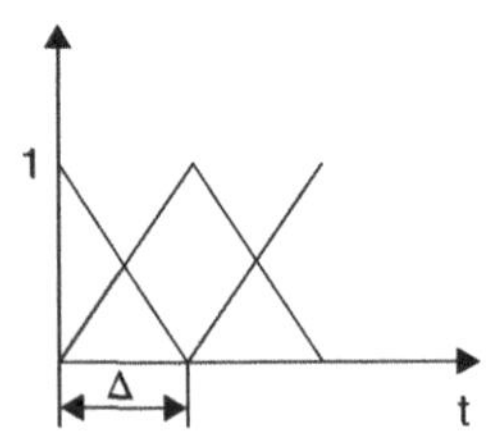

Compare solution to that obtained using MoM point matching. Discuss the results and differences.

P5.2 A rectangular PEC beam infinite in z-direction is covered on all its sides by a dielectric layer with thickness 1 cm and dielectric constant $\varepsilon_r = 4$. The cross section of the PEC beam is 1×4 in.^{2} and it is illuminated by a plane wave TM_z polarized propagating along y-axis at 10 GHz and with amplitude 1 V/m as shown in the sketch

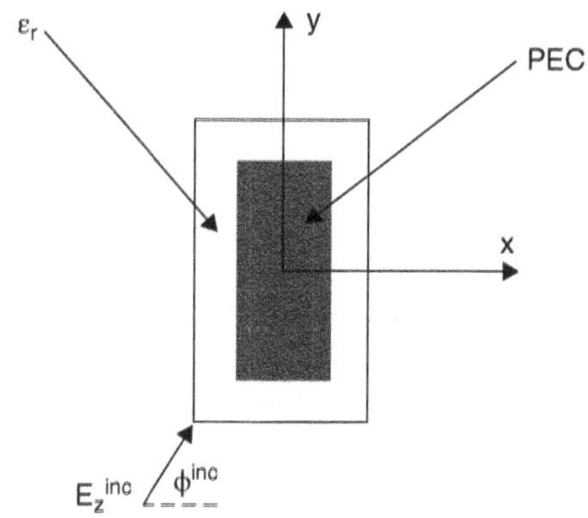

(a) Write the EFIE integral equation characterizing the problem.
(b) Formulate the numerical solution using point matching MoM volume version. Compute and plot the current distribution on the conductive beam for segmentation $\Delta_x \times \Delta_y = (\lambda/10)^2, (\lambda/30)^2$.
(c) Compute and plot the RCS $\sigma_{TM}(\phi)$ for subdivisions $\Delta_x \times \Delta_y = (\lambda/10)^2, (\lambda/30)^2$.
(d) Find the optimal dielectric layer thickness for minimum RCS $\sigma_{TM}(\pi/2)$.

P5.3 A rectangular PEC beam with cross section 3×8 in.^{2} is illuminated by a TM_z plane wave with amplitude 1V/m at 90 deg azimuth angle, as shown P5.1. The operational frequency is 10 GHz.

(a) Formulate the CFIE integral equation with the z-directed unknown current distribution on the beam. Assume coupling coefficient $\alpha = 0.2$.

(b) Formulate the numerical solution using MoM (point matching). Compute and plot the current distribution on the beam for subdivisions $\Delta = \lambda/30$.

(c) Compute and plot the scattering pattern from the beam.

(d) Compute and plot the IFR_e as a function of w/λ in the range [0.3–4.0] with w being the narrow width of the beam.

P5.4 Repeat problem P5.3 for TE_z polarization.

(a) Prove that the CFIE integral equation characterizing the problem is

$$\alpha E_t^{inc} - (1-\alpha)\eta_0 H_z^{inc} = (1-\alpha)\eta_0 J_t + (1-\alpha)\eta_0 \int_s J_t(t')\hat{n}(t')\cdot\nabla\frac{1}{4j}H_0^{(2)}(k_t|\underline{r}-\underline{r}'|)dt'$$
$$+ j\alpha\frac{\eta_0}{k_0}\hat{t}\cdot(\nabla_t\nabla_t\cdot + k_t^2)\int_s \hat{t}(t')J_t(t')\frac{1}{4j}H_0^{(2)}(k_t|\underline{r}-\underline{r}'|)dt'$$

(b) Formulate the numerical solution of the MFIE and CFIE (with $\alpha = 0.2$) equations using point-matching MoM and use the conjugate gradient iterative method for matrix inversion.

(c) Compute and plot the current distribution induced on the beam for $\Delta = \lambda/20$ segmentation.

(d) Compute and plot the RCS $\sigma_{\mathrm{TE}}(\phi)$ for both formulations MFIE and CFIE. Discuss differences.

P5.5 Consider a dielectric rectangular beam with dimensions 2×0.4 in.^{2} and dielectric constant $\varepsilon_r = 4.6$, as shown in the sketch:

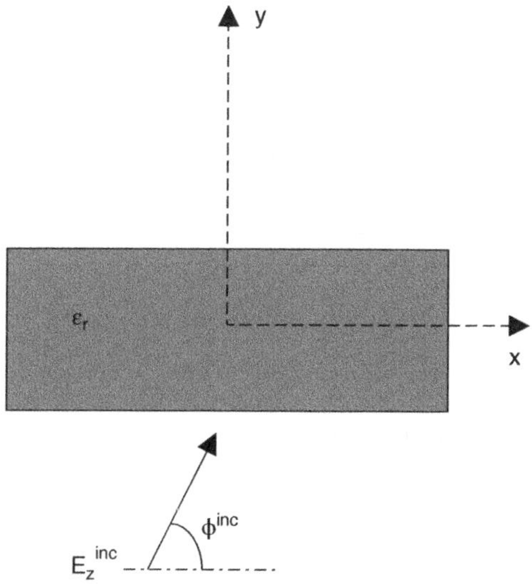

The beam is illuminated by a plane wave with amplitude 1V/m, at 5.6 GHz, TM polarized and propagating in the direction $\phi^{inc} = 90$ deg.

(a) Formulate the EFIE integral eq. of the problem.

(b) Formulate the numerical solution of the EFIE equation using point-matching MoM (volume formulation). Compute and plot the RCS $\sigma_{TM}(\phi)$ for cross section segmentation $\Delta_x \times \Delta_y = (\lambda/20)^2$.

On the beam axis, ($y = 0$) 8 PEC strips were inserted with width 0.062 in. and spacing 0.277 in. between centers, as shown in the sketch:

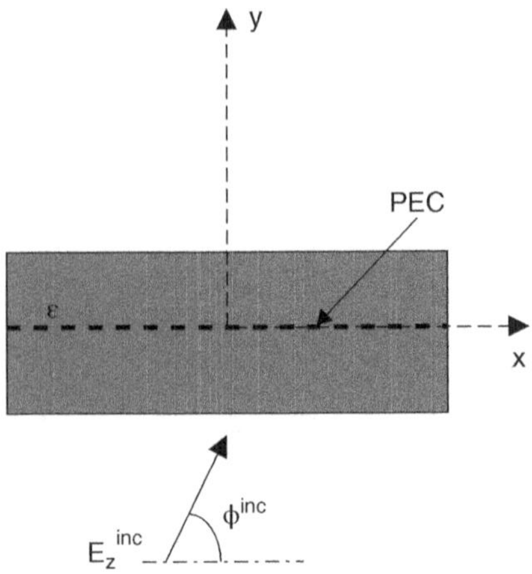

(c) Formulate the new EFIE equation characterizing the problem.

(d) Formulate the numerical solution using point-matching MoM (volume formulation). Compute and plot the RCS $\sigma_{TM}(\phi)$ for cross section segmentation $\Delta_x \times \Delta_y = (\lambda/20)^2$. Assume on the metal strips one basis function. Compare and discuss the results to (b) results.

6

Ground-Based Radomes

Large radar antennas are generally covered with radomes to protect them from extreme weather conditions and to enable their continuous precision operation without sacrificing their performance. The radomes are assembled from many panels connected together with metallic or dielectric beams because of their large size. In Chapter 2 and 3, the panel design process, which may include dielectric and FSS panels was discussed and analysis methods for optimum design were described. In Chapter 5, the scattering effect of a single dielectric or conductive beam with arbitrary cross section and for different incident angles and polarizations was analyzed. This chapter describes the problem of integration and systematical design of the beams geometry for minimum scattering and presents the effect of the radome on the radiation pattern of an antenna which is enclosed within a space frame structure, as shown in Fig. 6.1.

The radome design shown in Fig. 6.1 is based on the architectural concept often called a *geodesic dome*, in which a spherical structure is built from simple geometrical shapes. The geodesic dome inventor was W. Bauersfeld, who worked for Zeiss company in Germany and developed this structural concept for a planetarium built in 1922. Twenty years later, a US architect, B. Fuller, from Massachusetts, reinvented and popularized the design. The concept of the geodesic dome is to use an original triangle and divide it into multiple triangle divisions until the spherical surface of the dome is reached. The higher the subdivision used, the smoother and closer to spherical shape is the dome. On the other hand, a higher subdivision increases the number of beams needed for the construction and increases the optical blockage to the enclosed antenna in the radome. Therefore, there are trade-offs on how many subdivisions are used in the design. On the internet can be found

Radome Electromagnetic Theory and Design, First Edition. Reuven Shavit.

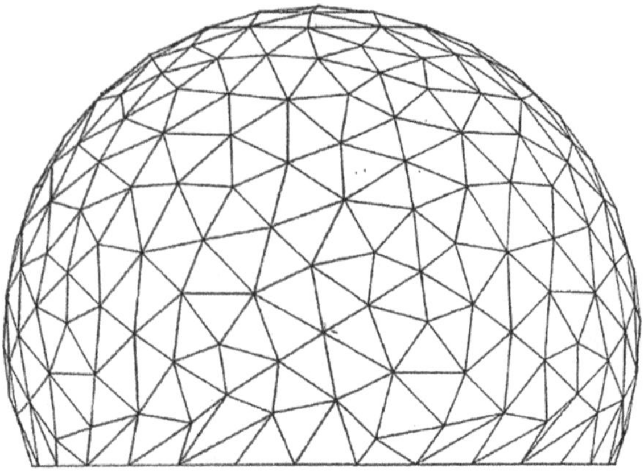

Figure 6.1 Geometry of a typical space frame radome.

many calculators to compute the beam lengths, center position, and space orientation, pending on the number of triangle divisions.

The spherical shape for the radome is chosen not only because of structural and manufacturing considerations but also because it provides the desired symmetry when the enclosed antenna performs scanning. From an electrical viewpoint, the beams and the window panels may be regarded as a collection of scatterers, which scatter the energy incident upon them from the transmitting antenna. In microwave applications, the window panels have little effect (especially for frequencies below X band); hence, the presence of the radome can be determined solely by evaluating the scattering properties of the beams and superimposing the effects in the far field. The radome is designed to have a quasi-random geometry in contrast to a regular geodesic dome design to reduce the cumulative effect of scattering from the beams.

The degree of randomness of the beam geometry can increase with a decrease in the beam length. However, such a decrease in the beam length also has associated negative effects, such as an increase in the total blockage due to the beams and decrease in the cost-effectiveness of the panel manufacturing process. Consequently, there is a trade-off issue when randomness is considered in the design process of the radome. The frequency bandwidth of the radome, in terms of transmission losses and sidelobe perturbation performance, is determined by the combined effect of the panels and its beams. The panels are usually thin low-loss plastic membranes for metal and dielectric space frames, or A-sandwich type panels for sandwich space frame radomes. The panels exhibit a wide frequency bandwidth, which normally does not limit the overall radome performance. The physical dimensions of the beams are determined by the stresses they have to withstand due to all the environmental and physical loads, including extremely high wind-loading. These beams may degrade the total system performance by introducing high levels of scattering, which would limit the operational frequency bandwidth of the radome. The analysis of the scattering effect of space-frame radomes was first conducted by Kay [1].

Reduction of the scattering effect from the beams entails two steps:

1. Minimizing the scattering level from the individual beams
2. Optimizing the radome geometry

Various methods to reduce the scattering effects from the beams, either as individuals or as an assembly in a space-frame radome and their effects on the radiation characteristics of an antenna enclosed in the radome, are reviewed in the following.

6.1 Scattering from an Individual Beam

Scattering from metal and dielectric beams vary with frequency and is contributed by the electric polarization currents induced in the beams. This effect can be characterized by the induced field ratio (*IFR*) [2, 3] and the beam scattering pattern as explained in Chapter 5. The *IFR* parameter determines the scattering level from a beam of infinite length compared to the scattering from an aperture with the beam dimensions. The *IFR* of an ideal invisible beam is equal to zero and achieving this would be the design goal for a good space-frame radome. A metal beam that causes only optical blockage has an *IFR* equal to 1 with a phase of 180 deg. Fig. 6.2 presented in [2, 4] shows the dependence of the *IFR* for a perfect electric conductor (PEC) circular cylinder, while Fig. 6.3, also from [2, 4], shows the same dependence for a dielectric circular cylinder with $\varepsilon_r = 5$ for parallel (TM case) and perpendicular (TE case) polarizations.

As one can observe, the *IFR* for the metallic cylinder varies monotonically for both polarizations and shows a decrease of the *IFR* amplitude for the parallel polarization and an increase for perpendicular polarization for an increase in the frequency. For high frequencies, the *IFR* tends to be 1 with phase of 180 deg (optical blockage) for both

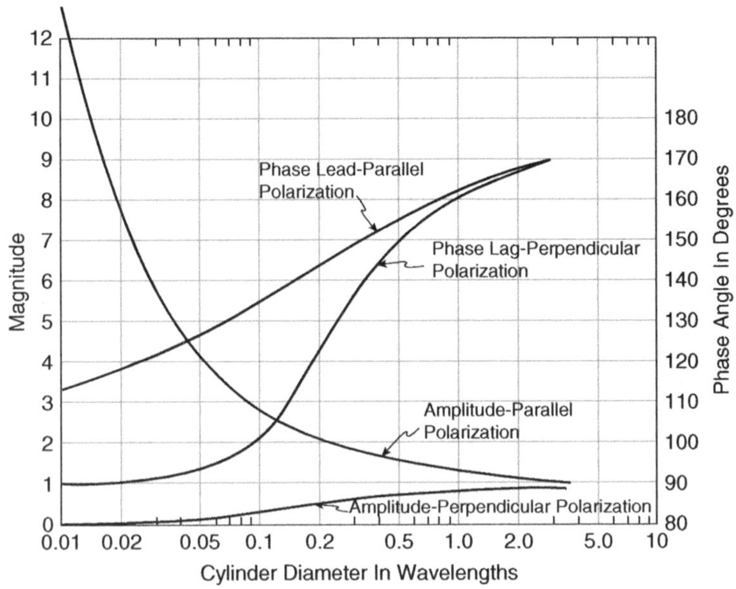

Figure 6.2 The *IFR* of a PEC cylinder as a function of its diameter in wavelengths [4].

parallel and perpendicular polarizations. On the other hand, the picture is totally different for the dielectric cylinder. As one can observe, the *IFR* is very low for both polarizations at low frequencies, where the radius of the cylinder is small compared to the wavelength, and, therefore, the use of dielectric beams at low frequencies (less than 1 GHz) is attractive. However, as the frequency increases, the induced polarization currents in the beam increase and, correspondingly, so does its *IFR*. Also, the *IFR* becomes oscillatory and the performance of the radome is not good over a broad frequency band. Similar results and trends are described in [5] and [6] for rectangular conductive and dielectric beams, respectively. Consequently, a dielectric space frame radome (DSF) is attractive for low frequencies (less than 1 GHz), while a metal space frame (MSF) radome can operate above 1 GHz over a wide frequency range. However, its minimal scattering losses are determined by the total optical blockage of the beams introduced in the antenna aperture. Another drawback of the scattering level for an MSF radome is an increase in the antenna sidelobe level above –25 dB for certain high-performance radar applications. Therefore, there is a real

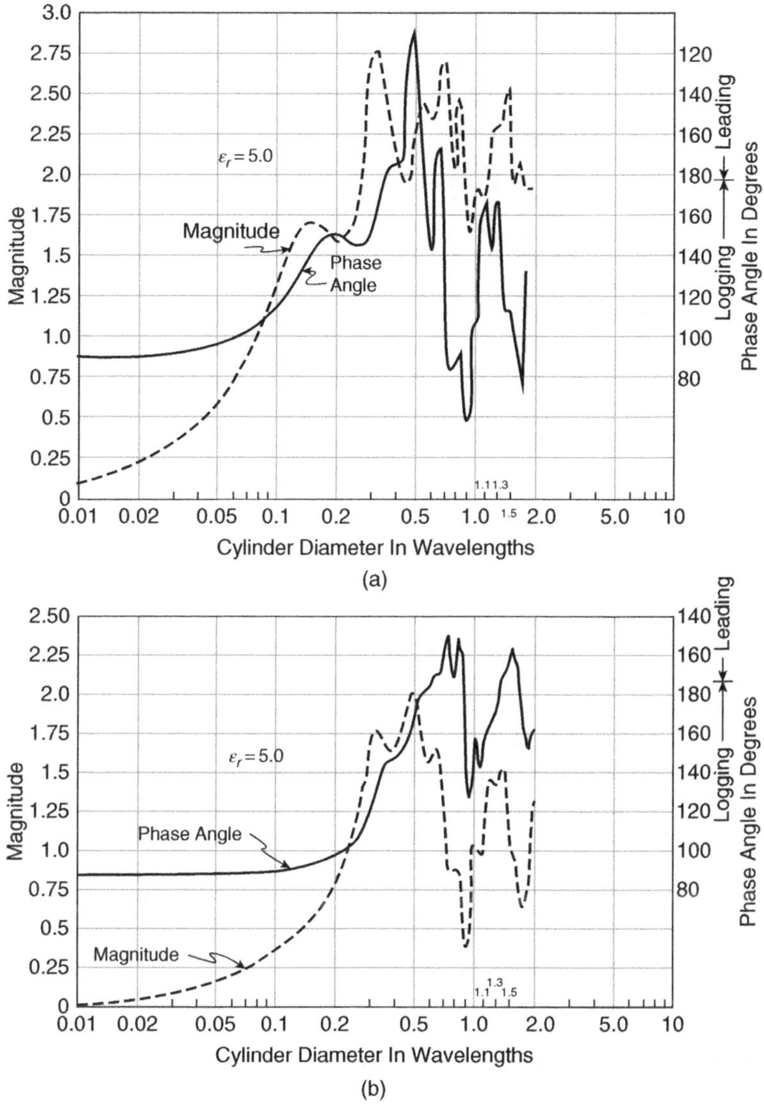

Figure 6.3 The *IFR* of a dielectric cylinder with $\varepsilon_r = 5$ as a function of its diameter in wavelengths [2]: (a) parallel polarization, (b) perpendicular polarization.

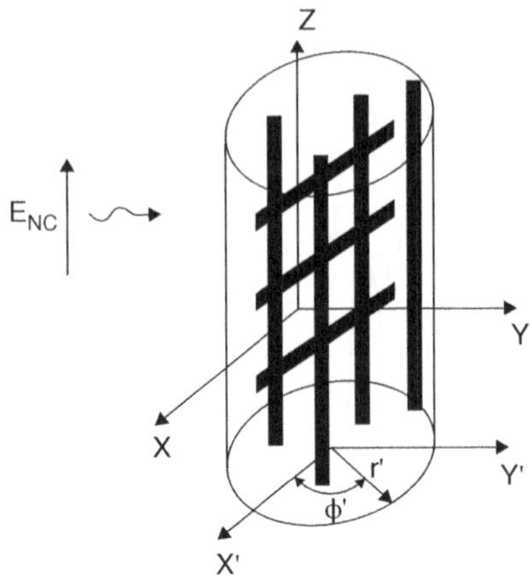

Figure 6.4 Basic geometry of a dielectric beam with tuning grid.

need to reduce the scattering level below the optical blockage for frequencies above 1 GHz.

Tuning techniques, such as insertion of conductive strips in the dielectric seams connecting the sandwich panels of the sandwich radome, are used to reduce the scattering level caused by the dielectric seams to provide a performance better than that of MSF or DSF radomes, as described in [7, 8], and in section 5.4. The currents induced in the conductive strips tend to offset the polarization currents induced in the seams and thereby reduce the tuned seams *IFR*. Fig. 6.4 shows the tuning mechanism with vertical and horizontal conductive strips.

The problem of computing the scattering from a dielectric beam with a rectangular cross section loaded with conductive strips has been addressed by Michielssen and Mittra [9, 10] to solve its combined field integral equation (CFIE) using the method of moments (MoM) and is described in Chapter 5.

6.2 Scattering Analysis of the Beams Assembly

The approach introduced by Kay in [1] is followed in this work to derive the expression for the perturbed far-field radiation pattern of the antenna enclosed in the radome. Applications of Kay [1] approach to MSF and DSF can be found in [11, 12], and to tuned and untuned sandwich radomes in [13, 14, 15]. Fig. 6.5 shows a schematic diagram of the scattering mechanism of the beams. The aperture of the antenna

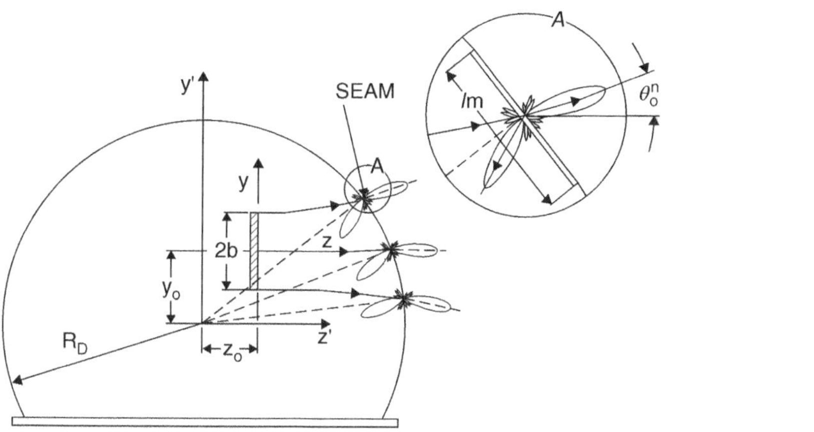

Figure 6.5 The geometry of the radome and the scattering mechanism.

is chosen to be elliptical, with its major axes $2a \times 2b$ and its center is assumed to be offset from the radome center $(0, y_0, z_0)$. The coordinate system for the scattering analysis is shown in Fig. 6.5.

Let the coordinates of the center of the mth beam be $(x_{\text{m}}, y_{\text{m}}, z_{\text{m}})$. In the analysis we made the assumption that the maximum forward scattering from a singular beam is in the direction of the incident field as shown in Fig. 6.5. This assumption enables to consider the total scattering from the illuminated beams as equal to the scattering from these beams projected on the antenna aperture. The angle of inclination of the projected beam on the antenna aperture (x-y plane) with respect to the y-axis is δ_{m}. The peak of the scattering pattern from the mth beam is determined by the incident angle (θ_0^m, ϕ_0^m) of the corresponding constituent ray of the Gaussian beam at its center, as shown in Fig. 6.5. In this derivation, we use the coordinate system (θ, ϕ), where θ is the elevation angle measured in the y-z plane and ϕ is the azimuthal angle measured from the x-axis in the x-y plane.

To determine the effect of perturbation due to the beam scattering, we initially compute the unperturbed far-field pattern of the antenna enclosed by the radome, and then superpose this result to the scattering pattern of the beams assembly.

In the scattering analysis we make the following assumptions that are valid approximations for the purpose of computing the perturbed far-field radiation pattern:

1. The field incident upon the radome due to the enclosed antenna can be represented by using a divergent Gaussian beam model. Further,

only the beams within the incident beam of the enclosed antenna contribute to the scattering.
2. The total scattered field from the beams is assumed to be a superposition of all individual beams. Mutual coupling and multiple scattering effects are negligible.
3. The field incident upon a particular radome beam member does not vary significantly over the extent of the member.
4. The field radiated by the mth beam is equivalent to that arising from a straight rectangular current strip projected on the antenna aperture with length l_m, width w_m, and constant transverse and axial current distributions.

To determine the perturbation due to the beam scattering, we initially compute the unperturbed far-field pattern of the antenna enclosed in the radome. In our analysis, we have assumed that the aperture distribution $f(x,y)$ [14] of the elliptical aperture of the antenna with dimensions $2a \times 2b$, is expressed as

$$f(x,y) = C + (1-C)\left[1 - \frac{x^2}{a^2} - \frac{y^2}{b^2}\right]^p \tag{6.1}$$

where C is the pedestal illumination and p is an exponential factor, which determines the shape of the aperture distribution. Given the aperture distribution $f(x,y)$, one can compute the far-field pattern $F(\theta,\phi)$ [16], which can be expressed in a closed form as

$$F(\theta,\phi) = \pi a^2 \cos\alpha \left[C\frac{2J_1(u)}{u} + (1-C)\frac{2^{p+1}\Gamma(p+1)J_{p+1}(u)}{u^{p+1}}\right] \tag{6.2}$$

where $\cos\alpha = b/a$, $J_p(u)$ is the Bessel function of pth order, $\Gamma(p+1)$ is the Gamma function and $u = ka\sin\theta(1-\sin^2\phi\sin^2\alpha)^{1/2}$. To evaluate the field incident upon the beams, the radiated near field from the elliptical aperture can be approximated by a Gaussian beam [17] with two independent waists $W_{el}(z)$ and $W_{az}(z)$ in elevation and azimuth, respectively. The waist $W(z)$ is given by [17],

$$W(z) = W_0\left[1 + (\lambda_0 z/\pi W_0^2)^2\right]^{1/2} \tag{6.3}$$

where, W_0 is the waist of the Gaussian beam at $z = 0$, and it can either be $W_{0,el}$ (elevation) or $W_{0,az}$ (azimuth). W_0 can be computed by equating the actual 3dB beamwidth of the aperture to the 3dB beamwidth of the Gaussian beam approximation. The radius of curvature $R(z)$ of the Gaussian beam is given by [17]:

$$R(z) = z\left[1 + (\pi W_0^2/\lambda_0 z)^2\right] \tag{6.4}$$

There are two radii of curvature, R_{az} and R_{el} in azimuth and elevation, respectively. Given $W(z)$ and $R(z)$, one can compute the field intensity $f(x,y,z)$ of the Gaussian beam at any distance z from the antenna aperture, by using:

$$f(x,y,z) = \left[\frac{W_{0,el}W_{0,az}}{W_{el}W_{az}}\right]^{1/2} e^{-\frac{x^2}{W_{az}^2}-\frac{y^2}{W_{el}^2}} e^{-jkz} e^{-j\frac{\pi}{\lambda}\left(\frac{x^2}{R_{az}}+\frac{y^2}{R_{el}}\right)} e^{j\tan^{-1}\left(\frac{\lambda_0 z}{\pi W_{0,el}W_{0,az}}\right)} \tag{6.5}$$

One can observe that the intensity of the electric field decays exponentially transverse to the propagation z-axis. Moreover, the intensity of the Gaussian beam drops by more than 30 dB at a lateral distance of $4W_{az}$ in x direction and $4W_{el}$ in y direction. Given this, the scattering effects from beams beyond a lateral distance of $4W_{az}$ (x direction) and $4W_{el}$ (y direction) can be ignored.

Using the approximations mentioned above, the far-field scattering pattern $I_m(\theta,\phi)$ from the mth beam can be written as,

$$I_m(\theta,\phi) = w_m l_m \frac{\sin A_m}{A_m}\frac{\sin B_m}{B_m} e^{jk(x_m u_m + y_m v_m + z_m d_m)} \tag{6.6}$$

where,

$$\begin{aligned} u_m &= \cos(\theta-\theta_0^m)\sin(\phi-\phi_0^m) \\ v_m &= \sin(\theta-\theta_0^m) \\ d_m &= \cos(\theta-\theta_0^m)\cos(\phi-\phi_0^m) \end{aligned} \tag{6.7}$$

and

$$\begin{aligned} A_m &= \frac{kw_m}{2}\sin\theta'\cos(\phi'+\delta_m) \\ B_m &= \frac{kl_m}{2}\sin\theta'\sin(\phi'+\delta_m) \\ \sin\theta' &= \left[\sin^2(\theta-\theta_0^m)+\cos^2(\theta-\theta_0^m)\sin^2(\phi-\phi_0^m)\right]^{1/2} \\ \sin\phi' &= \frac{\sin(\theta-\theta_0^m)}{\sin\theta'} \end{aligned} \tag{6.8}$$

The scattering from the beams is weighted by the antenna illumination function $f(x_m, y_m, z_m)$, given by (6.5), g_m, which is proportional to the *IFR*, and the scattering pattern $I_m(\theta,\phi)$ given by (6.6). Thus, the total scattered field from M beams can be written in the form

$$F_s(\theta,\phi) = \sum_{m=1}^{M} g_m I_m(\theta,\phi/\theta_0^m,\phi_0^m) f_m(x_m,y_m,z_m) \tag{6.9}$$

where [12]

$$g_m = g_{||}\cos^2\delta_m + g_{\perp}\sin^2\delta_m \tag{6.10}$$

and $g_{||}$, $g_{\perp}$ are the parallel and perpendicular components of the beam *IFR*. By combining the unperturbed pattern of the antenna $F(\theta,\phi)$ given by (6.2) and the scattered pattern from the beams $F_s(\theta,\phi)$ given by (6.9), one can compute the perturbed pattern $F'(\theta,\phi)$ by expressing it as

$$F'(\theta,\phi) = F(\theta,\phi) + F_s(\theta,\phi) \tag{6.11}$$

As an example of the scattering effect of the beams, we consider a vertically polarized antenna aperture 27 × 2 ft^2 operating at 1.09GHz, which is enclosed in a 56-ft diameter tuned sandwich radome. We assume that the aperture taper is 15 dB, i.e., $p = 2$. The measured parallel polarization *IFR* is –0.013 + j0.013 and the perpendicular polarization *IFR* is –0.12 + j0.001, for a beam width of 4 in. The Gaussian beam spreading results in an illuminated aperture on the radome surface of 29.6 × 13.2 ft^2, which indicates that more beams are illuminated than is evident from the projected 27 × 2 ft^2 aperture. Fig. 6.6 [13] shows the comparison between the computed scattering patterns with and without the inclusion of the divergence effect of the radiating aperture.

One can observe that the divergence causes the scattered energy from the beams to be more uniformly distributed over a wider sector. The total transmission loss (with illumination divergence) from

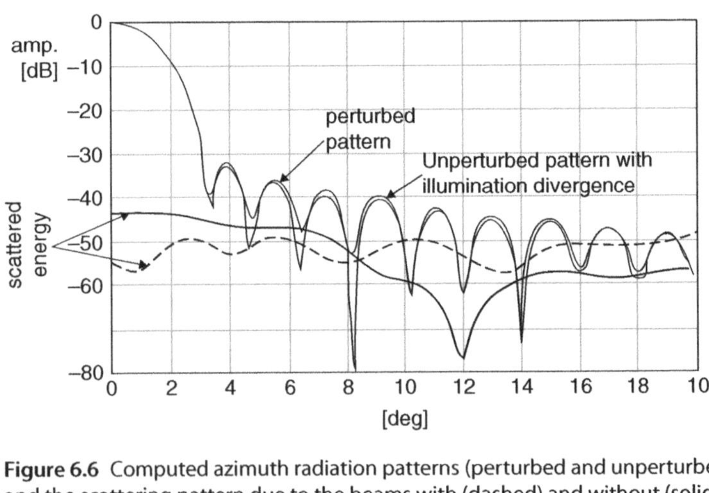

Figure 6.6 Computed azimuth radiation patterns (perturbed and unperturbed) and the scattering pattern due to the beams with (dashed) and without (solid) the divergence effect @ 1.09 GHz.

the radome improves from 0.06 dB to 0.02 dB, and the scattering level becomes more uniform.

Radiation parameters affected by the scattered energy from the beams
The scattered energy by the beams affects a number of radiation parameters [18], as for example:

1. Transmission loss
2. Sidelobe level increment of the enclosed antenna
3. Null depth in a difference pattern
4. Beamwidth change
5. Boresight error
6. Boresight error slope
7. Cross-polarization ratio
8. Antenna noise temperature

6.2.1 Transmission Loss

The transmission through the radome not only depends on the construction of the radome window, but also on the antenna look angle, the field distribution over the antenna aperture, the location of the antenna inside the radome and the radome/antenna size ratio. Using ray-tracing methods [1] the transmission through the radome can be determined by summing up the losses from all rays, which have different weight functions and look angles. The total loss through the radome, L_t is given by

$$L_t = L_w + L_j$$

where

L_w = transmission loss due to the radome wall in dB
L_j = transmission loss due to the beams scattering in dB

$$L_j = 10 \cdot \log |1 + \rho_s \cdot \mathrm{Re}\{IFR\}|^2 \tag{6.12}$$

ρ_s = physical blockage of projected beams on the antenna aperture

6.2.2 Sidelobe Level Increment

At each angle, the energy scattered by the radome and computed by the model presented in [1] interferes with the radiated energy, according to the amplitude and phase relationship between these two contributions. A sidelobe level (*SL*) in (dB) below the peak of the sum pattern is affected by the scattered energy (*SE*) in (dB) relative to the peak. The resulting sidelobe perturbation ΔSL(dB) can be approximated by

$$\Delta SL = 20\log\left(1 + 10^{(SE-SL)/20}\right) \tag{6.13}$$

6.2.3 Null Depth Increment

The scattered energy, whether it is co-polarized or cross-polarized, fills in the null of a monopulse system. The on-axis contribution of the scattered energy interacts with the weak energy in the null pattern radiated in this direction and affects the pattern according to the amplitude and phase relationship between the two sources. The null's depth increment in the worst case can be approximated by eq. (6.13), in which SL(dB) can be replaced by the null depth.

6.2.4 Beamwidth Change

The interference of the scattered energy with the main beam energy causes a $\Delta\theta$ change in the antenna beamwidth BW. Maximum effect occurs when both sides of the beam are affected in the same way, either with in-phase or out-of-phase combination. The approximated antenna beamwidth variation at the 3 dB points is described in [18] and is given by

$$\Delta\theta = \frac{2 \cdot BW \cdot 20\log\left(1 + 10^{\frac{SE+3}{20}}\right)}{12} \tag{6.14}$$

This formula assumes a phase addition of the antenna field and the radome scattering field at the −3 dB points.

6.2.5 Boresight Error

Boresight error (*BSE*) results from an imbalance between the two halves of the antenna. The source of the asymmetry may be a result of the difference in panel construction or the presence of more beams in front of one side of the antenna. The contribution of the beams is a function of their distribution in front of the antenna. An approximate formula [18], which estimates the beam-induced boresight error, reads:

$$BSE(rad) = \frac{0.27 \cdot \lambda_0 \cdot l \cdot w \cdot |\mathrm{Im}(IFR)|}{R^3} \tag{6.15}$$

where

w = the width of the beam
l = the average length of the beams
R = the radius of the antenna
λ_0 = the wavelength

This formula assumes that there is an entire beam on one side of the aperture, uncompensated by the presence of another beam on the opposite half. The actual imbalance is smaller in the large-antenna case.

6.2.6 Boresight-Error Slope

The rate of change of the boresight error is dependent on the change in the number of beams in front of the antenna. Following [18], it can be approximated by

$$Slope = BSE/\theta_r \tag{6.16}$$

where,

BSE = boresight error
θ_r = subtended angle of the radome's beam from the antenna center

6.2.7 Cross-Polarization Ratio

The way in which the radome affects cross-polarization is also the same way it affects the sidelobes; that is, the scattered energy vectorially adds to the field radiated by the antenna without the radome. The principal difference here is that the energy scattered into the cross-polarization field must be less than half of the co-polarized scattered energy; thus, the scattered-energy level used for the effect on cross-polarization energy is 3 dB lower for the change in the sidelobe level calculation.

6.2.8 Antenna Noise Temperature

The noise temperature contribution due to the radome includes three factors:

1. Noise temperature contribution, NT_1 due to absorption in the radome window
2. Noise temperature contribution, NT_2 due to reflection from the radome window
3. Noise temperature contribution, NT_3 due to scattering from the beams

The antenna noise temperature T_A described in Appendix C is given by

$$T_A \triangleq \frac{1}{4\pi}\int_{4\pi} D(\theta',\phi')T_s(\theta')d\Omega' \tag{6.17}$$

in which $D(\theta, \phi)$ is the antenna directivity (far-field approximation) as a function of the space coordinates (θ,ϕ) and $T_s(\theta)$ is the sky temperature at the antenna location. A typical function is shown in [20]:

In Fig. 6.7 we can see the sky temperature as a function of frequency with the elevation angle θ being the parameter. In general, the sky noise increases with frequency because the attenuation through the atmosphere increases with frequency. On the other hand, at a certain frequency increase in the elevation angle reduces the sky noise because the propagation path and the attenuation through the atmosphere is reduced as the elevation angle is increased. Moreover, at 22.5 GHz we can notice an increase in the noise sky due to the absorption line of the oxygen gas. The noise contribution due to absorption in the radome window can be approximated by $NT_1 = 300\ P_a$, where P_a is the amount of energy absorbed in the radome window. When computing the reflection contribution, it is assumed that half of this energy is reflected to the cold atmosphere and half to the warm earth (300°K). Therefore, $NT_2 = 150P_r$, where P_r is the amount of energy reflected by

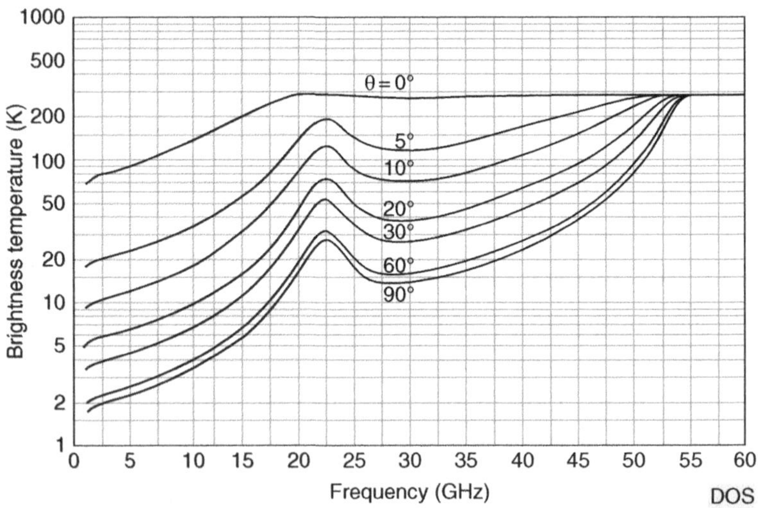

Figure 6.7 Sky temperature for clear air for 7.5 g/m^3 of water vapor concentration [20].

the radome. The noise temperature contribution due to scattering is dependent on the antenna elevation angle. Thus, for a typical 10deg. elevation angle, we assume that half of the scattered energy goes forward to the cold atmosphere and the other half backward. Furthermore, half of the backward energy is reflected by the antenna reflector back to the cold atmosphere. Under this assumption we obtain, $NT_3 = 75P_s$, where P_s is the amount of energy scattered by the radome beams.

6.3 Geometry Optimization

The total scattering effect of the antenna from all the beams in front of the aperture is computed by superposing all of the scattering effects from the individual beams. Optimization of the radome geometry takes into consideration factors such as minimization of the total beam length, increase in panel dimensions, minimization of the total number of parallel beams, and uniform beam density throughout the antenna/radome scan angles. Fig. 6.8 [13] shows the difference in the beams scattering level and its distribution for a radome geometry

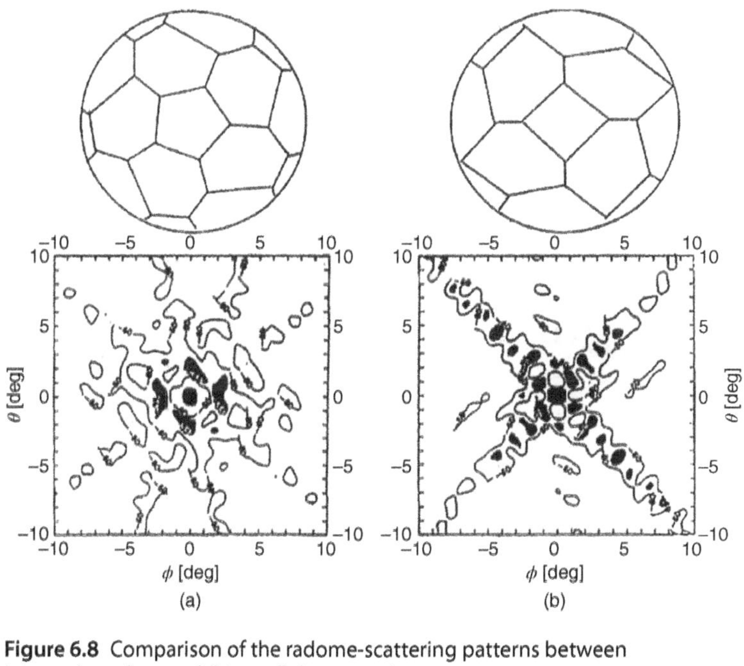

Figure 6.8 Comparison of the radome-scattering patterns between (a) quasi-random and (b) parallel geometries.

projected on a circular antenna aperture for the case of a parallel (bad) geometry, and its comparison with the distribution for the quasi-random (good) geometry.

One can observe that the scattered field for the case of the quasi-random geometry is more uniformly distributed, and has a lower intensity in comparison to that of the parallel geometry. Accordingly, the effect on the radiation characteristics of the antenna enclosed by the radome with quasi-random panelization is lower. Fig. 6.9 [13] shows the difference in the beams scattering level, and distributions for a quasi-random radome geometry with and without tuned beams.

The results in Fig. 6.9 demonstrate the significance of tuning the beams in a sandwich radome. Another important factor in the radiation characteristics of an enclosed antenna in a sandwich radome is the variability of the blockage by the beams, through the radome with the antenna aspect angle. In a well-designed radome, the blockage to the antenna is almost independent of the aspect angle of the antenna

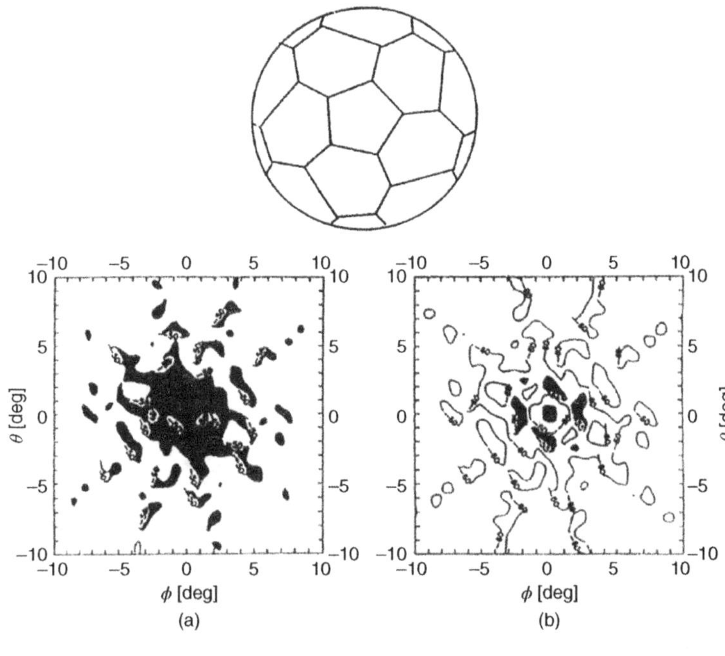

Figure 6.9 Comparison between the scattering patterns of a quasi-randomized geometry with (a) untuned and (b) tuned beams.

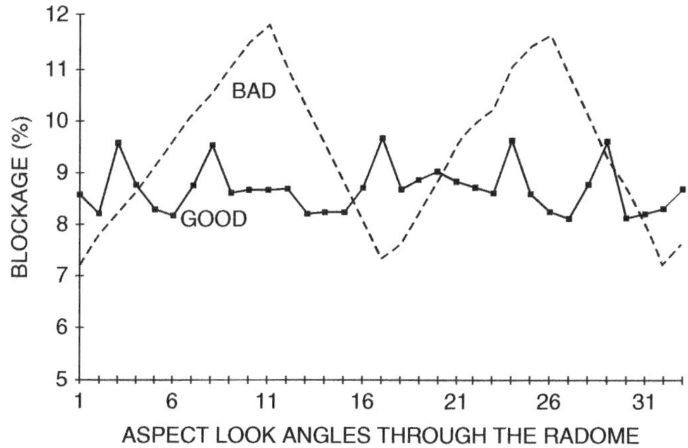

Figure 6.10 Beams blockage dependence on aspect angle for a "bad" and a "good" (uniform density) radome geometry.

relative to the radome, while large variability may be encountered in a badly designed radome. Fig. 6.10 [13] shows the variability of the beams blockage due to beams in a "good" and a "bad" sandwich radome design.

Scattering effects from the beams play an important role in the design of large ground-based radomes. To obtain a high-performance radome, one should consider both reducing the scattering from the individual beams and optimizing the beam geometry.

The scattering from the beams affects many radiation characteristics of the antenna and not just the transmission loss. In this chapter, we have reviewed the analytical and numerical tools to analyze these effects. The analysis shows that the DSF radome performs well for low frequencies, while the MSF radome has a wide frequency range, although its performance is limited. Finally, if we require a high-performance radome operating in a limited frequency range we should consider the use of a tuned sandwich radome.

6.4 Intermodulation Distortion in MSF Radomes

Large reflector antennas and metal space frame (MSF) radomes are fabricated from many panels and beams interconnected at joints with bolts and rivets. The contact at these junctions is not perfect and tiny gaps of air with the size of 10–100 angstroms are generated. These tiny gaps are filled, as time is passing, by metal oxide and metal-insulator-metal (MIM) junctions are generated. The MIM junctions constitute nonlinear circuit elements, due to the electron tunneling effect [21, 22] through the thin insulator gap, such that if illuminated by high RF power levels (larger than 10 mW/cm^2) they generate spurious signals. The electron tunneling effect is nonexistent if the gaps are out of the range of 10–100 angstroms. These nonlinear elements give rise to intermodulation products (IMP) signals above the receiver noise level, when subject to high-power microwave radiation by two or more transmitters used simultaneously at different frequencies, f_1 and f_2. In radar and satellite transceiver systems, the close proximity of the transmit and receive bands cause the intermodulation products (IMP) of certain transmitted frequencies to fall within the receive band. The third-order product, $(2f_1 - f_2)$ being the highest level is of most concern; however, higher-order products may also be troublesome, depending on transmit power levels, receiver sensitivities, and frequencies employed. These transmissions-generated IMP signals contaminate the receive band and may seriously degrade the signal-to-noise capability of the receiver.

The effect of the IMP in radar systems was first identified during World War II on systems with antennas that had rusty bolts, particularly antennas on ships at sea. At that time, it was referred as the "rusty-bolt" phenomenon. They did not really realize what it was at that time, but they knew that if they prevented the bolts from rusting, the IMP in the receiving system would be brought under control. The remedies proposed to reduce the IMP effect in such systems are:

- Cover suspicious junctures with highly conductive metal tape.
- Weld wherever possible to increase conductivity across possible air gaps.
- Use insulators with thickness larger than 100 angstroms to avoid the electron tunneling effect.

6.4.1 The IMP Effect in MSF Radomes

IMP signal level is an overall system design problem starting with the most fundamental consideration of the frequency spectrum allocations of the transmit and receive bands. If the frequencies and separations are properly chosen, IMP generation can be expected to be below practical system noise levels and out of the receiving band. In an MSF radome enclosed-antenna system, there are many contributors to IMP, including the following:

- Waveguide feed horn
- Subreflector
- Main reflector
- The MSF radome

In a well-designed system, the system engineer is well aware of the various contributions and the IMP power levels should be more than 200 dB below the level of the transmitted carriers in order to be discarded. In many cases, the IMP effect of the radome is negligible compared to the other sources. During the last years, various methods to eliminate the generated IMP have been investigated and implemented in MSF radomes:

- Separation of the MSF beams by insulating material
- Use of plastic sleeves and washers to isolate the radome hardware (bolts and nuts)
- Use of nylon nuts
- Special care of workmanship and cleanliness of manufacture to avoid sharp metal edges, micro-cracks in metal surfaces, metal scratches, metal burrs, imperfect welding, and impurities in plating

Fig. 6.11 shows a typical comparison between the measured third-order IMP signal levels of a standard MSF radome and a treated MSF radome [22]. The test data was acquired for two transmitting frequencies f_1 and f_2 illuminating the radome's junctions with power levels of 300 mW/cm^2.

One can observe that a reduction of up to 30 dB in the IMP distortion can be obtained using the treatment of the MSF radome.

An important method to measure the level and source origin of the IMP signals is the RF imaging technique [23]. Using this method, the IMP signals are measured in the near-field of the radome junction and are back-projected on a plane crossing the problematic junctions and panels. This process enables us to locate the suspicious regions that generate IMP signals and treat them appropriately. However, imaging has its limitations in terms of ultimate resolution that can be obtained $\approx\lambda/2$.

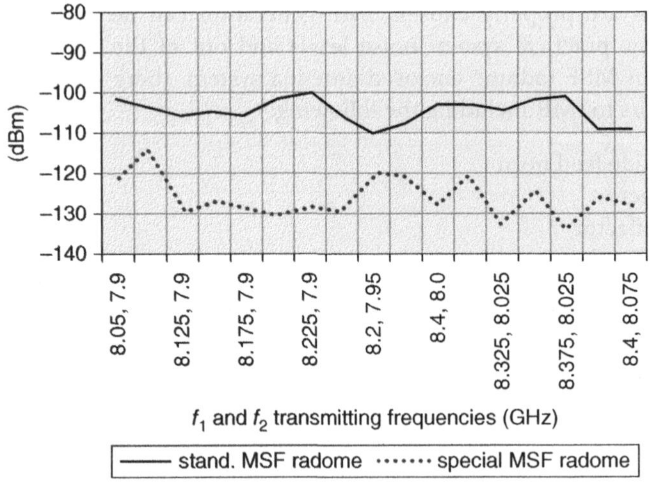

Figure 6.11 Third IMP signals power levels for a standard MSF radome and a special treated MSF radome [22].

References

1 Kay, AF. Electrical design of metal space frame radomes. *IEEE Trans. Antennas Propagat.*, 13(2), 188–202, 1965.
2 Kennedy, PD. An analysis of the electrical characteristics of structurally supported radomes. Ohio State University, Columbus, 1958.
3 Rusch, WVT, Appel-Hansen, J, Klein, CA, and Mittra, R. Forward scattering from square cylinders in the resonance region with application to aperture blockage. *IEEE Trans. Antennas Propagat.*, 24)(2), 182–189, 1976.
4 Vitale, JA. Large radomes. In *Microwaves Scanning Antennas*. New York: Academic Press, 1966.
5 Andreasen, MG. Scattering by Conducting Rectangular Cylinders. *IEEE Trans. Antennas Propagat.*, 12(6), 746–754, 1964.
6 Richmond, JH. Scattering by a Dielectric Cylinder of Arbitrary Cross Section Shape. *IEEE Trans. Antennas Propagat.*, 13(3), 334–341, 1965.
7 Chang, KC, and Smolski, AP. The effect of impedance matched radomes on SSR antenna systems. In *IEEE Conf. Proc. Radar 87*, London, 1987.

8 Smith, FC, Chambers, B, and Bennett, JC. Improvement in the electrical performance of dielectric space frame radomes by wire loading. In *ICAP 89*, Coventry, UK, 1989.

9 Michielssen, E, and Mittra, R. RCS reduction of dielectric cylinders using a simulated annealing approach. In *IEEE APS*, Dallas, Texas, 1990.

10 Michielssen, E, and Mittra, R. TE plane wave scattering by a dielectric cylinder loaded with perfectly conducting strips. In *IEEE APS*, Dallas, Texas, 1990.

11 Katashaya, SR, and Evans, BG. Depolarization properties of metal space frame radomes. *Int'l Journal of Satellite Comm.*, 2, 61–72, 1984.

12 Chang, KC, and Smolski, AP. A radome for air traffic control SSR radar systems. In *IEEE Conf. Proc. Radar 87*, London, UK, 1987.

13 Shavit, R, Smolski, AP, Michielssen, E, and Mittra, R. Scattering analysis of high performance large sandwich radomes. *IEEE Trans. Antennas Propagat.*, 40(2), 126–133, 1992.

14 Vironc, G, Tascone, R, Addamo, G, and Peverini, O. A. A design strategy for large dielectric radome compensated joints. *IEEE Antennas and Wireless Propagat. Lett.*, 28(7), 1257–1268, 2009.

15 Mishra, S., Sarkar, M, and Daniel, A. Optimization of radome wall and joint for X-band reflector antenna using floquet modal analysis. *Journal of Electromagnetic Waves and Applications*, 28(7), 1257–1268, 2014.

16 Cornbleet, S. Microwave optics. New York: Academic Press, 1976.

17 Goldsmith, P. F. Quasi-optical techniques at millimeter and submillimeter wavelengths. In *Infrared and Millimeter Waves*. New York, New York Academic, 1982.

18 Skolnik, M. Radar handbook. New York: John Wiley & Sons, 1990.

19 Balanis, CA. Antenna theory—analysis and design. Hoboken, NJ: John Wiley & Sons, 2005.

20 Flock, WL, and Smith, EK. Natural radio noise—A mini-review. *IEEE Trans. Antennas Propag.*, 32(7), 762–767, 1984.

21 Higa, WH. Spurious signals generated by electron tunneling on large reflector antennas. *Proceedings of the IEEE*, 63(2), 306–313, 1975.

22 Ray, CE, and Exum, NL. IM Product Abatement in the DSCS Heavy Terminal. In *AP-S International Symposium*, Vancouver, 1985.

23 Aspden, PL, Anderson, AP, and Bennett, JC. Microwave holographic imaging of intermodulation product sources applied to reflector antennas. In *ICAP*, London, 1989.

Problems

P6.1 A 10-ft circular aperture antenna with uniform field distribution, vertical polarization, and operating at 5.6 GHz is enclosed in a spherical space frame radome. The beams' cross section is 0.4×2 in.^{2}, with the narrow side facing the antenna aperture. The sketch shows the radome's beam projections on the antenna aperture.

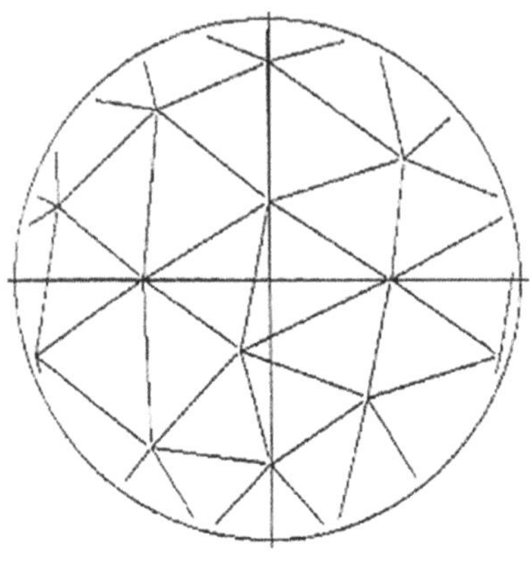

Assume that the radome is a metal space frame with $IFR_{e,h} = -1+j0$.

(a) Compute and plot the far-field radiation pattern of the antenna without radome, the scattering pattern and the antenna radiation pattern with radome (neglect insertion loss due to radome panels). What is the insertion loss and BSE

(b) Compute the insertion loss and BSE using the approximate eqs. (6.12) and (6.15).

P6.2 Repeat P6.1, but assume that the radome's beams are made of dielectric material with $\varepsilon_r = 4.6$, $\tan\delta = 0.02$ and find the $IFR_{e,h}$ of a single beam using your code for numerical solution, or use for this purpose a commercial EM simulator.

(a) Compute and plot the far-field scattering pattern and the antenna radiation pattern with and without the radome (neglect insertion loss due to radome panels). What is the insertion loss and the BSE in this case?
(b) Compute the insertion loss and BSE using the approximate eqs. (6.12) and (6.15).

P6.3 Repeat P6.2, but assume that at the center of the dielectric beam were inserted 8 conductive (PEC) parallel strips with width 0.062 in. and center-to-center spacing of 0.277 in. Find the $IFR_{e,h}$ of a single tuned beam using your code for numerical solution or use for this purpose a commercial EM simulator.
(a) Compute and plot the far-field scattering pattern and the antenna radiation pattern with and without the radome (neglect insertion loss due to radome panels). What is the expected insertion loss and the BSE?
(b) Compute the insertion loss and BSE using the approximate eqs. (6.12) and (6.15).

P6.4 Repeat P6.3, but with the wide side of the beam facing the antenna aperture.

P6.5 Discuss the differences in the radome performance if the space frame is made of metal, dielectric, or tuned dielectric.

P6.6 For problem P6.1, compute the antenna noise temperature of the antenna and antenna with radome for an antenna pointing at 20 deg elevation.

7

Measurement Methods

Radomes in general and large ground-based radomes in particular are composed of two main components: panels and beams. Therefore, to closely monitor the design and fabrication process, these two components need to be individually characterized and measured. Moreover, at the end of the design process, the performance of the entire radome is also evaluated by measuring the far-field radiation pattern of the enclosed antenna covered by the radome for different antenna orientations, dual polarizations, and in the entire frequency band. Usually, a comparison is made to the radiation pattern of the antenna without the radome. The panel characterization procedure is mainly focused to determine the transmission and reflection through the panel and is described in section 7.1. This measurement determines the accuracy of the simulation and the assumptions made for the electrical parameters of the materials used in the multilayer structure of the panel.

Accurate characterization of the scattering properties of dielectric and metal beams is essential in the electromagnetic analysis of large sandwich and dielectric/metal space frame radomes assembled from a large number of beams and panels. The fundamental basis for such an analysis is the knowledge of the scattering characteristics from an individual beam located and oriented at a specified point in space and illuminated by a plane wave propagating in a particular direction. Computation of the scattering parameters of an arbitrary cross section beam is described in Chapter 5 and their measurement is described in section 7.2. Rusch [1] suggested an experimental procedure to measure the *IFR* value in the far field for a cylinder with arbitrary cross section, as described in section 7.2.1. This procedure lacks the capability to measure the scattering pattern and lacks the ability to filter out the reflections from surrounding objects, which

Radome Electromagnetic Theory and Design, First Edition. Reuven Shavit.

compromise the measurement accuracy. The near-field measurement method described in section 7.2.2 bridges this gap and enables us to evaluate both *IFR* and the scattering pattern. The focused beam system described in section 7.2.3 enables us to measure both *IFR* and the scattering pattern and, in addition, is immune to errors caused by reflections in the measurement setup system.

7.1 Panel Measurements

Fig. 7.1 shows the test setup to measure the transmittance (amplitude and phase) through a sandwich panel sample. It consists of transmitting and receiving antennas and the panel crossing their line of sight. The illuminated panel is located in the far field of the transmitting and receiving antennas. The system is calibrated without the panel and in the next step the panel is introduced crossing the line connecting the two antennas. Before the test begins, it is important to verify that the panel size and the distances between the transmitting and receiving antennas to the panel are appropriate to minimize illumination of the panel edges and avoid diffraction that may add measurement noise and affect the measurement accuracy.

Usually, the measurement is performed for dual polarizations (parallel and perpendicular), at different incident angles and in the entire frequency band. The experimental data is usually compared to the simulation results performed with the methods described in Chapters 2 and 3 for validation purposes. This type of measurement can be also

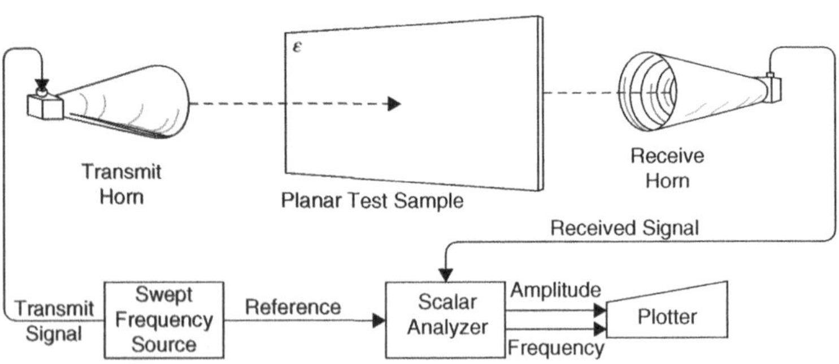

Figure 7.1 Panel measurement setup.

used to determine the electrical parameters (ε_r, tan δ) of the materials utilized in the construction of the sandwich panel, if a one-layer and homogeneous panel is considered. In this case the transmittance data (amplitude and phase), the thickness of the sample and the frequency of operation are used to determine the electrical parameters of the tested panel. The measured transmittance data is compared to the simulated data and the electrical parameters (ε_r, tan δ) are varied until minimum mean square error (MMSE) is obtained between the measured and simulated transmittance data. The evaluation of the material electrical parameters based on the measured data is described in [2].

7.2 Characterization of Forward-Scattering Parameters

Accurate characterization of the scattering properties of dielectric and metal beams is essential in the electromagnetic analysis of large sandwich and dielectric/metal space frame radomes assembled from a large number of beams or panels. These beams or panel joints introduce scattering effects, which modify the radiation characteristics of the antenna enclosed in the radome. The total scattering effect of the radome may be evaluated as the summation of the scattered fields from all beams or panel joints illuminated by the antenna (in a transmit mode analysis), as described in Chapter 6. The fundamental basis for such a computation is the knowledge of the scattering characteristics from an individual beam located and oriented at a specified point in space and illuminated by a plane wave propagating in a particular direction. Therefore, the ability to compute and measure the scattering characteristics of an arbitrary shaped cylinder is essential to the entire scattering analysis. Kay [3] was the first to introduce the concepts of *IFR* (induced field ratio) and scattering pattern to characterize the scattering effect of the beams on the MSF radome electromagnetic performance. The *IFR* is equal to the ratio of the maximum forward far-field scattering field to the hypothetical forward maximum radiated far field from a 2D aperture with a width equal to the optical projected width of the beam on the incident plane wave front [4]. Low values of *IFR* are indicative of low scattering effect in the forward direction. In his analysis Kay [3] disregarded the cross-polarized scattering field from the beams and neglected the variation of the co-polarized scattering pattern for obliquely incident plane waves. This missing information can be recovered using the suggested experimental methods. The full scattering analysis for an arbitrary shaped dielectric beam illuminated by an oblique incident plane wave was first formulated by Rojas [5]. The analysis shows the existence of cross-polarized scattering field for oblique incident angles, absent at normal incidence.

Analytical computations of the scattering characteristics for cylinders can be performed for canonical cross sections such as circular and elliptical, but for arbitrarily shaped cylinders, numerical computations using techniques such as the MoM [6] and the FEM [7] are required and are described in Chapter 5. These numerical techniques are in many cases numerically intensive. Consequently, in many practical instances to reduce development cycle times, control manufacturing processes, and verify numerical predictions, an accurate method to determine experimentally the *IFR* and the scattering patterns (both co-polarized and cross-polarized) of the cylinder is required.

7.2.1 Far-Field Probing

The *IFR* measurement of a cylindrical beam using the far-field probing measurement technique was developed by Rusch [1] and is described in the following. Fig. 7.2 describes the geometry of the proposed measurement setup:

H_1 is the transmitting horn, and H_2 is the receiving horn. The illuminated arbitrarily shaped cylinder is located in the far field of the transmitting and receiving antennas. The cylinder, mounted on a positioner track, moves perpendicularly to the line connecting the transmitting and receiving horns. The position of the cylinder along the x-axis and the amplitude and phase of the received signal are recorded. The system is calibrated before the cylinder movement, and the recorded signal $R_{z,x}(x)$ is referenced to this value. The subscripts z and x stand for vertical and horizontal polarizations of the incident field, respectively. Neglecting edge currents on the top and bottom of the cylinder, the scattered field is given by [1]:

$$\underline{E}^s(0, y_0, 0) = -\frac{j\omega\mu}{4\pi}\hat{z}\int_{-L/2}^{L/2}\oint_{S_l} J_{sz}\frac{e^{-jk_0 r}}{r}\,dl\,dz \tag{7.1}$$

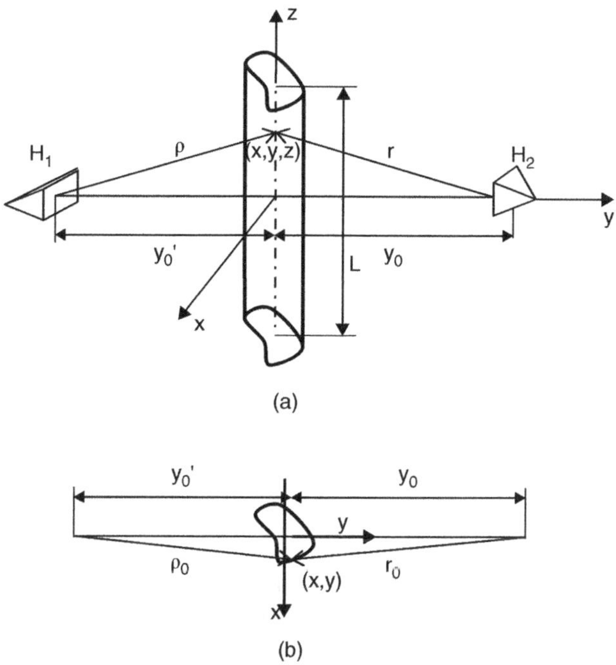

Figure 7.2 The geometry of the proposed far-field probing measurement setup: (a) side view; (b) top view.

where its parameters as shown in Fig. 7.2 are:

$$\rho = \sqrt{x^2 + (y_0' + y)^2 + z^2} = \sqrt{\rho_0^2 + z^2}$$
$$r = \sqrt{x^2 + (y_0 - y)^2 + z^2} = \sqrt{r_0^2 + z^2}$$
$$\rho_0 = \sqrt{x^2 + (y_0' + y)^2}$$
$$r_0 = \sqrt{x^2 + (y_0 - y)^2} \tag{7.2}$$

and J_{sz} is the induced current on the cylinder. Evaluating the z-integral in eq. (7.1) using saddle-point technique [8] yields (assuming the cylinder to be long in terms of the wavelengths and neglecting the end effects due to low illumination):

$$\underline{E}^s(0, y_0, 0) = -\frac{j\omega\mu}{4\pi} e^{-j\pi/4} \sqrt{\frac{2\pi}{k_0}} \hat{z} \oint_{S_l} \frac{J_{sz}}{(e^{-jk_0\rho_0}/\rho_0)} \frac{e^{-jk_0(\rho_0 + r_0)}}{\sqrt{\rho_0 r_0(\rho_0 + r_0)}} dl \tag{7.3}$$

If the cylinder is in the far field of the horns and vice versa (with respect to maximum width of the cylinder), the following approximations can be made. In phase expressions

$$\begin{aligned}\rho_0 + r_0 &\cong y_0 + y_0' \\ \rho_0 &\cong y_0' + y\end{aligned} \tag{7.4}$$

and in amplitude approximations,

$$\begin{aligned}\rho_0 &\cong y_0' \\ r_0 &\cong y_0\end{aligned} \tag{7.5}$$

Then

$$\underline{E}^s(0, y_0, 0) \cong \hat{z}\frac{e^{j\pi/4} w E_0' e^{-jk_0(y_0+y_0')}}{\sqrt{\lambda y_0 y_0'(y_0 + y_0')}} IFR_e \tag{7.6}$$

in which

$$IFR_e = \frac{\eta_0}{2wE_0'(e^{-jk_0 y_0'}/y_0')}\oint_{S_t} J_{sz} e^{jk_0 y}\, dl \tag{7.7}$$

where w is the optical blockage of the cylinder on the incident field wavefront. E_0' is the amplitude of the electric field generated by the transmitting horn H_1. Accordingly, the total field at the receiving port of H_2 horn in the presence of the cylinder is

$$\underline{E}^T(0, y_0, 0) = \underline{E}^{inc}(0, y_0, 0)\left[1 + we^{j\pi/4} IFR_e \sqrt{\frac{y_0 + y_0'}{\lambda y_0 y_0'}}\right] \tag{7.8}$$

Moreover,

$$\underline{E}^{inc}(0, y_0, 0) = \frac{E_0' e^{-jk_0(y_0+y_0')}}{y_0 + y_0'}\hat{z} \tag{7.9}$$

is the incident field from H_1 to H_2 in the absence of the cylinder. The amplitude change $\Delta\alpha$ and the phase change $\Delta\phi$ in the incident field on H_2 is measured when the cylinder is inserted between H_1 and H_2. Thus, $\underline{E}^{inc}(0, y_0, 0)e^{-\Delta\alpha}e^{-j\Delta\phi}$ denotes the total field $\underline{E}^T(0, y_0, 0)$, which is then resolved from (7.8) to yield

$$IFR_e = (e^{-\Delta\alpha}e^{-j\Delta\phi} - 1)\frac{e^{-j\pi/4}}{w}\sqrt{\frac{\lambda y_0 y_0'}{y_0 + y_0'}}. \tag{7.10}$$

An identical expression can be derived to determine IFR_h.

Fig. 7.3 shows the comparison of $IFR_{e,h}$ as a function of w/λ computed numerically with MoM as described in Chapter 5 and processed

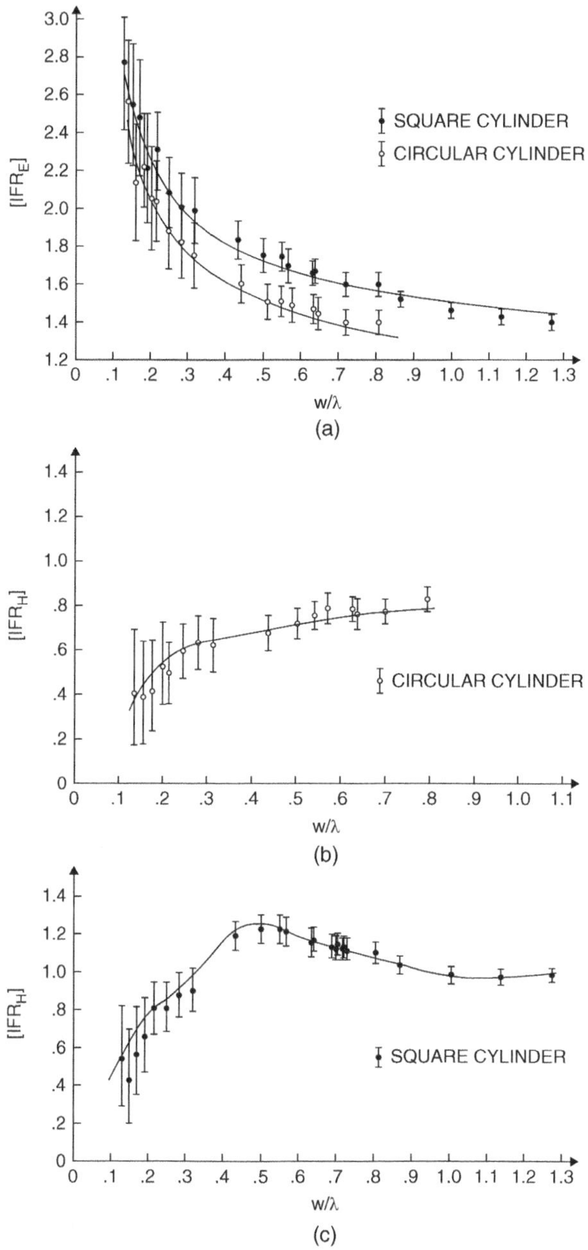

Figure 7.3 Comparison of $IFR_{e,h}$ as a function of w/λ computed and processed through measurements using the far-field probing technique: (a) IFR_e for circular and square PEC cylinder; (b) IFR_h for circular PEC cylinder; (c) IFR_h for square PEC cylinder [1].

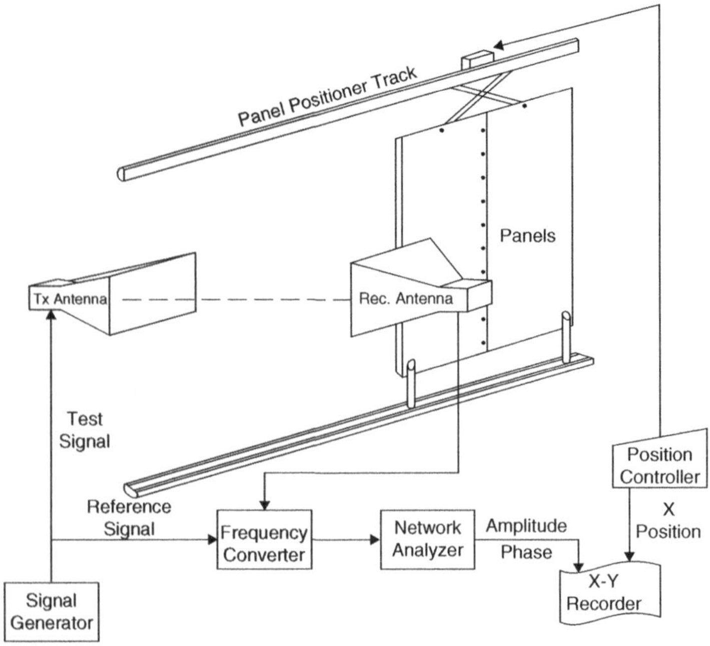

Figure 7.4 The measurement setup for panel/seam test.

through measurements using the far-field probing technique, for circular and square PEC cylinders as described in [1]. One can observe a nice agreement between the simulations and the experimental results.

A practical implementation of the setup presented in Fig. 7.2, for beams as well as a seam connecting two adjacent sandwich panels, that avoids multiple reflections measurement errors on the system axis is shown in Fig. 7.4.

The two panels, with a seam in-between, are moved on a track in front of a pair of receiving and transmitting horns. The amplitudes and phases variation of the receiving signal are recorded throughout the panel movement. Using the measured phase and amplitude perturbation in the seam region, one can compute the *IFR* of the seam by using the formula given in eq. (7.10). Typical recordings of signals (amplitude and phase) for both the parallel (TM case) and perpendicular (TE case) polarizations for untuned (dielectric beam) and tuned (dielectric beam with conductive strips), as described in section 5.4, are shown in Fig. 7.5.

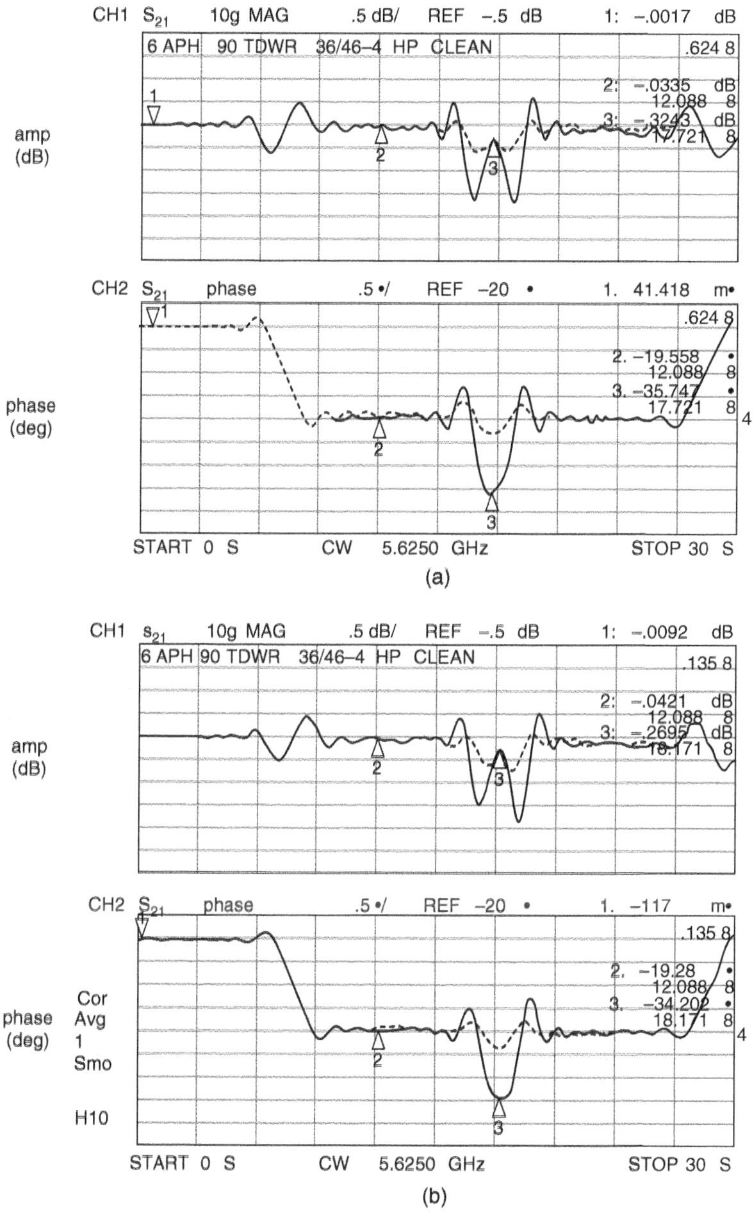

Figure 7.5 Measured *TL* and *IPD* for tuned (- - - -) and untuned (———) seam: (a) vertical polarization; (b) horizontal polarization.

One can observe the reference signal before the panels cross the boresight line between the transmitting and receiving horns; the diffraction effect when the panels begin to cross this line; the insertion phase delay (*IPD*) and the transmission loss (*TL*) of the panel; and the phase and amplitude perturbations in the area of the seams. The recorded signal is symmetric with respect to the seam axis. The perturbation of the amplitude and phase in the seam region is significantly reduced for the tuned case, resulting in a lower forward scattering level (*IFR*). Moreover, one can observe a significant reduction in the *IPD* introduced by the seam for the tuned case.

7.2.2 Near-Field Probing

Rusch [1] developed the *IFR* concept and suggested an experimental procedure to measure its value in the far field for an arbitrarily shaped cylinder. This procedure lacks the capability to measure the beam-scattering pattern. In this section, a combined experimental and numerical procedure based on near-field probing to determine the scattering characteristics of a composite beam is presented in [10]. The accuracy of the *IFR* value is comparable to that measured in the far field by Rusch's method, but in addition, it provides the scattering pattern. A similar measurement technique based on the standard near-field probing technique, but for a stationary cylinder in front of a large phased array and a moving probe, was suggested in [11].

The schematic configuration of the measurement setup is shown in Fig. 7.6.

The transmitting antenna is a horn with aperture dimensions $A \times B$ and linear polarization (parallel or perpendicular to the cylinder axis). The illuminated arbitrarily shaped cylinder is located in the far field of the transmitting antenna. An open waveguide with aperture dimensions $a \times b$ probes the near field in the vicinity of the cylinder along x-axis at a distance z_0. The distance z_0 is large enough ($>3\lambda$) to avoid multiple reflections between the probe and the cylinder. The probe is stationary and the cylinder, mounted on a positioner track, moves in front of the probe. The position of the cylinder along the x-axis and the amplitude and phase of the received signal are recorded. The system is calibrated before the cylinder movement, and the recorded signal $R_{y,x}(x,z_0)$ is referenced to this value. The subscripts y and x stand for vertical and horizontal polarizations of the incident field, respectively. The advantage of the suggested procedure instead of the

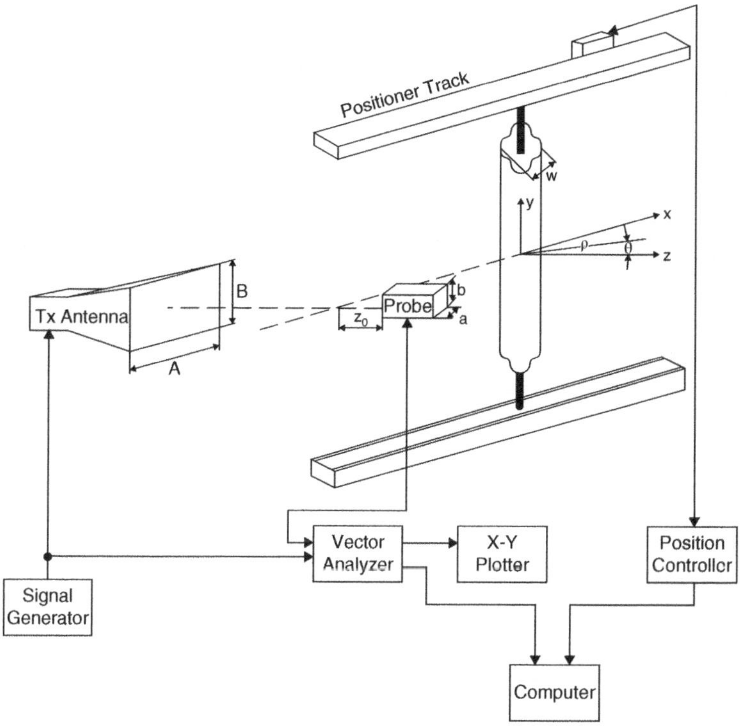

Figure 7.6 Schematic configuration of the near-field measurement setup.

standard procedure in which the probe moves in front of the cylinder, is that it keeps the RF cables stationary and does not introduce measurement errors throughout their motion as encountered in the standard near-field procedure. Throughout the cylinder movement the incident field angle varies, however, for close proximity of the probe to the cylinder surface, the error is minimal and affordable.

Without loss of generality, we can assume that the incident field is y directed. Moreover, we assume that the cylinder is infinite in y direction; therefore, all fields are only x and z dependent. The measured scattered field $E_y^s(x,z)$ by the cylinder can be expressed in terms of its plane wave spectrum $E_y^s(k_x)$ and the plane-wave spectrum of the probe $\widetilde{P}_y(k_x)$ [10]:

$$E_y^s(x,z) = \int_{-\infty}^{\infty} \widetilde{P}_y(k_x)\widetilde{E}_y^s(k_x)e^{-jk_x x}e^{-jk_z z}dk_x \tag{7.11}$$

where $k_z = \sqrt{{k_0}^2 - k_x^2}$ and k_0 is the propagation constant in free space. The Fourier inversion of (7.11) gives the scattered plane wave spectrum of the cylinder in terms of the measured scattered near-field $E_y^s(x, z_0)$

$$\widetilde{E}_y^s(k_x) = \frac{e^{jk_z z_0}}{2\pi} \widetilde{P}_y^{-1}(k_x) \int_{S_a} E_y^s(x, z_0) e^{jk_x x} dx \tag{7.12}$$

where S_a is the aperture domain along x-axis on which the near-field is probed. Given the plane wave spectrum $\widetilde{E}_y^s(k_x)$, we can compute the scattered far-field $E_\phi^s(\rho, \theta)$, for $\rho \to \infty$ using the steepest descent path method [8]. In the far-field limit $k_x = k_0 sin\theta$ and $k_z = k_0 cos\theta$ in which θ is shown in Fig. 7.6. The result of such derivation is

$$E_\phi^s(\rho, \theta) = 2\pi \sqrt{\frac{jk_0}{2\pi\rho}} e^{-jk_0\rho} \cos\theta \widetilde{E}_y^s(k_0 \cos\theta) \tag{7.13}$$

The plane-wave spectrum of the open waveguide probe is computed in [8], with the result being

$$\widetilde{P}_y(\theta) = \cos\theta \frac{\pi^2 \cos\left(\frac{\pi a}{\lambda} \sin\theta\right)}{\pi^2 - 4\left(\frac{\pi a}{\lambda} \sin\theta\right)^2} \tag{7.14}$$

Substitution of (7.14) into (7.12) and substituting the result into (7.13) yields the scattered far-field pattern $F_y^s(\theta)$ of the cylinder in terms of the near-field data for vertical polarization:

$$F_y^s(\theta) = \frac{\pi^2 - 4\left(\frac{\pi a}{\lambda} \sin\theta\right)^2}{\pi^2 \cos\left(\frac{\pi a}{\lambda} \sin\theta\right)} \int_{S_a} E_y^s(x, z_0) e^{jk_0 x \sin\theta} dx \tag{7.15}$$

By a similar derivation, the appropriate plane wave spectrum of the open waveguide probe for perpendicular polarization using Yaghian model [12] is

$$\widetilde{P}_x(\theta) = \frac{1 + \frac{\beta}{k} \cos\theta}{1 + \frac{\beta}{k}} \frac{\sin\left(\frac{\pi b}{\lambda} \sin\theta\right)}{\frac{\pi b}{\lambda} \sin\theta} \tag{7.16}$$

in which β is the propagation constant in the open waveguide for the dominant TE_{10} mode, we obtain the scattering pattern $F_x^s(\theta)$ for horizontal polarization

$$F_x^s(\theta) = \frac{1 + \frac{\beta}{k}}{1 + \frac{\beta}{k} \cos\theta} \frac{\frac{\pi b}{\lambda} \sin\theta}{\sin\left(\frac{\pi b}{\lambda} \sin\theta\right)} \int_{S_a} E_x^s(x, z_0) e^{jk_0 x \sin\theta} dx \tag{7.17}$$

The integrals in (7.15) and (7.17) are Fourier type and can be evaluated efficiently by the FFT algorithm. The approximate forward scattering pattern, $F_a(\theta)$, is equal to the radiation pattern from an equivalent aperture with the cylinder shadow width, w and with constant field distribution. Its analytical form is [13]

$$F_a(\theta) = \frac{\sin\left(\pi \frac{w}{\lambda} \sin\theta\right)}{\pi \frac{w}{\lambda} \sin\theta} \tag{7.18}$$

The $IFR_{e,h}$ is defined as the ratio of the forward scattered far-field $E^s_{y,x}(\rho, 0)$ to the hypothetical far-field $E^r(\rho, 0)$ radiated in the forward direction by the plane wave E^{inc} in the reference aperture of width w equal to the shadow of the geometrical cross section of the cylinder on the incident wavefront [4]. The hypothetical field $E^r(\rho, \theta)$ can be obtained from eqs. (7.12) and (7.13) with E^{inc} as the aperture field. Following these steps, we obtain the following:

$$E^r(\rho, 0) = E^{inc} w \sqrt{\frac{jk_0}{2\pi\rho}}\, e^{jk_0 z_0} e^{-jk_0\rho} \tag{7.19}$$

The ratio of (7.13) for $\theta = 0$ and (7.19) gives the $IFR_{e,h}$ of the cylinder in terms of the near-field data:

$$IFR_{e,h} = \frac{1}{w} \int_{S_a} \frac{E^s_{y,x}(x, z_0)}{E^{inc}} dx \tag{7.20}$$

Fig. 7.7 shows typical recorded near-field signals (amplitude and phase) for both vertical polarization (VP) and horizontal polarization (HP) in the course of a metal circular cylinder with 0.75 in. diameter motion. The operating frequency is 10 GHz and the distance between the probe and cylinder is $z_0 = 3.5$ in.

The abscissa represents the relative distance along x-axis from the line connecting the transmitting horn and the probe. One can notice the level of the reference signal, $R_{y,x}(x,z_0)$ before the cylinder crosses this line and the diffraction effect, when the cylinder crosses this line. The recorded signal is symmetric with respect to the boresight line. Based on the near-field data accumulated as shown above, one can compute the *IFR* and the scattering pattern. The result of such calculation is illustrated in Fig. 7.8. For comparison are also shown the approximate scattering patterns described in eq. (7.18) and the analytical scattering pattern for the circular cylinder derived in [14]. For vertical polarization (VP), the analytical scattering pattern is

$$E^s_z \sim \sum_{n=-\infty}^{\infty} \frac{J_n(kw/2)}{H^{(2)}_n(kw/2)} e^{jn\theta} \tag{7.21}$$

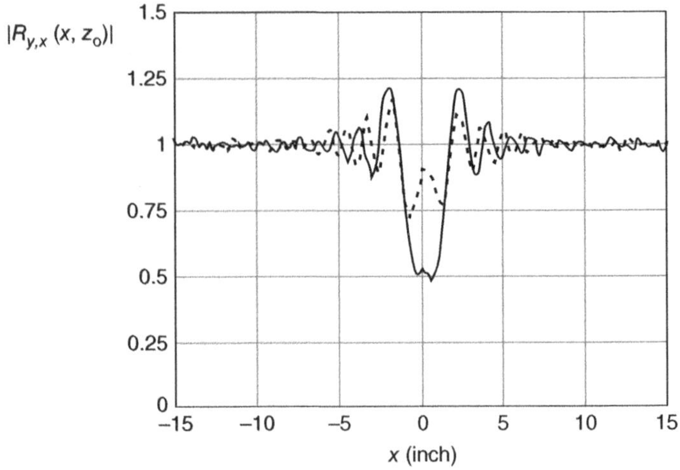

(a)

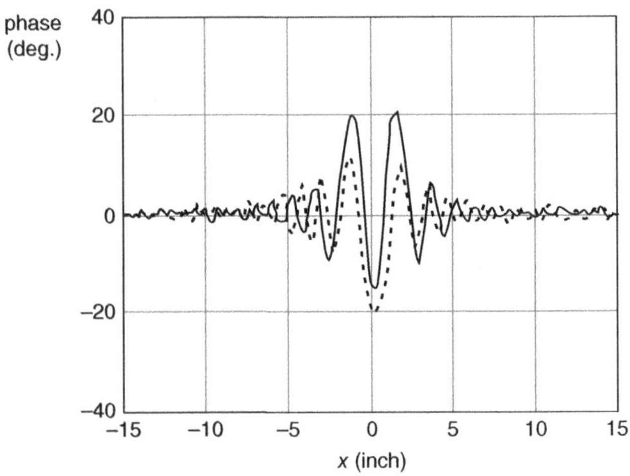

(b)

Figure 7.7 Recorded signal throughout the 0.75″ diameter metal cylinder motion (——) VP, (– – – ·) HP: (a) amplitude; (b) phase.

and for horizontal polarization (HP), the analytical scattering pattern is

$$H_z^s \sim \sum_{n=-\infty}^{\infty} \frac{J_n'(kw/2)}{H_n^{(2)\prime}(kw/2)} e^{jn\theta} \tag{7.22}$$

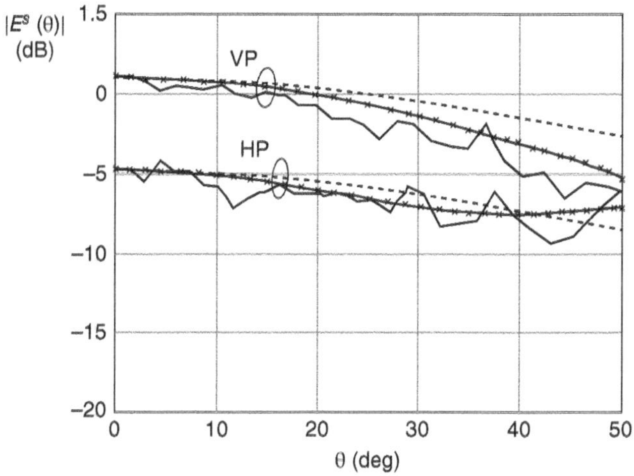

Figure 7.8 Scattering pattern of a metal cylinder with diameter 0.75″ (—) based on near-field data, (-★-★-) analytical solution eq. (7.21), (- - - -) approximate form eq. (7.18).

in which $J_n\,(x)$ and $H_n^{(2)}\,(x)$ are the Bessel function of nth order and Hankel function of second type and n^{th} order, respectively and $J_n'\,(x)$, $H_n^{(2)\prime}$ denote their derivatives.

In the forward scattering direction ($\theta = 0$) the amplitude of the scattered field is equal to $E^s(0) = 20\log|IFR{\cdot}w/\lambda|$. In the computation based on the near-field data $N = 256$ sampling points were used and the spacing between sampling points was less than the Nyquist criterion $\lambda/2$. One can observe a good agreement between the computed pattern based on near-field data and the analytical solution.

Table 7.1 compares the *IFR* values at $\theta = 0$ deg based on the accumulated near-field data to the *IFR* far-field values measured by Rusch's method [1] and the exact analytical value. The metallic beam was circular with 0.75 in. diameter, and the plastic beam was made from plastic with $\varepsilon_r = 4.2$ and dimensions 0.66×2.25 in.^{2} with the incident wave illuminating normally its wide side. The operating frequency was 10 GHz.

The metallic beam *IFR* analytical value was computed using the derivation in [14]. The plastic beam *IFR* computed value was computed using MoM, as explained in Chapter 5. We can observe a good agreement among the measured *IFR* values by both methods and

Table 7.1 Comparison of *IFR* values for metallic and plastic beams

	Beam Type	Computed	Measured Near Field	Measured Far Field
Vertical Polarization	Metallic circular beam-0.75" diam.	−1.31 + j0.55	−1.37 + j0.64	−1.25 + j0.52
	Plastic rectangular beam-0.66"×2.25"	−0.73 − j0.34	−1.03 − j0.81	−0.97− j0.18
Horizontal Polarization	Metallic circular beam-0.75" diam.	−0.68 − j0.33	−0.72 − j0.26	−0.69 − j0.36
	Plastic rectangular beam-0.66"×2.25"	−1.71 + j0.93	−2.25 + j0.41	−1.8 + j0.49

the computed values. The proposed near-field approach verifies well with the exact scattering characteristics computed analytically and numerically by the MoM.

7.2.3 Focused-Beam System

In this section, a combined experimental and numerical procedure to determine the scattering characteristics of arbitrarily shaped cylinders using a focused-beam system is presented. In the proposed system, the *IFR* is computed based on measurement data and the scattering pattern is directly measured in contrast to the method suggested in [1], which requires post processing of the measured data. The focused-beam system has a unique feature that minimizes measurement errors due to specular and diffuse reflections from adjacent objects, in contrast to open measurement systems.

The unique characteristics of the focused beam measurement system provides a tightly constrained field distribution at the beam waist, normally midway between the transmitting and receiving ends, which facilitates the separation of the incident field from the scattered field by a cylinder inserted in such a system at the beam waist. This feature significantly improves the accuracy of the measured scattering parameters of the tested cylinder. Moreover, this new method also provides a reduction of the measurement errors associated with specular and diffuse reflections from scatterers in the vicinity of the test facility in contrast to the measurement errors present in previous systems [1].

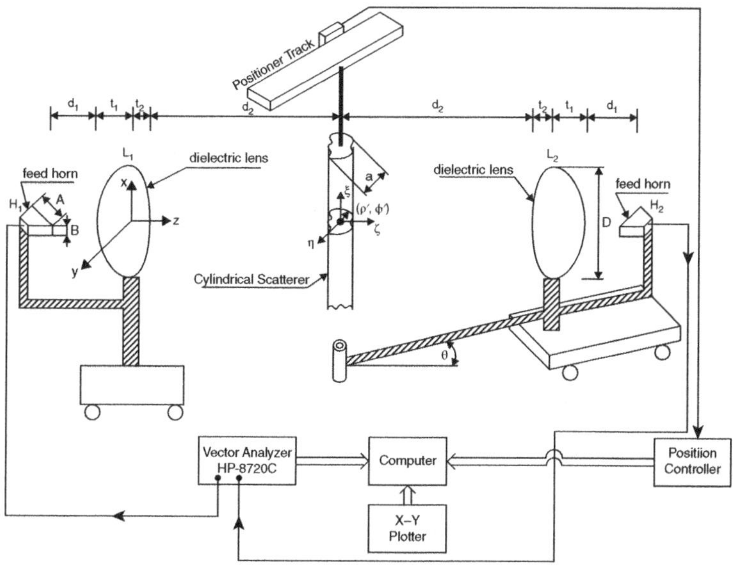

Figure 7.9 Schematic configuration of the focused beam system.

The scattering analysis presented is that of an arbitrarily shaped cylinder cross section measured in a focused beam system for oblique incidence. The analysis presented extracts the *IFR* and the co-polarized/cross-polarized scattering patterns of the cylinder, which is a unique feature of this method.

A schematic form of the focused beam measurement system is shown in Fig. 7.9.

The system is comprised of two identical axially symmetric dielectric lenses L_1 and L_2 with aperture diameters D, and two identical linearly polarized feed horns H_1 and H_2. Each lens is designed so that its two foci are located at distances f_1 and f_2 from the lens surface on opposite sides of the lens. The distance of the feed horn phase center from the lens surface is d_1. In principle, the system is configured such that $f_1 = d_1$; however, due to movement of the feed horn phase center with frequency, and the fact that the location of the Guassian beam waist on the other side of the lens will be frequency dependent, this requirement is not perfectly achieved at all frequencies.

The energy radiated by the transmitting feed horn H_1 is captured by the lens L_1, and is focused on the opposite side of the lens to the common focal point of the two lenses at a distance d_2 from the lens surface, coupled to the receiving lens L_2 and refocused into the feed horn H_2. Accordingly, feed horn H_1 and lens L_1 will be designated as the transmitting antenna, while feed horn H_2 and lens L_2 will be designated as the receiving antenna. The focal plane of the system is defined as the plane passing through the internal (common) focal point of the two lenses and perpendicular to the system axis. This plane also represents the approximate location of the beam waists of the Gaussian beams of the two antenna systems. In the measurement to be described, the center of the cylindrical scatterer is located in this focal plane, and its axis makes an angle θ_0 with the system axis, as shown in Fig. 7.9.

In the analysis, we make the assumption that the propagation mechanism of the focused beam system can be described by fundamental Gaussian beams [16]. Accordingly, the divergent Gaussian beam of the transmitting feed horn H_1 (cf. Fig. 7.9) is transformed by the transmiting lens to a convergent Gaussian beam with minimum waists w_{0x} and w_{0y} in the internal focal plane. Due to symmetry considerations, the convergent Gaussian beam is transformed into a divergent Gaussian beam captured by the receiving lens, and subsequently transformed into a convergent Gaussian beam captured by the receiving feed horn H_2. In the internal focal plane, we obtain a minimum Gaussian beam waist with constant phase distribution.

Initially, before the cylinder is moved into the focal plane the co-polarized coupling between the two focused beams is recorded (amplitude and phase) to provide a reference value. The tested cylinder is then positioned in the focal plane, inclined an angle θ_0 relative to the system axis and the coupling between the two focused beams normalized to the no cylinder case, $R(y) = \Delta\alpha\, e^{j\Delta\phi}$ is recorded ($\Delta\alpha$ is the relative amplitude and $\Delta\phi$ is the relative phase). Then, the receiving lens L_2 and its feeding horn H_2 are rotated in predetermined angular increments on an arch with its center of rotation located in the focal plane. For each angular point two readings are taken with and without the cylinder. The procedure is repeated for the cross-polarized case with the receiving feed horn H_2 rotated 90 deg around its axis [17].

The x and z coordinates of the lens two hyperbolic contours shown in Fig. 7.9 can be computed by geometrical optics and are described in [15]:

$$x_{1,2} = \sqrt{(n^2-1)(z_{1,2}+t_{1,2})^2 + 2(n-1)f_{1,2}(x_{1,2}+t_{1,2})} \tag{7.23}$$

in which the subscripts 1,2 designate the two opposite side contours of the lens, $t_{1,2}$ denote the lens thicknesses as shown in Fig. 7.9 which can be computed by [15]

$$t_{1,2} = \frac{1}{n+1}\left[\sqrt{f_{1,2}^2 + \frac{(n+1)D^2}{4(n-1)}} - f_{1,2}\right] \tag{7.24}$$

and $n=\sqrt{\varepsilon_r}$ is the refraction index of the lens material. In analogy to the analysis of thin dielectric lenses [18], we define for each lens an equivalent focal length f given by $1/f=1/f_1+1/f_2$. Goldsmith [16] has shown that a feed horn radiation characteristics, can be approximated by a fundamental Gaussian beam with minimum waists w_{0e} and w_{0h}, representing the radiation in E and H planes. The waists w_{0e} and w_{0h} are related to the feed horn 10 dB beamwidths BW_e and BW_h by the approximate formula $w_{0e,h}=0.68\lambda/BW_{e,h}$. The feed horn was designed such that w_{0e} and w_{0h} are close in value. Accordingly, we can define an equivalent quasi-circular minimum waist of the feed horn by $w_{of} = \sqrt{w_{0e}w_{0h}}$. In addition, we made the assumption [16] that the two lenses act as phase transformers, each providing a phase advancement approximately proportional to the square of the distance r of a ray from the axis of propagation:

$$\Delta\phi = \pi r^2 / \lambda f \tag{7.25}$$

This phase advancement transforms the diverging Gaussian beam of H_2 feed horn (cf. Fig. 7.9) to a converging Gaussian beam with minimum waist of w_{0l}, located a distance d_2 from the lens surface. By symmetry, this converging Gaussian beam is transformed into an outward Gaussian beam captured by the second lens and transformed into an inward Gaussian beam captured by H_1 feed horn, as shown in Fig. 7.9. Accordingly, at half the distance between the lenses, we obtain a minimum Gaussian beam waist w_{0l} with constant phase distribution.

System Analysis without the Scatterer

The angular dependence of the co-polarized transmission loss between the two antennas is proportional to the coupling between the inward and outward Gaussian beams in the focal plane, as shown in [16]. The co-polarized electric field distribution of the inward Gaussian beam in the focal plane ($z = 0$) is described in [16] by $f^t_{copol} e^{-\frac{x^2}{w_{0x}^2} - \frac{y^2}{w_{0y}^2}}$, in which $f^t_{copol} = const.$ In a similar fashion, the cross-polarized term can be described by $f^t_{xpol}(y)e^{-\frac{x^2}{w_{0x}^2}-\frac{y^2}{w_{0y}^2}}$ for an $\hat{x}$-polarized electric field. Accordingly, the total electric field distribution of the inward Gaussian beam in the focal plane can be described by

$$\underline{f}^t(x,y) = (f^t_{copol}\hat{x} + f^t_{xpol}(y)\hat{y})e^{-\frac{x^2}{w_{0x}^2}-\frac{y^2}{w_{0y}^2}} \tag{7.26}$$

where f^t_{copol} and $f^t_{xpol}(y)$ are determined by the co-polarized and cross-polarized characteristics of the feed horn H_1 and the lens L_1. In most cases, $f^t_{copol} >> f^t_{xpol}$. The electric field distribution of the outward Gaussian beam in the focal plane, $\underline{f}^r(x,y)$ can be expressed in a similar fashion, but with the addition of a linear progressive phase due to the nonalignment of this beam with the system axis (cf. Fig. 7.9).

$$\underline{f}^r(x,y) = (f^r_{copol}\hat{x} + f^r_{xpol}(y)\hat{y})e^{-\frac{x^2}{w_{0x}^2}-\frac{y^2}{w_{0y}^2}}e^{-jky\sin\theta} \tag{7.27}$$

Due to the symmetry of the system, it can be assumed that $f^t_{copol} = f^r_{copol} = f^t_0$ and $f^t_{xpol} = f^r_{xpol}$. The coupling factor, $C_c(\theta)$ between the two beams in the focal plane is computed by performing a two-dimensional integral in the focal plane over the scalar product of the two electric field distributions, $\underline{f}^t(x,y)$ and $\underline{f}^r(x,y)$ as described in [16]. This exercise results in a co-polarized coupling factor, $C^c_{copol}(\theta)$ and a cross-polarized coupling factor, $C^c_{xpol}(\theta)$. Using separation of variables, the integration over x can be performed analytically and the result absorbed in $f_0{}^t$. Thus, the simplified co-polarized coupling factor, $C^c_{copol}(\theta)$ is given by

$$C^c_{copol}(\theta) = \int_{-\infty}^{\infty} \underline{f}^t \cdot \underline{f}^{r*}\, dy \cong |f^t_0|^2 \int_{-\infty}^{\infty} e^{-\frac{2y^2}{w_{0y}^2}} \cdot e^{jky\sin\theta} dy \tag{7.28}$$

Evaluation of the integral results in

$$C^c_{copol}(\theta) = C_0 e^{-\frac{1}{2}\left(\frac{w_{0y}\pi}{\lambda}\sin\theta\right)^2}; \quad C_0 = |f^t_0|^2 w_{0y}\sqrt{\pi/2} \tag{7.29}$$

The co-polarized coupling factor, $C^c_{copol}(\theta)$ can be either computed through (7.29) or directly measured. In a similar fashion, the cross-polarized coupling factor, $C^c_{xpol}(\theta)$ can be computed by

$$C^c_{xpol}(\theta) = 2\mathrm{Re}\left\{\int_{-\infty}^{\infty} f^t_{copol} f^{t*}_{xpol}(y) e^{-\frac{2y^2}{w_{0y}^2}} e^{jky\sin\theta} dy\right\} \tag{7.30}$$

or directly measured, if the receiving feed horn H_2 is rotated by 90 deg around its axis.

System Analysis with the Scatterer

In this case, the cylinder's scattered field $\underline{f}^s(x, y)$ in the focal plane is added to the inward Gaussian beam and the total field is coupled to the outward Gaussian beam of the receiving end. In a similar fashion to the assumption made in (7.26), the scattered field can be approximated by

$$\underline{f}^s = \begin{cases} (f^s_{copol}(y)\hat{x} + f^s_{xpol}(y)\hat{y})e^{-\frac{x^2}{w_{0x}^2}-\frac{y^2}{w_{0y}^2}} & ; \quad |y| \leq \frac{a}{2} \ and \ |x| < \infty \\ 0 & \quad elsewhere \end{cases} \tag{7.31}$$

in which $f^s_{copol}(y)$ and $f^s_{xpol}(y)$ are the co-polarized and cross-polarized scattered electric fields in the focal plane ($z = 0$) for *uniform plane wave* illumination and a is the projected width of the cylinder in the focal plane. $f^s_{copol}(y)$ and $f^s_{xpol}(y)$ are tapered by the exponential term due to the Gaussian beam characteristics in the focal plane. In the co-polarized case, both feed horns at the transmit and receive ends are aligned, and we can neglect the cross-polarized contribution since $f^t_{copol} >> f^t_{xpol}$. The coupling factor in the focal plane, $C^t_{copol}(\theta)$ between the total field in the vicinity of the cylinder, $(f^t_{copol} + f^s_{copol}(y))e^{-\frac{x^2}{w_{0x}^2}-\frac{y^2}{w_{0y}^2}}$ and the outward Gaussian beam, $f^r_{copol}e^{-\frac{x^2}{w_{0x}^2}-\frac{y^2}{w_{0y}^2}}e^{-jky\sin\theta}$ is computed by performing a two-dimensional integral in the focal plane over the scalar product of the two electric field distributions, as described in [16]. Using separation of variables, the integration over x can be performed analytically and the result absorbed in both f^t_{copol} and $f^s_{copol}(y)$. Accordingly, the co-polarized coupling can be described by

$$C^t_{copol}(\theta) = \int_{-\infty}^{\infty} (f^t_{copol} + f^s_{copol}(y)) \cdot f^{r*}_{copol} e^{-\frac{2y^2}{w_{0y}^2}} e^{jk_0 y \sin\theta} dy \tag{7.32}$$

If we normalize both sides by C_0 (the value recorded at $\theta = 0$ deg without any scatterer) and substitute (7.28) into (7.32), we obtain

$$C^t_{copol}(\theta) - C^c_{copol}(\theta) = f^{t*}_0 \int_{-a/2}^{a/2} f^s_{copol}(y)e^{-\frac{2y^2}{w_{0y}^2}} e^{jk_0 y \sin\theta} dy \tag{7.33}$$

All quantities on the LHS of (7.33) can be either analytically computable or directly measurable. Moreover, if we recall [19] that the scattering radiation pattern, $E^s_{copol}(\theta)$ for uniform plane wave illumination can be computed by

$$E^s_{copol}(\theta) = \int_{-a/2}^{a/2} f^s_{copol}(y)e^{jk_0 y \sin\theta} dy \tag{7.34}$$

and compare $E^s_{copol}(\theta)$ expression with the RHS of (7.33), one can recognize that the RHS of (7.33) is the co-polarized scattering pattern of a *tapered* scattered field distribution in the focal plane. Thus, $E^s_{copol}(\theta)$ for uniform plane wave illumination can be reconstructed from the data in (7.33), either by applying an FFT algorithm [17] or by using Gabor basis functions to express $f^s_{copol}(y)$ as described in [20].

The cross-polarized coupling factor is obtained if the feed horn at the receiving end is rotated 90 deg around its axis. In a similar fashion to the co-polarized case, the cross-polarized coupling factor, $C^t_{xpol}(\theta)$ between the co-polarized total field in the vicinity of the cylinder and the cross-polarized field of the outward Gaussian beam of the receiving end can be expressed by

$$C^t_{xpol}(\theta) = C^c_{xpol}(\theta) + \int_{-a/2}^{a/2} (f^s_{copol}(y)\cdot f^{r^*}_{xpol}(y) + f^s_{xpol}(y)\cdot f^{r^*}_{copol}(y))e^{-\frac{2y^2}{w^2_{0y}}} e^{jk_0 y \sin\theta} dy \tag{7.35}$$

Eq. (7.35) can be simplified if the assumption that $f^s_{copol}(y) = Const = f^s_0$ is made on the scatterer surface in the focal plane ($z = 0$). In this case the co-polarized scattered field, f_0^s can be evaluated from (7.33) for $\theta = 0$ deg to be

$$\frac{f^s_0}{f^t_0} = \left(\frac{C^t_{copol}(0)}{C_0} - 1\right)\frac{1}{erf\left(\frac{a}{\sqrt{2}w_{0y}}\right)} \tag{7.36}$$

in which $erf(x) = 2/\sqrt{\pi}\int_0^x e^{-t^2}dt$. Substitution of (7.36) into (7.35) yields

$$C^t_{xpol}(\theta) - C^c_{xpol}(\theta) - \frac{f^s_0}{f^t_0}\int_{-a/2}^{a/2} f^{r^*}_{xpol}(y)e^{-\frac{2y^2}{w^2_{0y}}} e^{jk_0 y \sin\theta} dy = \int_{-a/2}^{a/2} f^s_{xpol}(y)e^{-\frac{2y^2}{w^2_{0y}}} e^{jk_0 y \sin\theta} dy \tag{7.37}$$

All quantities involved on the LHS of (7.37) are either directly measured, such as $C^t_{xpol}(\theta)$, $C^c_{xpol}(\theta)$, $f^r_{xpol}(y)$, or computable. Similar considerations to the co-polarized case lead us to recognize that the RHS of (7.37) is proportional to the cross-polarized scattering pattern of a *tapered* cross-polarized scattered field distribution in the focal plane, with $\exp(-2y^2/w^2_{0y})$ being the tapering factor. The correct cross-polarized scattering pattern, $E^s_{xpol}(\theta)$ can be reconstructed from the data in (7.37), applying an FFT algorithm in a similar fashion to the procedure described in the co-polarized case.

The IFR_e (co-polarized TM case) of an arbitrary shaped cylinder for normal incidence is given by Rusch [1]

$$IFR_e = -\frac{\eta_0}{2aE_0}\int_s J_{sz'} e^{jk\rho' \sin\phi'}\, dl \tag{7.38}$$

in which $\eta_0 = \sqrt{\frac{\mu_0}{\varepsilon_0}} \cong 120\pi\,[\Omega]$ is the characteristic impedance of free space (μ_0 and ε_0 are the permeability and permittivity of free space), $J_{sz'}$ is the axial induced current distribution on the cylinder and E_0 is the intensity of the incident electric field. The extension of (7.38) to the oblique incidence case would be

$$IFR_e = -\frac{\eta_0 \sin\theta_0}{2aE_0}\int_s J_{sz'} e^{jk_0\rho' \sin\theta_0 \sin\phi'}\, dl \tag{7.39}$$

Similarly, for the TE case the IFR_h would be,

$$IFR_h = \frac{\sin\theta_0}{2aH_0}\int_s H_{z'}(\hat{z}\cdot\hat{n}) e^{jk_0\rho' \sin\theta_0 \sin\phi'}\, dl \tag{7.40}$$

in which $H_{z'}$ is the total axial magnetic field at the surface of the cylinder and $\hat{n}$ is the unit vector normal to the cylinder surface. An alternative representation [15] valid for both TM and TE cases of the *IFR* derived in terms of the respective scattered electric field, $f^s_{copol}(y)$ in the focal plane is

$$IFR_{e,h} = \left(10^{\frac{\Delta\alpha}{20}} e^{j\Delta\phi} - 1\right)\sqrt{\frac{\pi}{2}}\,\frac{w_{0y}}{a}\,\frac{\displaystyle\int_{-a/2}^{a/2} f^s_{copol}(y)dy}{\displaystyle\int_{-a/2}^{a/2} f^s_{copol}(y)e^{-\frac{2y^2}{w_{0y}^2}}\, dy} \tag{7.41}$$

As a first-order approximation, we can assume that $f^s_{copol}(y) = const$ in the projected width. In this specific case, we obtain a very simple and compact expression for the $IFR_{e,h}$

$$IFR_{e,h} = \left(10^{\frac{\Delta\alpha}{20}} e^{j\Delta\phi} - 1\right)\frac{1}{erf\left(\frac{a}{\sqrt{2}w_{0y}}\right)} \tag{7.42}$$

in which $erf(x) = \frac{2}{\sqrt{\pi}}\int_0^x e^{-t^2}\, dt$ is the error function.

Fig. 7.10 shows a picture of a typical focused beam system of L-3 Communications-ESSCO, Concord, Massachusetts [17] testing beams for oblique incident angles.

An AEL horn antenna model H-1498 operating in the frequency range 2–18 GHz was chosen as feed. The horn has almost a constant

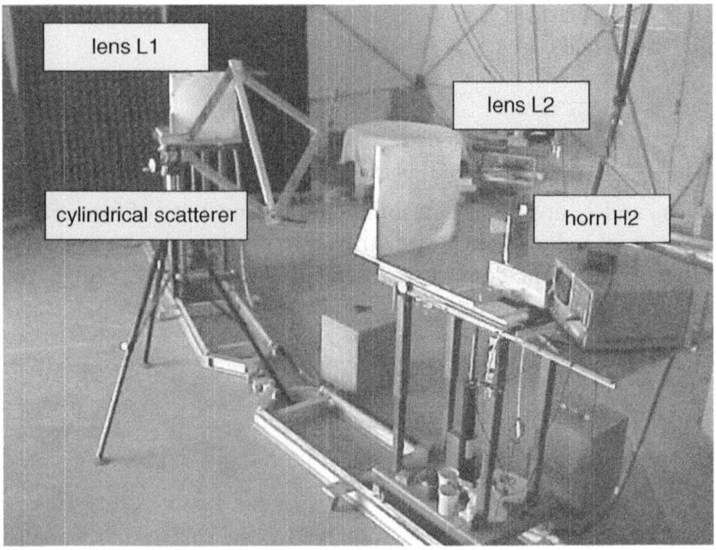

Figure 7.10 Focused beam system of L-3 Communications-ESSCO, Concord, Massachusetts.

10 dB beamwidth in both E and H planes over the entire frequency bandwidth. Based on the beamwidth data the equivalent feed horn Gaussian beam minimum waists, w_{0e} and w_{0h} were computed. The diameter of the lens was chosen to be 55.9 cm and it was manufactured from material with dielectric constant 2.3. The lens focal distances, f_1 and f_2 were chosen as 53.3 cm and 203.2 cm, respectively. Due to the finite diameter of the lens and the variation of the feed horn phase center location with frequency we had to determine, d_2 for each test frequency. At the operating frequency 12 GHz, the Gaussian beam waist size w_{0x} and w_{0y} in the focal plane were measured to be 6.54 cm and 6.78 cm, respectively, at –8.7 dB points (the field is e^{-1} of its value on axis).

Two rectangular cylinders (metal and dielectric) used in a typical metal and dielectric space frame radomes were chosen for testing the accuracy of the system. The cross section of the rectangular metal cylinder was 1.37 × 5.16 cm and that of the dielectric cylinder was 1.19 × 5.69 cm with $\varepsilon_r = 5.0$. The cylinders were tested (co-polarized and cross-polarized) at 12 GHz for both vertical (VP) and horizontal (HP) polarizations and for four oblique incident angles on their broadside.

Due to physical constraints of the measurement setup, the angular extent of the focused beam was θ_{max}=70 deg. The algorithm assumes that the cylinder is infinite in length; therefore, to avoid any error caused by its finite length, it is chosen to be longer than $4w_{0x}/\sin\theta_0$. Such a choice makes the finite length of the cylinder irrelevant, since for this case the top and bottom of the cylinder are almost not illuminated (–34 dB down). Accordingly, for practical reasons this bound makes the method less attractive for small inclination angles, θ_0.

Initially, direct measurements of the co-polarized $f^t_{copol}(y)$ and cross-polarized $f^t_{xpol}(y)$ field distributions without the scatterer were performed in the focal plane at 12 GHz using a small feed horn. Fig. 7.11 shows the co-polarized and cross-polarized amplitude and phase distribution in the focal plane for vertical polarization.

The comparison of the measured co-polarized data with the computed results using the assumption made in eq. (7.26) shows a nice agreement. One can observe that the maximum cross-polarized level (measured) due to the lens, L_1 and feed horn, H_1 is less than –25 dB. This information was stored in the computer for further processing.

Fig. 7.12 shows a typical recorded signal (amplitude and phase) for both co-polarized and cross-polarized measurements throughout the

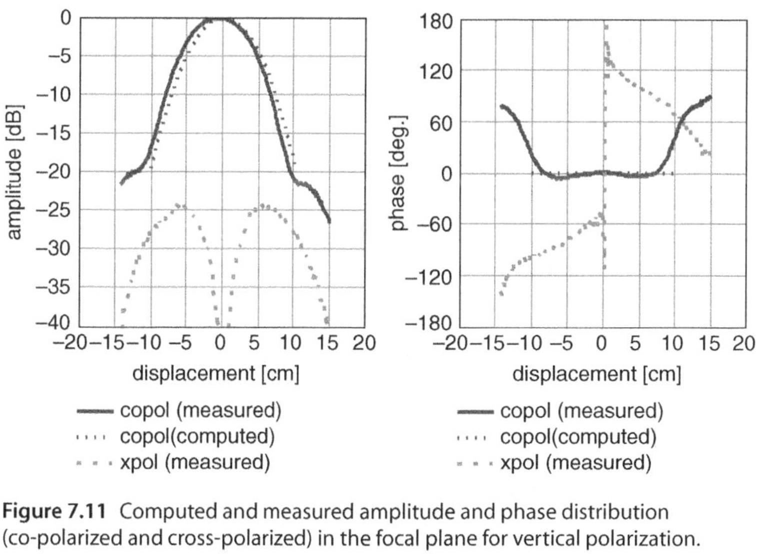

Figure 7.11 Computed and measured amplitude and phase distribution (co-polarized and cross-polarized) in the focal plane for vertical polarization.

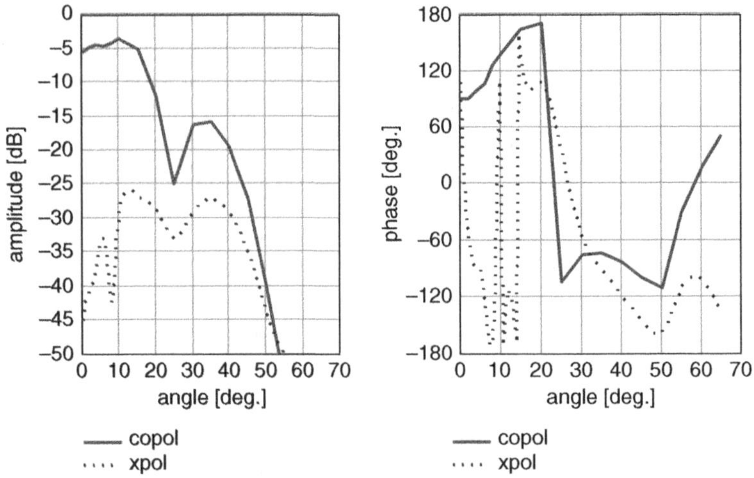

Figure 7.12 Recorded signal (amplitude and phase) for co-polarized and cross-polarized with the dielectric cylinder inclined 30 deg ($\theta_0 = 60$ deg) in the focal plane, illuminated on its broadside at 12 GHz with vertical polarization.

angular movement of the receiving end (horn H_2 and lens L_2) with the dielectric cylinder inclined 30 deg ($\theta_0 = 60$ deg).

The beam is illuminated by a vertical polarized wave on its broadside. The co-polarized recorded data (with and without the scatterer) was plugged in eq. (7.33) and processed with an FFT algorithm to evaluate the tapered and corrected field distributions of the scatterer in the focal plane, $f^{s}_{copol}(y)$, as described in [15]. The next step was to compute the reconstructed co-polarized radiation pattern of the scatterer using eq. (7.33). The cross-polarized recorded data (with and without the scatterer) was plugged in eq. (7.37), and the cross-polarized scattered radiation pattern was computed. Fig. 7.13 shows the comparison between the computed scattering pattern (co-polarized and cross-polarized) by FEM and the reconstructed scattering pattern (co-polarized and cross-polarized) using the recorded measured data in the focused beam system.

One can observe that a nice agreement in the main beam and in the sidelobe level is obtained, when we compare the computed and the reconstructed patterns. The ripple in the reconstructed pattern at $\theta \approx 23$ deg is probably caused by corrupted phase measurement data amplified by the reconstruction algorithm. One can observe that the cross-polarized scattering pattern for oblique incident

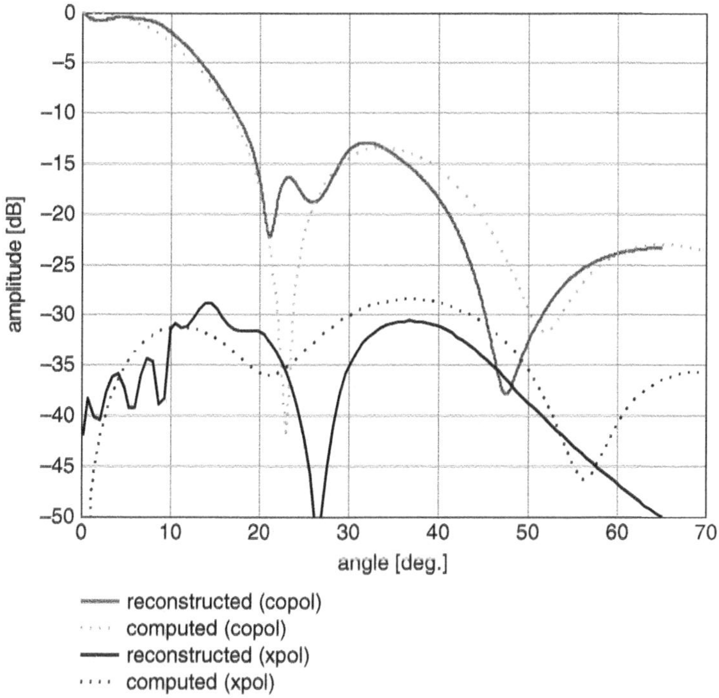

Figure 7.13 Comparison among computed (FEM) and reconstructed scattering radiation patterns (co-polarized + cross-polarized) of the dielectric cylinder inclined 30 deg ($\theta_0 = 60$ deg) in the focal plane, illuminated on its broadside at 12 GHz with vertical polarization.

angles is significant and cannot be ignored in the scattering analysis of large space frame radomes. Moreover, it is interesting to note that the cross-polarized level in the direction of the incident field vanishes. These observations confirm Rojas [5] computational results, which predicted a significant cross-polarized scattering level of dielectric cylinders for oblique incident angles. This cross-polarized level vanishes completely for normal incidence on both metal and dielectric beams.

In addition, the data from all measurements taken for both the dielectric and metal cylinders was processed to obtain the IFR_e and IFR_h based on eq. (7.41) in the 2D case. The results obtained were compared to the IFR values computed numerically by MoM [6] for an infinite length metallic cylinder (2D case) using two codes

Table 7.2 Comparison between measured and computed IFR values

Beam type	Incident angle, θ_0	Computed (MoM/FEM)		Measured	
		IFR_e	IFR_h	IFR_e	IFR_h
Metal cylinder (a=5.16 cm)	90^0	$1.14\angle 172.6°$	$0.98\angle -176.6°$	$1.06\angle 172.4°$	$1.14\angle -177.5°$
	75^0	$1.15\angle 172.5°$	N/A	$1.1\angle 176.1°$	$1.2\angle -178.1°$
	60^0	$1.15\angle 171.9°$	N/A	$1.12\angle 178.5°$	$1.3\angle 179.5°$
	45^0	$1.16\angle 170.2°$	N/A	$1.24\angle -178.3°$	$1.26\angle 178.5°$
Dielectric cylinder (a=5.69 cm)	90^0	$1.91\angle 163.8°$	$1.76\angle 167.1°$	$1.85\angle 159.0°$	$1.90\angle 166.9°$
	75^0	$1.93\angle 161.7°$	$1.79\angle 165.0°$	$1.86\angle 157.9°$	$1.94\angle 164.1°$
	60^0	$1.99\angle 157.0°$	$1.86\angle 162.2°$	$1.93\angle 154.5°$	$2.06\angle 160.6°$
	45^0	$2.02\angle 148.6°$	$1.79\angle 159.0°$	$1.93\angle 148.9°$	$1.91\angle 158.3°$

named ANYMFRE and ANYMFRH for TE and TM polarizations, respectively, and by FEM [21] for a dielectric cylinder (2D case) using a code named FEFD. All three codes were developed at the University of Illinois at Champaign–Urbana, Illinois. Computation results of the *IFR* for four oblique incident angles at 12 GHz are shown in Table 7.2.

One can observe a nice agreement between the numerical computation (MoM and FEM) and the processed measurement data acquired in the focused beam system. It is interesting to note that the *IFR* variation with the incident angle is minor for both IFR_e and IFR_h of the dielectric and metal cylinders. This observation justifies to a certain extent the original assumption made by Kay [3] in his co-polarized scattering analysis of the metal space frame radome in which he neglected the variation of the *IFR* for different oblique incident angles and considered only the effect of the projected beams on the antenna surface in the radome enclosure.

References

1 Rusch, WVT, Appel-Hansen, J, Klein, CA, and Mittra, R. Forward scattering from square cylinders in the resonance region with application to aperture blockage. *IEEE Trans. Antennas Propagat.*, 24(2), 182–189, 1976.

2 Afsar, MN, Birch, JR, and Clarke, RN. The measurement of the properties of materials. *Proceedings of the IEEE*, 74(1), 183–199, 1986.

3 Kay, AF. Electrical design of metal space frame radomes. *IEEE Trans. Antennas Propagat.*, 13(2), 188–202, 1965.
4 Kennedy, PD. An analysis of the electrical characteristics of structurally supported radomes. Columbus: Ohio State University, 1958.
5 Rojas, RG. Scattering by an Inhomogeneous Dielectric/Ferrite Cylinder of Arbitrary Cross-Section Shape-Oblique Incidence Case. *IEEE Trans. Antennas Propagat.*, 36(2), 238–246, 1988.
6 Richmond, JH. Scattering by a dielectric cylinder of arbitrary cross section shape. *IEEE Trans. Antennas Propagat.*, 13(3), 334–341, 1965.
7 Wu, RB, and Chen, CH. Variational reaction formulation of scattering problem for anisotropic dielectric cylinders. *IEEE Trans. Antennas Propagat.*, 34(5), 640–645, 1986.
8 Felsen, LB, and Marcuvitz, N. Radiation and scattering of waves. Englewood Cliffs, NJ: Prentice-Hall, 1973.
9 Shavit, R, Smolski, AP, Michielssen, E, and Mittra, R. Scattering analysis of high performance large sandwich radomes. *IEEE Trans. Antennas Propagat.*, 40(2), 126–133, 1992.
10 Shavit, R, Cohen, A, and Ngai, E. Characterization of forward scattering parameters from arbitrarily shaped cylinders by near-field probing. *IEEE Trans. Antennas Propagat.*, 43(6), 585–589, 1995.
11 Grimm, KR, and Hoffman, JB. Measurement of small radar target forward scattering by planar near-field scanning. In *Symp. on Precision Electromagnetics Measurements*, Gaithersburg, MD, 1986.
12 Yaghjian, AD. An overview of near-field antenna measurements. *IEEE Trans. Antennas Propagat.*, 34(1), 30–45, 1986.
13 Balanis, CA. Antenna theory—Analysis and design. Hoboken, NJ: Wiley, 2005.
14 Harrington, RF. Time-harmonic electromagnetic fields. New York: McGraw-Hill, 1961.
15 Shavit, R, Wells, T, and Cohen, A. Forward-scattering analysis in a focused-beam system. *IEEE Trans. Antennas Propagat.*, 46(4), 563–568, 1998.
16 Goldsmith, PF. Quasi-optical techniques at millimeter and submillimeter wavelengths. In *Infrared and Millimeter Waves.* New York: New York Academic, 1982.
17 Shavit, R, Sangiolo, J, and Monk, T. Scattering analysis of arbitrarily shaped cylinders in a focused beam system-oblique incidence

case. *IEEE Proc. Microwave Antennas Propagat.*, 148(2), 73–78, 2001.
18 Lo, YT, and Lee, SW. Antenna handbook: Theory, applications, and design. New York: Van Nostrand Reinhold, 1988.
19 Elliott, RS. Antenna theory and design. Englewood Cliffs, NJ: Prentice-Hall, 1981.
20 Shavit, R. Improved scattering analysis of arbitrary-shaped cylinders in a focused beam system using Gabor representation. *IEEE Proc. Microw. Antennas Propagat.*, 146(3), 193–196, 1999.
21 Jin, J. The finite element method in electromagnetics. New York: John Wiley & Sons, 1993.

Problems

P7.1 Design a focused Gaussian beam setup operating in the frequency band 8–12 GHz based on two lenses with diameter 22 in. made of PTFE with $\varepsilon_r = 2.3$ and focal distances $f_1 = 21$ in. and $f_2 = 80$ in. The lenses are fed by two small horns.
(a) Plot the lenses contour and calculate their thicknesses.
(b) Compute the aperture dimensions of the feeding horns to illuminate the lens edges at –20 dB relative to peak.
(c) Compute the equivalent Gaussian beam waists of the feeding horns in *E* and *H* planes.

P7.2 Compute the coupling (insertion loss) between the two feed horns of the focused beam system described in P7.1.

P7.3 A dielectric beam with cross section 2×0.48 in.^{2} is positioned at the center of the line connecting the two lenses of the focused beam setup described in P7.1. Making the assumption that the near field scattering from the beam can be represented by an elliptical Gaussian beam with the narrow waist equal to the beam width facing the receiving lens and the wider waist equal to the incident field Gaussian beam at the beam location. Find the coupling between the scattering beam and the lens at the receiving end of the system.

P7.4 Compute the *IFR* of the beam described in P7.3.

Appendices

A

Vector Analysis

A.1 Coordinate Transformations

The major three coordinate systems to be considered are: rectangular (x,y,z), cylindrical (ρ,ϕ,z), and spherical (r,θ,ϕ), and they are shown in Fig. A.1.

$$\begin{aligned} x &= \rho\cos\phi = r\sin\theta\cos\phi \\ y &= \rho\sin\phi = r\sin\theta\sin\phi \\ z &= r\cos\theta \end{aligned} \tag{A.1}$$

The corresponding unit vectors in the three coordinate systems are provided in the following tables:

Rectangular to cylindrical

	$\hat{x}$	$\hat{y}$	$\hat{z}$
$\hat{\rho}$	$cos\phi$	$sin\phi$	0
$\hat{\phi}$	$-sin\phi$	$cos\phi$	0
$\hat{z}$	0	0	1

Radome Electromagnetic Theory and Design, First Edition. Reuven Shavit.

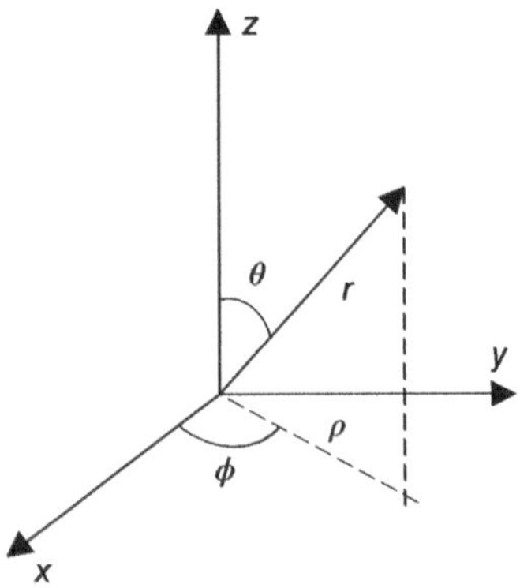

Figure A.1 Rectangular, cylindrical, and spherical coordinates.

Rectangular to spherical

	$\hat{x}$	$\hat{y}$	$\hat{z}$
$\hat{r}$	$\sin\theta\cos\phi$	$\sin\theta\sin\phi$	$\cos\theta$
$\hat{\theta}$	$\cos\theta\cos\phi$	$\cos\theta\sin\phi$	$-\sin\theta$
$\hat{\phi}$	$-\sin\phi$	$\cos\phi$	0

Cylindrical to spherical

	$\hat{\rho}$	$\hat{\phi}$	$\hat{z}$
$\hat{r}$	$\sin\theta$	0	$\cos\theta$
$\hat{\theta}$	$\cos\theta$	0	$-\sin\theta$
$\hat{\phi}$	0	1	0

Antennas and radomes are mounted on positioners in antenna far-field range to measure their radiation patterns. In general, there are two types of positioners: azimuth (A) over elevation (E) and elevation (e) over azimuth (a). Their angular scan are related to the spherical coordinates (θ,ϕ) in the following manner.

A.1.1 Azimuth over Elevation Positioner

Initially the positioner is rotated in azimuth an angle A to z' around the y-axis and then in elevation an angle E to z'' around the x'-axis.

$$\sin E = \sin\theta \sin\phi$$
$$\tan A = \tan\theta \cos\phi$$
$$\cos\theta = \cos E \cos A$$
$$\tan\phi = \frac{\tan E}{\sin A} \quad \text{(A.2)}$$

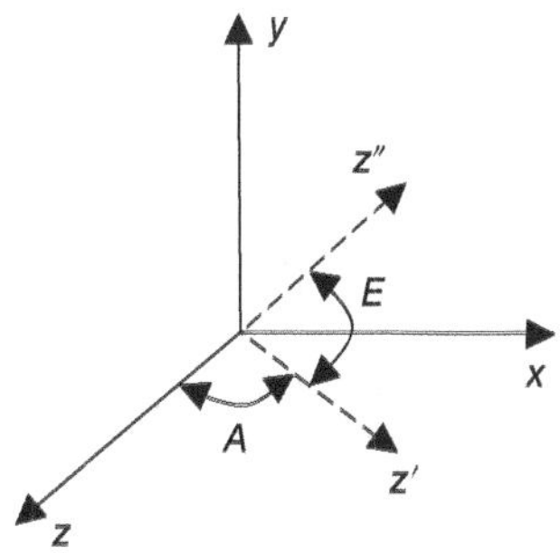

A.1.2 Elevation over Azimuth Positioner

Initially, the positioner is rotated in elevation an angle e to z' around x-axis and then in azimuth an angle a to z'' around the y-axis.

$$\tan e = \tan\theta \sin\phi$$
$$\sin a = \sin\theta \cos\phi$$
$$\cos\theta = \cos e \cos a$$
$$\tan\phi = \frac{\sin e}{\tan a} \quad \text{(A.3)}$$

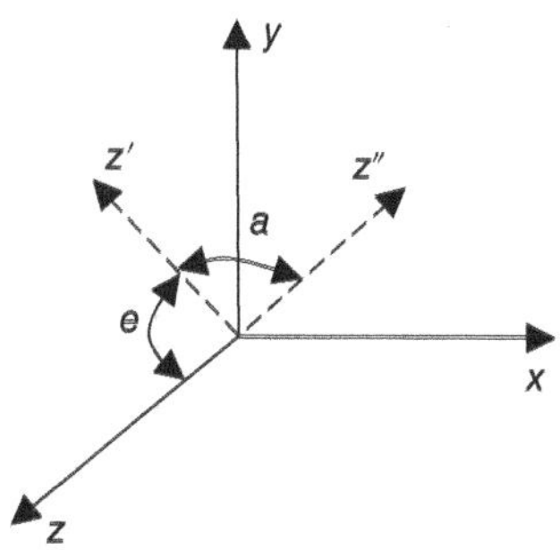

A.2 Vector Differential Operators

Rectangular coordinates:

$$\nabla f = \hat{x}\frac{\partial f}{\partial x} + \hat{y}\frac{\partial f}{\partial y} + \hat{z}\frac{\partial f}{\partial z} \tag{A.4}$$

$$\nabla \cdot \underline{A} = \frac{\partial A_x}{\partial x} + \frac{\partial A_y}{\partial y} + \frac{\partial A_z}{\partial z} \tag{A.5}$$

$$\nabla \times \underline{A} = \hat{x}\left(\frac{\partial A_z}{\partial y} - \frac{\partial A_y}{\partial z}\right) + \hat{y}\left(\frac{\partial A_x}{\partial z} - \frac{\partial A_z}{\partial x}\right) + \hat{z}\left(\frac{\partial A_y}{\partial x} - \frac{\partial A_x}{\partial y}\right) \tag{A.6}$$

$$\nabla^2 f = \frac{\partial^2 f}{\partial x^2} + \frac{\partial^2 f}{\partial y^2} + \frac{\partial^2 f}{\partial z^2} \tag{A.7}$$

$$\nabla^2 \underline{A} = \hat{x}\nabla^2 A_x + \hat{y}\nabla^2 A_y + \hat{z}\nabla^2 A_z \tag{A.8}$$

Cylindrical coordinates:

$$\nabla f = \hat{\rho}\frac{\partial f}{\partial \rho} + \hat{\phi}\frac{1}{\rho}\frac{\partial f}{\partial \phi} + \hat{z}\frac{\partial f}{\partial z} \tag{A.9}$$

$$\nabla \cdot \underline{A} = \frac{1}{\rho}\frac{\partial(\rho A_\rho)}{\partial \rho} + \frac{1}{\rho}\frac{\partial A_\phi}{\partial \phi} + \frac{\partial A_z}{\partial z} \tag{A.10}$$

$$\nabla \times \underline{A} = \hat{\rho}\left(\frac{1}{\rho}\frac{\partial A_z}{\partial \phi} - \frac{\partial A_\phi}{\partial z}\right) + \hat{\phi}\left(\frac{\partial A_\rho}{\partial z} - \frac{\partial A_z}{\partial \rho}\right) + \hat{z}\frac{1}{\rho}\left(\frac{\partial(\rho A_\phi)}{\partial \rho} - \frac{\partial A_\rho}{\partial \phi}\right) \tag{A.11}$$

$$\nabla^2 f = \frac{1}{\rho}\frac{\partial}{\partial \rho}\left(\rho\frac{\partial f}{\partial \rho}\right) + \frac{1}{\rho^2}\frac{\partial^2 f}{\partial \phi^2} + \frac{\partial^2 f}{\partial z^2} \tag{A.12}$$

$$\nabla^2 \underline{A} = \nabla(\nabla \cdot \underline{A}) - \nabla \times \nabla \times \underline{A} \tag{A.13}$$

Spherical coordinates:

$$\nabla f = \hat{r}\frac{\partial f}{\partial r} + \hat{\theta}\frac{1}{r}\frac{\partial f}{\partial \theta} + \hat{\phi}\frac{1}{r\sin\theta}\frac{\partial f}{\partial \phi} \tag{A.14}$$

$$\nabla \cdot \underline{A} = \frac{1}{r^2}\frac{\partial(r^2 A_r)}{\partial r} + \frac{1}{r\sin\theta}\frac{\partial(\sin\theta A_\theta)}{\partial \theta} + \frac{1}{r\sin\theta}\frac{\partial A_\phi}{\partial \phi} \tag{A.15}$$

$$\begin{aligned}\nabla \times \underline{A} = {} & \hat{r}\frac{1}{r\sin\theta}\left(\frac{\partial}{\partial \theta}(A_\phi \sin\theta) - \frac{\partial A_\theta}{\partial \phi}\right) + \hat{\theta}\frac{1}{r}\left(\frac{1}{\sin\theta}\frac{\partial A_r}{\partial \phi} - \frac{\partial(rA_\phi)}{\partial r}\right) \\ & + \hat{\phi}\frac{1}{r}\left(\frac{\partial(rA_\theta)}{\partial r} - \frac{\partial A_r}{\partial \theta}\right)\end{aligned} \tag{A.16}$$

$$\nabla^2 f = \frac{1}{r^2}\frac{\partial}{\partial r}\left(r^2\frac{\partial f}{\partial r}\right) + \frac{1}{r^2 \sin\theta}\frac{\partial}{\partial \theta}\left(\sin\theta\frac{\partial f}{\partial \theta}\right) + \frac{1}{r^2 \sin^2\theta}\frac{\partial^2 f}{\partial \phi^2} \quad \text{(A.17)}$$

$$\nabla^2 \underline{A} = \nabla(\nabla \cdot \underline{A}) - \nabla \times \nabla \times \underline{A} \quad \text{(A.18)}$$

Vector identities:

$$\underline{A} \cdot \underline{B} = |A||B| \cos\theta \quad \text{(A.19)}$$

$$|\underline{A} \times \underline{B}| = |A||B| \sin\theta \quad \text{(A.20)}$$

$$\underline{A} \cdot \underline{B} \times \underline{C} = \underline{A} \times \underline{B} \cdot \underline{C} = \underline{C} \times \underline{A} \cdot \underline{B} \quad \text{(A.21)}$$

$$\underline{A} \times \underline{B} = -\underline{B} \times \underline{A} \quad \text{(A.22)}$$

$$\underline{A} \times (\underline{B} \times \underline{C}) = (\underline{A} \cdot \underline{C})\underline{B} - (\underline{A} \cdot \underline{B})\underline{C} \quad \text{(A.23)}$$

$$\nabla(fg) = g\nabla f + f\nabla g \quad \text{(A.24)}$$

$$\nabla \cdot (f\underline{A}) = \underline{A} \cdot \nabla f + f\nabla \cdot \underline{A} \quad \text{(A.25)}$$

$$\nabla \cdot (\underline{A} \times \underline{B}) = (\nabla \times \underline{A}) \cdot \underline{B} - (\nabla \times \underline{B}) \cdot \underline{A} \quad \text{(A.26)}$$

$$\nabla \times (f\underline{A}) = (\nabla f) \times \underline{A} + f\nabla \times \underline{A} \quad \text{(A.27)}$$

$$\nabla \times (\underline{A} \times \underline{B}) = \underline{A}\nabla \cdot \underline{B} - \underline{B}\nabla \cdot \underline{A} + (\underline{B} \cdot \nabla)\underline{A} - (\underline{A} \cdot \nabla)\underline{B} \quad \text{(A.28)}$$

$$\nabla(\underline{A} \cdot \underline{B}) = (\underline{A} \cdot \nabla)\underline{B} + (\underline{B} \cdot \nabla)\underline{A} + \underline{A} \times (\nabla \times \underline{B}) + \underline{B} \times (\nabla \times \underline{A}) \quad \text{(A.29)}$$

$$\nabla \cdot \nabla \times \underline{A} = 0 \quad \text{(A.30)}$$

$$\nabla \times (\nabla f) = 0 \quad \text{(A.31)}$$

$$\nabla \times \nabla \times \underline{A} = \nabla\nabla \cdot \underline{A} - \nabla^2 \underline{A} \quad \text{(A.32)}$$

$$\int_v \nabla \cdot \underline{A} dv = \oint_s \underline{A} \cdot \underline{ds} \quad \textit{3D divergence theorem} \quad \text{(A.33)}$$

$$\int_s \nabla \cdot \underline{A} ds = \oint_c \underline{A} \cdot \hat{n} dl \quad \textit{2D divergence theorem} \tag{A.34}$$

$$\int_s T\nabla \cdot \underline{A} ds = \int_c T\underline{A} \cdot \hat{n} dl - \int_s \nabla T \cdot \underline{A} ds \tag{A.35}$$

$$\int_s (\nabla \times \underline{A}) \cdot \underline{ds} = \oint_c \underline{A} \cdot \underline{dl} \quad \textit{Stokes theorem} \tag{A.36}$$

$$\int_v \nabla \times \underline{A} dv = -\oiint_s \underline{A} \times \underline{ds} \tag{A.37}$$

$$\int_v \nabla w dv = \oiint_s w\underline{ds} \tag{A.38}$$

$$\int_s \hat{n} \times \nabla w ds = \oint_c w\underline{dl} \tag{A.39}$$

B

Dielectric Constants and Loss Tangent for Some Radome Materials

Organic radome materials (8.5 GHz)

Material	Dielectric Constant	Loss Tangent
Thermal plastics		
Lexan	2.86	0.006
Teflon	2.10	0.0005
Noryl	2.58	0.005
Kydox	3.44	0.008
Laminates		
Epoxy E glass cloth	4.4	0.016
Polyester-E glass cloth	4.1	0.015
Polyester-quartz cloth	3.7	0.007
Polybutadiene	3.83	0.015
Fiberglass laminate polybenzimidazole resin	4.9	0.008
Quartz-reinforced polyimide	3.2	0.008
Duroid 5650 (loaded Teflon)	2.65	0.003

Ceramic radome materials (8.5 GHz)

Material	Density [g/cm^3]	Dielectric Constant	Loss Tangent
Aluminium oxide	3.32	7.85	0.0005
Alumina hot-pressed	3.84	10.0	0.0005
Beryllium oxide	2.875	6.62	0.001
Boron nitride, hot-pressed	2.13	4.87	0.0005
Boron nitride, pyrolitic	2.14	5.12	0.0005

Material	Density [g/cm³]	Dielectric Constant	Loss Tangent
Magnesium aluminate (spinel)	3.57	8.26	0.0005
Magnesium aluminum silicate (cordierite ceramic)	2.44	4.75	0.002
Magnesium oxide	3.30	9.72	0.0005
Pyroceram 9606		5.58	0.0008
Rayceram 8		4.72	0.003
Silicon dioxide	2.20	3.82	0.0005
Silica-fiber composite (AS-3DX)	1.63	2.90	0.004
Slip-cast fused silica	1.93	3.33	0.001
Silicon nitride	2.45	5.50	0.003

C

Basic Antenna Theory

C.1 Vector Potentials

In antenna problems, one is interested in determining the fields at remote points from the sources. The sources are assumed to be oscillating harmonically with time ($e^{j\omega t}$) at an angular frequency ω and are confined to a volume V. For mathematical convenience and for better physical insight it is customized to define actual electric current and charge sources $\underline{J}$, ρ, and equivalent magnetic current and charge sources $\underline{J}_m$, ρ_m. The introduction of the magnetic sources based on equivalence and Schellkunoff theorems [1] lead to the definition of two sets of Maxwell equations:

$$\begin{aligned} \nabla \times \underline{E}_1 &= -j\omega \underline{B}_1 \\ \nabla \times \underline{H}_1 &= j\omega \underline{D}_1 + \underline{J} \\ \nabla \cdot \underline{D}_1 &= \rho \\ \nabla \cdot \underline{B}_1 &= 0 \\ \nabla \cdot \underline{J} &= -j\omega\rho \end{aligned} \quad ; \quad \begin{aligned} \nabla \times \underline{E}_2 &= -j\omega \underline{B}_2 - \underline{J}_m \\ \nabla \times \underline{H}_2 &= j\omega \underline{D}_2 \\ \nabla \cdot \underline{D}_2 &= 0 \\ \nabla \cdot \underline{B}_2 &= \rho_m \\ \nabla \cdot \underline{J}_m &= -j\omega\rho_m \end{aligned} \tag{C.1}$$

in which $\underline{E}_1$, $\underline{D}_1$, $\underline{H}_1$, $\underline{B}_1$ are the electric field, displacement vector, magnetic field, and magnetic induction related to the electric sources and $\underline{E}_2$, $\underline{D}_2$, $\underline{H}_2$, $\underline{B}_2$ are the corresponding vectors related to the equivalent magnetic sources. Correspondingly, the total fields will be

$$\begin{aligned} \underline{E} &= \underline{E}_1 + \underline{E}_2 \\ \underline{H} &= \underline{H}_1 + \underline{H}_2 \end{aligned} \tag{C.2}$$

In addition to Maxwell equations, we can define additional constitutive relations between the electric and displacement vectors, the magnetic and the magnetic induction vectors, and the induced currents and electric fields, such that

$$\begin{aligned} \underline{D} &= \varepsilon \underline{E} \\ \underline{B} &= \mu \underline{H} \\ \underline{J} &= \sigma \underline{E} \end{aligned} \tag{C.3}$$

in which ε is the permittivity, μ is the permeability, and σ is the conductivity. The objective in the solution of the electric and magnetic fields is to obtain a scalar differential equation, which can be solved analytically or numerically. This objective can be achieved if two vector potentials are introduced, the magnetic vector potential $\underline{A}$ related to the electric sources and the electric vector potential $\underline{F}$, related to the equivalent magnetic sources. They are defined by

$$\begin{aligned}\underline{H}_1 &\triangleq \nabla \times \underline{A} \\ \underline{E}_2 &\triangleq -\nabla \times \underline{F}\end{aligned} \tag{C.4}$$

Substitution of (C.4a) and (C.3b) into (C.1a) results in

$$\nabla \times (\underline{E}_1 + j\omega\mu\underline{A}) = 0 \rightarrow \underline{E}_1 = -j\omega\mu\underline{A} - \nabla\Phi \tag{C.5}$$

in which Φ is the scalar electric potential. Substitution of (C.5) and (C.3a) into (C.1b) yields

$$\begin{aligned}\nabla \times \nabla \times \underline{A} &= k^2\underline{A} - j\omega\varepsilon\nabla\Phi + \underline{J} \\ \nabla(\nabla \cdot \underline{A}) - \nabla^2\underline{A} &= k^2\underline{A} - j\omega\varepsilon\nabla\Phi + \underline{J}\end{aligned} \tag{C.6}$$

in which $k \triangleq \omega\sqrt{\mu\varepsilon}$. The vector potential $\underline{A}$ is single valued only if $\nabla \times \underline{A}$ and $\nabla \cdot \underline{A}$ are defined. Accordingly, if we use *Lorentz gauge* [1] and define $\nabla \cdot \underline{A} \triangleq -j\omega\varepsilon\Phi$, eq. (C.6b) simplifies to

$$\nabla^2\underline{A} + k^2\underline{A} = -\underline{J} \tag{C.7}$$

Equation (C.7) is a vector differential equation, which can be decomposed into three scalar differential equations for A_x, A_y, A_z components in a rectangular coordinate system with J_x, J_y, J_z current sources. Similar derivation for $\underline{E}_2$, $\underline{D}_2$, $\underline{H}_2$, $\underline{B}_2$, equivalent magnetic current sources, and using *Lorentz gauge* $\nabla \cdot \underline{F} \triangleq -j\omega\mu\Phi_m$ (Φ_m is the magnetic scalar potential) for the electric vector potential $\underline{F}$ results in the vector differential equation

$$\nabla^2\underline{F} + k^2\underline{F} = -\underline{J}_m \tag{C.8}$$

Solution of the differential equations (C.7) and (C.8) enables us to evaluate the vector potentials $\underline{A}$ and $\underline{F}$. Knowledge of these vectors enables us to compute the electric and magnetic fields of the system:

$$\begin{aligned}\underline{E} &= -j\omega\mu\left[\underline{A} + \frac{1}{k^2}\nabla(\nabla \cdot \underline{A})\right] - \nabla \times \underline{F} \\ \underline{H} &= -j\omega\varepsilon\left[\underline{F} + \frac{1}{k^2}\nabla(\nabla \cdot \underline{F})\right] + \nabla \times \underline{A}\end{aligned} \tag{C.9}$$

As a first step in the evaluation of the vector potentials, $\underline{A}$ and $\underline{F}$ for a given source distribution is their evaluation for an elementary dipole. This solution is also called the Green's function. Using the superposition principle enables us to evaluate the total vector potentials in the

Figure C.1 Elementary dipole at origin.

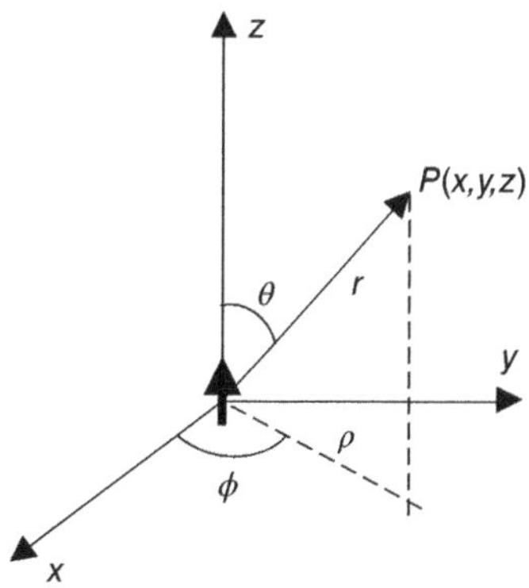

system through integration. Without loss of generality, we can consider the solution of an elementary dipole z-directed and positioned at the origin, as shown in Fig. C.1.

In this case, eq. (C.7) simplifies to

$$\nabla^2 A_z + k^2 A_z = -\delta(x)\delta(y)\delta(z) \tag{C.10}$$

Due to the symmetry of the problem $A_z = A_z(r)$, and it is convenient to solve it in spherical coordinates (r,θ,ϕ). As a first step, we solve its homogeneous version:

$$\frac{1}{r^2}\frac{\partial}{\partial r}\left[r^2\frac{\partial A_z}{\partial r}\right] + k^2 A_z = 0 \tag{C.11}$$

If we define $A_z \triangleq \frac{g(r)}{r}$, eq. (C.11) transforms to

$$\frac{1}{r^2}\frac{d}{dr}\left[r^2\left(\frac{g'(r)}{r} - \frac{g(r)}{r^2}\right)\right] + k^2\frac{g(r)}{r} = 0$$

$$\frac{1}{r^2}\frac{d}{dr}[rg'(r) - g(r)] + k^2\frac{g(r)}{r} = 0$$

$$\frac{1}{r^2}[\cancel{g'(r)} + rg''(r) - \cancel{g'(r)}] + k^2\frac{g(r)}{r} = 0$$

$$g'' + k^2 g = 0 \tag{C.12}$$

The differential equation of $g(r)$ is of second order, and its solution is $g(r) = e^{\pm jkr}$. Accordingly,

$$A_z(r) = C\frac{e^{\pm jkr}}{r} \tag{C.13}$$

Taking in consideration that in this problem we encounter only outward waves, we keep only the solution of $A_z(r) = C\frac{e^{-jkr}}{r}$. Evaluation of the coefficient C is possible if we integrate both sides of (C.10) over a spherical volume and take the sphere radius to zero:

$$\int_{V\to 0}(\nabla^2 A_z + k^2 A_z)dv = -\int_{V\to 0}\delta(x)\delta(y)\delta(z)dv \tag{C.14}$$

Thus,

$$\int_{V\to 0} k^2 A_z dv = \int_{V\to 0} k^2 C\frac{e^{-jkr}}{r}r^2\sin\theta\, d\theta\, d\phi\, \mathrm{dr} \to 0$$

$$\int_{V\to 0}\nabla^2 A_z dv = \int_{V\to 0}\nabla\cdot(\nabla A_z)dv = \oint_{S|_{V\to 0}}\nabla A_z\cdot \underline{ds} = \hat{r}\cdot\int_0^{\pi}\int_0^{2\pi}\nabla A_z r^2\sin\theta\, d\theta\, d\phi\Big|_{r\to 0}$$

$$4\pi r^2\frac{d}{dr}\left(C\frac{e^{-jkr}}{r}\right)\Big|_{r\to 0} = -4\pi C = -1$$

$$C = \frac{1}{4\pi}$$

$$A_z(r) = \frac{e^{-jkr}}{4\pi r} \tag{C.15}$$

The result in (C.15) can be extended to currents in y, z directions and distributions in a volume V as shown in Fig. C.2. The vector $\underline{r}'$ represents the vector to a source point and $\underline{r}$ the vector to the observation point.

$$\underline{A}(r) = \frac{1}{4\pi}\int_V \underline{J}(\underline{r}')\frac{e^{-jk|\underline{r}-\underline{r}'|}}{|\underline{r}-\underline{r}'|}dv' \tag{C.16}$$

where

$$|\underline{r}-\underline{r}'| = \sqrt{(x-x')^2+(y-y')^2+(z-z')^2} \tag{C.17}$$

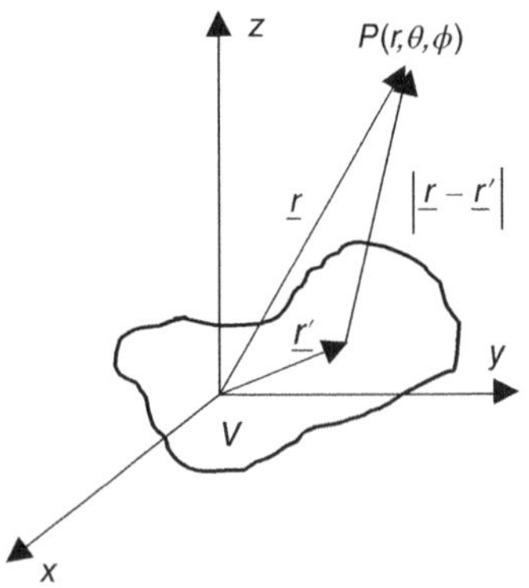

Figure C.2 Source distribution in volume V and observation point at $P(r,\theta,\phi)$.

In a similar fashion, we can derive the expression for the electric potential vector $\underline{F}$ to be

$$\underline{F}(r) = \frac{1}{4\pi}\int_V \underline{J}_{-m}(\underline{r}')\frac{e^{-jk|\underline{r}-\underline{r}'|}}{|\underline{r}-\underline{r}'|}dv' \tag{C.18}$$

Substitution of (C.16) and (C.18) into (C.9) yields the electric and magnetic fields radiated by an electric and magnetic source distributions in an infinite medium.

C.2 Far-Field Approximation

The integrals in (C.16) and (C.18) can be simplified if we restrict our observation to large distances (far-field approximation) from the radiating sources. To obtain an expression for these cases, we can expand $|\underline{r}-\underline{r}'|$ by a Newton series:

$$\begin{aligned}|\underline{r}-\underline{r}'| &= \sqrt{(r\sin\theta\cos\phi - x')^2 + (r\sin\theta\sin\phi - y')^2 + (r\cos\theta - z')^2}\ldots\\ &\approx r - \underbrace{(x'\sin\theta\cos\phi + y'\sin\theta\sin\phi + z'\cos\theta)}_{L} + \ldots\\ &+\frac{1}{2r}\underbrace{[x'^2 + y'^2 + z'^2 + (x'\sin\theta\cos\phi + y'\sin\theta\sin\phi + z'\cos\theta)^2]}_{\Psi} + o\left(\frac{1}{r^2}\right)\end{aligned} \tag{C.19}$$

Substituting (C.19) into (C.16) and if we keep only the first three items of the expansion, we obtain the Fresnel approximation:

$$\underline{A} \sim \frac{1}{4\pi}\frac{e^{-jkr}}{r}\int_V \underline{J}e^{jkL-jk\frac{\Psi}{2r}}dv' \tag{C.20}$$

On the other hand, if we keep only the first two items, we obtain Fraunhofer approximation (far-field approximation):

$$\underline{A} \sim \frac{1}{4\pi}\frac{e^{-jkr}}{r}\int_V \underline{J}e^{jkL}dv' \tag{C.21}$$

Next step is evaluation of (C.9) for far-field approximation. For this task, we need to perform the differential operations required in eq. (C.9) and keep items up to accuracy $o\left(\frac{1}{r}\right)$, such that

$$\begin{aligned}&\nabla\left(\frac{e^{-jk|\underline{r}-\underline{r}'|}}{|\underline{r}-\underline{r}'|}\right) \overset{Fraunhoffer}{\sim} \nabla\left(\frac{e^{-jkr+jkL}}{r}\right) = -jk\frac{e^{-jkr+jkL}}{r}\hat{r}\\ &\nabla\cdot\underline{A} = \frac{1}{4\pi}\int_V \underline{J}\cdot\nabla\left(\frac{e^{-jkR}}{R}\right)dv' \overset{ff}{\sim} -jk\hat{r}\cdot\underline{A}\end{aligned} \tag{C.22}$$

Substitution of (C.22) into (C.9) enables to evaluate the far-field approximation of $\underline{E}_1$:

$$\underline{E}_1 \overset{ff}{\sim} -j\omega\mu\left[\underline{A} + \frac{1}{k^2}(-jk)^2\hat{r}(\hat{r}\cdot\underline{A})\right] = -j\omega\mu[\underline{A} - \hat{r}(\hat{r}\cdot\underline{A})] = -j\omega\mu\underline{A}_T$$
$$\underline{A}_T \triangleq A_\theta\hat{\theta} + A_\phi\hat{\phi} \tag{C.23}$$

Similarly, we can approximate the far-field approximation of the electric vector potential $\underline{F}$ and the electric field $\underline{E}_2$ to obtain:

$$\underline{F} \overset{ff}{\sim} \frac{1}{4\pi}\frac{e^{-jkr}}{r}\int_V \underline{J}_{-m} e^{jkL} dv'$$
$$\underline{E}_2 = -\nabla\times\underline{F} = -\frac{1}{4\pi}\int_V \nabla\times\left(\underline{J}_{-m}\frac{e^{-jkR}}{R}\right)dv'$$
$$= -\frac{1}{4\pi}\int_V \left[\frac{e^{-jkR}}{R}\underbrace{\nabla\times\underline{J}_{-m}}_{=0} - \underline{J}_{-m}\times\nabla\left(\frac{e^{-jkR}}{R}\right)\right]dv'$$
$$\underline{E}_2 \overset{ff}{\sim} -\frac{1}{4\pi}jk\int_V \underline{J}_{-m}\times\hat{r}\frac{e^{-jkr+jkL}}{r}dv' = \hat{r}\times\frac{1}{4\pi}jk\frac{e^{-jkr}}{r}\int_V \underline{J}_{-m}e^{jkL}dv' = jk(\hat{r}\times\underline{F}) = jk\hat{r}\times\underline{F}_T$$
$$\underline{F}_T \triangleq F_\theta\hat{\theta} + F_\phi\hat{\phi} \tag{C.24}$$

Addition of the far-field approximations of $\underline{E}_1$ and $\underline{E}_2$ gives the total electric and magnetic fields:

$$\underline{E} \overset{ff}{\sim} -j\omega\mu\underline{A}_T + jk(\hat{r}\times\underline{F}_T)$$
$$\underline{H} \overset{ff}{\sim} \frac{1}{\eta}\hat{r}\times\underline{E}^{ff} \tag{C.25}$$

Evaluation of the transverse components $\underline{A}_T$ and $\underline{F}_T$ is possible by multiplying (C.21) and (C.24a) by the angular unit vectors $\hat{\theta}$ and $\hat{\phi}$ expressed in rectangular coordinates to obtain the following:

$$\mathrm{A}_\theta = \int_V [J_x\cos\theta\cos\phi + J_y\cos\theta\sin\phi - J_z\sin\theta]e^{jkL}dv'$$
$$\mathrm{A}_\phi = \int_V [-J_x\sin\phi + J_y\cos\phi]e^{jkL}dv'$$
$$\mathrm{F}_\theta = \int_V [J_{mx}\cos\theta\cos\phi + J_{my}\cos\theta\sin\phi - J_{mz}\sin\theta]e^{jkL}dv'$$
$$\mathrm{F}_\phi = \int_V [-J_{mx}\sin\phi + J_{my}\cos\phi]e^{jkL}dv'$$
$$L = x'\sin\theta\cos\phi + y'\sin\theta\sin\phi + z'\cos\theta$$
$$\underline{A}_T = \frac{1}{4\pi}\frac{e^{-jkr}}{r}(\mathrm{A}_\theta\hat{\theta} + \mathrm{A}_\phi\hat{\phi})$$
$$\underline{F}_T = \frac{1}{4\pi}\frac{e^{-jkr}}{r}(\mathrm{F}_\theta\hat{\theta} + \mathrm{F}_\phi\hat{\phi}) \tag{C.26}$$

C.3 Directivity and Gain

The antenna directivity is defined as the ratio of the radiation intensity in a specific direction in space to the average radiation intensity of the radiating element [1], i.e.,

$$D(\theta,\phi) = \frac{|\underline{E}(\theta,\phi)|^2}{\frac{1}{4\pi}\int_{\Omega} |\underline{E}(\theta',\phi')|^2 \sin\theta' d\theta' d\phi'} \tag{C.27}$$

The maximum directivity D_m is defined in the maximum radiation direction (θ_0, ϕ_0):

$$D_m = \frac{4\pi}{\int_{4\pi} F(\theta',\phi')^2 \sin\theta' d\theta' d\phi'} \tag{C.28}$$

in which $F(\theta',\phi') = \left|\frac{\underline{E}(\theta',\phi')}{\underline{E}_{max}}\right|$ is the antenna normalized radiation pattern. The antenna gain G is related to the maximum directivity through the antenna efficiency e, such that $G = eD_m$. For a pencil beam antenna with 3 dB beamwidths BW_e and BW_h in the E and H planes, it is easy to show that an approximate relation between the directivity and beamwidths is [1]:

$$D_m \approx \frac{4\pi}{BW_e BW_h} \tag{C.29}$$

with BW_e and BW_h units being in radians. Another important parameter characterizing receiving antennas is the antenna effective aperture $A_r(\theta,\phi)$. The directivity and the effective aperture are related through [1]:

$$A_r(\theta,\phi) = \frac{\lambda^2}{4\pi} D(\theta,\phi) \tag{C.30}$$

C.4 Antenna Noise Temperature

The antenna noise temperature T_A [°K] in Kelvin degrees is the temperature of a hypothetical resistor at the input of an ideal noise-free receiver that would generate the same output noise power P_n per unit bandwidth as that at the antenna output at a specified frequency. This noise can be written in the form [2]

$$P_n = KT_A \Delta f \tag{C.31}$$

in which $K = 1.38 \cdot 10^{-23} \left[\frac{Joule}{{}^0K}\right]$ is the Boltzman coefficient and Δf is the frequency bandwidth. In other words, antenna noise temperature is a parameter that describes how much noise an antenna produces in a given environment. This temperature is not the physical temperature of the antenna. Antenna noise temperature has contributions from several sources, such as galactic radiation, earth heating, the sun, and antenna losses. These contributions can be represented by the sky temperature $T_s(\theta)$, which is dependent on the elevation angle θ and is minimal at zenith and maximum at $\theta = 0$ deg (horizon). The antenna noise temperature depends on the antenna coupling to all the noise sources in its environment, as well as on noise generated within the antenna. Thus, the noise received by the antenna in an angular steradian segment $d\Omega$ at the elevation angle θ can be computed through

$$dP_n = KT_s(\theta)\Delta f\alpha, \quad \alpha = \frac{|E(\theta,\phi)|^2 d\Omega}{\int_{4\pi} |E(\theta',\phi')|^2 d\Omega'} \tag{C.32}$$

Integration over the entire space and use of (C.27) to describe α yields the total noise received by the antenna:

$$P_n = K\Delta f \frac{1}{4\pi} \int_{4\pi} D(\theta',\phi')T_s(\theta')d\Omega \tag{C.33}$$

in which $D(\theta',\phi')$ is the antenna directivity. Using (C.33) into (C.31) yields the expression for the antenna noise temperature

$$T_A = \frac{1}{4\pi} \int_{4\pi} D(\theta',\phi')T_s(\theta')d\Omega \tag{C.34}$$

C.5 Basic Array Theory

The general results for far-field approximation obtained in (C.26) can be extended to an array of antenna elements. Fig. C.3 shows the general geometry of the array.

The global coordinate system is *(x,y,z)* and the observation point is *P(x,y,z)*. (ξ, η, ζ) is the source coordinate system, (ξ_i, η_i, ζ_i) is the local source coordinate system of the ith element, and (x_i, y_i, z_i) are the coordinates of the ith element local coordinate system origin. The number of elements in the array is $N + 1$. The relation between the global source

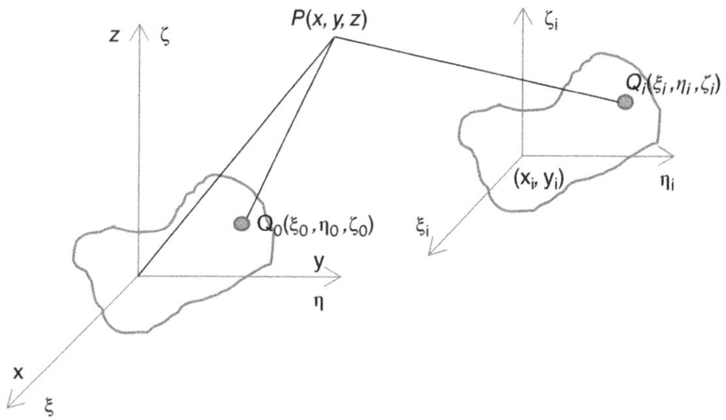

Figure C.3 The general array geometry with local and global coordinate systems.

coordinate system and the local coordinate system can be described by the following relations:

$$\begin{cases} \xi = \xi_i + x_i \\ \eta = \eta_i + y_i \\ \zeta = \zeta_i + z_i \end{cases} \quad ; \quad i = 0,1,\ldots,N \tag{C.35}$$

We assume that all elements in the array are identical in their geometry and space orientation. Accordingly, they have the same current distribution and differ only in their amplitude and phase. Thus, the current distribution of the ith element is related to the current distribution of the element at the origin by the relationship

$$\underset{\to}{J}(x_i + \xi_i, y_i + \eta_i, z_i + \zeta_i) = \frac{I_i}{I_0}\underset{\to}{J}(\xi_0, \eta_0, \zeta_0) \tag{C.36}$$

in which $\frac{I_i}{I_0}$ is a complex number. Substitution of (C.35) and (C.36) into (C.26) results in the following:

$$\mathrm{A}_\phi(\theta,\phi) = \overbrace{\int_{V_0}\left[-J_{x0}\sin\phi + J_{y0}\cos\phi\right] e^{jk(\xi_0 \sin\theta\cos\phi + \eta_0 \sin\theta\sin\phi + \zeta_0\cos\theta)} dv_0}^{=\mathrm{A}_e^{\phi}(\theta,\phi)} \cdot \underbrace{\sum_{i=0}^{N}\frac{I_i}{I_0}e^{jk(x_i \sin\theta\cos\phi + y_i\sin\theta\sin\phi + z_i cos\theta)}}_{=\mathrm{A}_a(\theta,\phi)}$$

$$\mathsf{A}_\theta(\theta,\phi) = \overbrace{\int_{V_0} [J_{x0}\cos\theta\cos\phi + J_{y0}\cos\theta\sin\phi - J_{z0}\sin\theta] e^{jk(\xi_0\sin\theta\cos\phi+\eta_0\sin\theta\sin\phi+\zeta_0\cos\theta)} dv_0}^{=\mathsf{A}_e^\theta(\theta,\phi)}$$
$$\cdot \underbrace{\sum_{i=0}^{N} \frac{I_i}{I_0} e^{jk(x_i\sin\theta\cos\phi+y_i\sin\theta\sin\phi+z_i cos\theta)}}_{=\mathsf{A}_a(\theta,\phi)} \tag{C.37}$$

which can be written in the short notation by,

$$\begin{aligned}\mathsf{A}_\theta &= \mathsf{A}_e^\theta \mathsf{A}_a \\ \mathsf{A}_\phi &= \mathsf{A}_e^\phi \mathsf{A}_a\end{aligned} \tag{C.38}$$

A_e^θ and A_e^ϕ represent the single element radiation properties of the element and A_a is the array factor. The far-field electric and magnetic fields of the array can be computed using (C.25). Usually in antenna array design the radiation pattern of the element is chosen to be closer to isotropic as possible and the array factor A_a determines the radiation characteristics of the entire array. In case of a 2D array positioned in the x-y plane with $2N_x$+1 rows of elements, each row parallel to the y-axis, with interspacing d_x between rows and $2N_y$+1 elements in each row with common spacing d_y between elements in the row, and if the current of the mn^{th} element is I_{mn}, the array factor is

$$\mathsf{A}_a = \sum_{m=-N_x}^{N_x} \sum_{n=-N_y}^{N_y} \frac{I_{mn}}{I_{00}} e^{jk\sin\theta(md_x\cos\phi+nd_y\sin\phi)} \tag{C.39}$$

D

Conjugate Gradient Algorithm

The conjugate gradient algorithm [3] is used for the solution of a matrix equation using an iterative procedure of orthogonal projections to minimize the error until the solution is reached. The algorithm essence is generation of orthogonal vectors and evaluation of their coefficients to obtain the required solution. It starts with the matrix equation under consideration,

$$\mathbf{Ax} = \mathbf{b} \tag{D.1}$$

in which $\mathbf{A}$ is a $N \times N$ nonsingular matrix, $\mathbf{b}$ is a vector $N \times 1$ and $\mathbf{x}$ is the unknown vector $N \times 1$. Let us define an internal product and the norm,

$$\begin{aligned} \langle \mathbf{x},\ \mathbf{y} \rangle &= \mathbf{x}^{\dagger}\mathbf{y} \\ \|\mathbf{x}\| &= \sqrt{\langle \mathbf{x},\ \mathbf{x} \rangle} \end{aligned} \tag{D.2}$$

where † denotes conjugate Hermitian. The iterative solution can be represented by

$$\mathbf{x}_n = \mathbf{x}_{n-1} + \alpha_n \mathbf{p}_n \tag{D.3}$$

in which $\mathbf{x}_{n-1}$ is the solution at the $n-1$ step, $\mathbf{p}_n$ is the direction vector, and α_n the coefficient that determines how much we advance in $\mathbf{p}_n$ direction. Initially, an educated guess is chosen and the residue $\mathbf{r}_0$ is determined by

$$\mathbf{r}_0 = \mathbf{Ax}_0 - \mathbf{b} \tag{D.4}$$

Initially, the direction vector is chosen to be

$$\mathbf{p}_1 = -\mathbf{A}^{\dagger}\mathbf{r}_0 \tag{D.5}$$

At each step, the direction coefficient α_n is calculated by

$$\alpha_n = -\frac{\langle \mathbf{Ap}_n, \mathbf{r}_{n-1} \rangle}{\|\mathbf{Ap}_n\|^2} \tag{D.6}$$

The residue at each step is calculated by

$$\mathbf{r}_n = \mathbf{r}_{n-1} + \alpha_n \mathbf{Ap}_n \tag{D.7}$$

and the direction vector for the next step is calculated by

$$\mathbf{p}_{n+1} = -\mathbf{A}^{\dagger}\mathbf{r}_n + \beta_n \mathbf{p}_n \tag{D.8}$$

in which

$$\beta_n = \frac{\|\mathbf{A}^\dagger \mathbf{r}_n\|^2}{\|\mathbf{A}^\dagger \mathbf{r}_{n-1}\|^2} \tag{D.9}$$

At each step, the residual norm is computed by

$$N_n = \frac{\|\mathbf{r}_n\|}{\|\mathbf{b}\|} \tag{D.10}$$

The algorithm stops when N_n goes below a certain threshold. Each iteration of the algorithm includes four products of matrices and vectors, but in practice, only two products are performed: $\mathbf{A}^\dagger \mathbf{r}_n$, $\mathbf{A}\mathbf{p}_n$, and the results are kept for the next iteration.

References

1 Elliott, RS. Antenna theory and design. Englewood Cliffs, NJ: Prentice Hall, 1981.
2 Balanis, CA. Antenna theory—analysis and design. Hoboken, NJ: John Wiley & Sons, 2005.
3 Peterson, AF, Ray, SL, and Mittra, R. Computational methods for electromagnetics. New York: IEEE Press, 1998.

Index

Printed and bound by CPI Group (UK) Ltd, Croydon, CR0 4YY

17/09/2026

14964296-0001